Introduction to Technical Writing

Process and Practice

Second Edition

Introduction to Technical Writing

Process and Practice

Second Edition

Lois Johnson Rew
San Jose State University

St. Martin's Press   New York

Senior editor: Catherine Pusateri
Development editor: Denise Quirk
Manager, publishing services: Emily Berleth
Project management: Omega Publishing Services, Inc.
Text design: Gene Crofts
Cover design: Jeheber & Peace, Inc.
Cover photo: Slide Graphics of New England, Inc.

Library of Congress Catalog Card Number: 92-50047

7 6 5 4 3
f e d c b a

For information, write:
St. Martin's Press, Inc.
175 Fifth Avenue
New York, NY 10010

ISBN: 0-312-06781-X

Acknowledgments

Acknowledgments and copyrights are continued at the back of the book on page 615, which constitutes an extension of the copyright page.

Chapter 2

Scenarios B and D from "Some *Practical* Exercises in Engineering Ethics," J. Wujek, 1989. © 1989, 1990 by J. Wujek. Used by permission of J. Wujek.

Chapter 6

Figure 6-1 reprinted by permission. Pages 79–81 from *A Pocket Pal: A Graphics Arts Production Handbook*, 12th edition. 1981.

Chapter 9

Mayonnaise and the Origin of Life, Harold J. Morowitz. Reprinted with permission of Ox Bow Press, P.O. Box 4045, Woodbridge, CT 06525. Copyright © 1985 by Harold J. Morowitz.

Chapter 12

Exercise: This material is reproduced, with permission, from Underwriters Laboratories Inc. *Standard for Safety for Microwave Cooking Appliances*, UL 923, Copyright by Underwriters Laboratories Inc. Publications Stock, 333 Pfingsten Road, Northbrook, Illinois 60062 U.S.A.

Exercise: Copyright Allen and Unwin, "Combine Harvester and Thresher," from *The Way Things Work*, Vol. 1, 1967.

Chapter 16

Figure 16-4 body of letter reprinted with permission of Bechtel Corporation.

For Bob

Preface

This second edition of *Introduction to Technical Writing: Process and Practice* is about workplace writing: the memos and letters, proposals and reports that professionals must write to carry on their work and to communicate with others in the scientific and technical communities. Like the first edition, this text builds on skills students have learned in freshman composition—prewriting, writing, and revising—and applies those skills to the demands of the working world. Also like the first edition, this text emphasizes the process of creating technical documents and provides numerous occasions to practice newly learned skills.

Features of the Second Edition

However, technical writing, like technology itself, has changed, and this second edition reflects those changes with the following features:

- **A new chapter on considering legal and ethical responsibilities.** This chapter is included as part of the writing process because as legal and ethical issues become increasingly important, students need to grapple with these issues before they write, as they are writing, and after they complete a document.
- **A new chapter on writing for international and multicultural audiences.** This chapter introduces students to the importance of these audiences and to the challenges writers face in meeting reader needs. Political and economic changes in the world—including the rise of Japan, the dissolution of the Soviet Union, and the expanding world market—make it imperative that students understand cultural and language issues that have an impact on effective writing.
- **A new chapter on designing documents.** Based on current research in cognition and readability, this chapter provides guidelines for students to design effective documents with whatever tools and options are available to them.
- **An expanded discussion of the collaborative process.** The Introduction, "Technical Writing and the Workplace," contains a section on the advantages and problems of collaboration in the workplace. In addition, the book includes 20 exercises and 25 writing assignments that can be done collaboratively. The

sample proposal and feasibility study are both examples of collaborative projects.

- **A new section on corporate culture.** The Introduction defines the concept of corporate culture and discusses its importance for students seeking their first professional positions.
- **Information about workplace writing from 20 professionals in a variety of technical fields.** I interviewed professionals from across the United States for this new edition. Their comments on the writing process and the forms of technical writing introduce most chapters in the text. Whatever their titles, all these professionals write as an important part of their jobs. By recounting their real-world experiences, they help students understand the value of what they are learning.
- **Guidance for preparing seven types of short reports.** The chapter on memorandums and informal reports is expanded to include information on occurrence reports, trip and conference reports, work activity reports, meeting minutes, investigation (field and laboratory) reports, information reports, and archival reports.
- **New writing and graphics examples that deal with current concerns of technical writers.** The proposal chapter now includes examples of both informal and formal proposals; the recommendation-report chapter contains a new feasibility study; and the chapter on research and project reports highlights portions of one scientific report and one industry report.
- **New writing topics and techniques in the contemporary workplace.** These include information on grant proposals; storyboarding; portfolio preparation and interviews; manuals, quick reference cards, and online instruction; laboratory reports; indexing; nonsexist language; and electronic communication.
- **Highlighted checklists at the end of each major section or chapter.** These checklists summarize the preceding information and allow students to apply what they have learned to whatever document they are writing.

Features Retained from the First Edition

With all its changes, the second edition of *Introduction to Technical Writing* retains these characteristics from the first edition:

- **A friendly and supportive tone, clear definitions of terms, and jargon-free explanations.** Students can read and understand text information on their own, freeing classroom time for discussion, application, and workshop.
- **A three-part structure covering the process, techniques, and forms of technical writing.** This structure provides flexibility in designing a syllabus. You can emphasize long or short writing assignments, reports or instructions, or any combination, and the text will work well for you and your students.
- **Emphasis on practice, with more than 125 exercises and 28 writing assignments.** No other text gives students so many chances to practice what

they are learning. In many chapters, exercises immediately follow the explanation of theory so students can apply what they have learned.

- **Emphasis on planning, scheduling, and organization.** As students begin to write in their chosen professions, prewriting tasks become increasingly important. *Introduction to Technical Writing* shows students how to plan and organize both long and short documents.
- **Stress on review, revision, and editing.** Students learn how to review and revise their own or others' documents. An extensive editing chapter and the handbook provide explanations of likely trouble spots and exercises to apply editing skills.
- **Chapter organization that promotes understanding.** Each chapter previews the contents with a list of significant topics, defines the relevant terms, divides each topic into manageable portions, and includes checklists that summarize and apply section and chapter content.
- **A wealth of examples.** The examples throughout the book include student and professional work. Both are understandable to general readers; that is, they are technical without being incomprehensible.

Organization of the Book

Introduction to Technical Writing begins with an extensive introductory chapter called "Technical Writing and the Workplace." This chapter explains the key terms *technical writing* and *technical editing* and the associated job responsibilities; discusses the importance of technical writing in any professional's career; defines and explains corporate culture; stresses the importance of collaboration and explains the ways professionals collaborate in producing documents; and introduces the 20 professionals whose experiences and observations are woven throughout the book. The chapter closes by previewing the contents and explaining how students can best use the text's information.

The text itself is divided into three parts plus a handbook. The eight chapters of Part One explain a general process students and professionals can follow in writing workplace documents. The process chapters elaborate on prewriting, writing, and revising tasks but are nonprescriptive, recognizing that individual writers approach the process in different ways and that the process itself is recursive. With the current emphasis on the legal and ethical responsibilities of professionals, those considerations are included in the process.

Part One divides the process into these eight chapters:

1. Planning
2. Considering Legal and Ethical Responsibilities
3. Gathering Information
4. Organizing Information
5. Writing the Draft
6. Making Information Accessible
7. Reviewing and Revising the Document
8. Editing

Each chapter includes long-term and short-term writing assignments, exercises, and checklists. These chapters on the writing process can be assigned in sequence or in combination with other chapters covering writing techniques and forms of writing.

Part Two includes five chapters, two of them new, that explain specific techniques writers use in a variety of documents. These chapters are as follows:

9. Defining and Comparing
10. Describing Objects and Places
11. Illustrating with Graphics
12. Designing the Document
13. Documenting Your Sources
14. Writing for International and Multicultural Audiences

Because these chapters cover discrete topics, they can be assigned in any sequence and paired with a variety of writing assignments. Each chapter includes checklists, exercises, and writing assignments. You could assign these chapters for major writing tasks, as support reading, as reference, or for oral reports.

Part Three covers the 10 major forms in which technical writing most often appears, with *form* meaning the way the document is organized and presented. *Introduction to Technical Writing* recognizes that certain forms have evolved because they successfully fulfill specific document purposes and meet specific reader needs. Students can learn from this book how workplace writing differs from academic writing and how they can apply general composition skills to the demands of their profession. Each chapter defines a form, shows how it is organized, provides several examples, and explains how to adapt the general process explained in Part One to this particular form. Part Three consists of the following 10 chapters:

15. Memorandums and Informal Reports
16. Business Letters
17. Résumés and Job-application Letters
18. Instructions and Manuals
19. Procedures
20. Abstracts and Summaries
21. Proposals
22. Comparative, Feasibility, and Recommendation Reports
23. Research and Project Reports
24. Speeches and Oral Presentations

The appendix contains a handbook that is a reference source for the most common student questions on punctuation, abbreviations, number use, capitalization, and spelling. Students can find answers to their questions and can also complete exercises that will help them apply their new knowledge. The handbook alone contains 15 exercises.

The text's three-part structure will give you flexibility in both reading and writing assignments, and the instructor's manual suggests possible syllabi for 10-week and 15-week courses. Depending on your students' majors and the demands of your local workplace, you can choose long or short assignments; reports, correspondence, or instructions; or any combination of writing types. You may want to assign a long-term project like a proposal, report, or manual that can include many shorter writing assignments: memos, letters, status reports, definition, description, documentation, graphics, document design, and preparation of a long document.

Ancillary Material

Ancillaries to the text include the instructor's manual, transparencies, and computer software: *Exercises for Revision and Design* and the *St. Martin's Hotline for Technical Writing*. The instructor's manual provides tips on course planning, syllabi for 15-week and 10-week terms, suggested approaches for teaching each chapter, and answers to the exercises. It also includes a pretest for assessing student skill in punctuation. A packet of transparencies is available upon adoption for classroom use in highlighting key concepts in the text. The disk of *Exercises for Revision and Design* allows students with computers to edit sample documents by applying what they learn about purpose and organization from reading the text. The *St. Martin's Hotline for Technical Writing* is an online handbook, compatible with most word-processing programs, that provides immediate answers to questions of style, grammar, and usage. Upon adoption of the text, instructors are authorized to make copies of the software for their students. If instructors prefer, they may order copies of the software for students to purchase through the bookstore.

Acknowledgments

In writing this second edition of *Introduction to Technical Writing*, I have benefited from the advice and help of many people—in industry, in the classroom, and in colleges and universities across the country. My thanks to all of you for your expertise, encouragement, and support.

For their in-depth review of the first edition and their excellent suggestions for revision, I am grateful to Penny Hirsch, Northwestern University; D'Wayne Hodgin, University of Idaho; and J. Fred Reynolds, Old Dominion University. I also wish to thank Erik A. Thelen, University of Wisconsin—Milwaukee, for his insightful review of the chapters new to this edition.

In addition, the following instructors helped me by responding to a questionnaire about the first edition and by telling me how they used the text in their teaching. My thanks to George Braine, University of South Alabama; Cynthia Butler, Florida Institute of Technology; Marjorie T. Davis, Mercer University; Melody DeMeritt, California Polytechnic State University; Donna Dowdney, DeAnza College; John W. Fiero, University of Southwestern Louisiana; Hedda A. Fish, San Diego State University; Kenneth E. Gadomski, University of South Alabama; Gary C. Huested, Florida Institute of Technology; Sharon Irvin, Florida Institute of Technology; Colin K. Keeney, University of Wyoming; Sherry Little, San Diego State University; Paula Ložar, DeAnza College; Gretchen Nordstrom, University of Wyoming; Ruth Perkins, Chemeketa Community College; Michael Piotrowski, University of Toledo; Beverly J. Reed, California Polytechnic State University; Judith Rehm, Embry-Riddle Aero University; Anne M. Rosenthal, San Jose State University; Ann Martin Scott, University of Southwestern Louisiana; Skaidrite Stelzer, University of Toledo; John S.

Terhes, Chemeketa Community College; Mary Beth Van Ness, University of Toledo; C. Wambeam-Carpenter, University of Wyoming; and Elizabeth White, California Polytechnic State University. The following DeAnza College students also responded to a questionnaire: Susan K. Boyd, Patricia F. Johnston, Carol Moore, and Elizabeth Plowman.

At San Jose State University, my colleagues Virginia De Araujo, John Galm, Nettye Goddard, Hans Guth, Denise Murray, and Patricia Nichols encouraged me throughout the process. In the office, Mark Bussmann and my graduate assistant Patrick Nolan supported my work with great good cheer. Judy Reynolds from Clark Library gave valuable research guidance, and the staff in Government Documents gave freely of their time and materials. Aviation professors Eugene Little and Richard Le Clair, engineering professors Patrick Pizzo and Joseph Wujek, and Martin Schulter of Disabled Student Services provided subject-matter expertise. Technical writing instructors Cindy Baer, Carol Gerich, Marie Highby, Diane McBurnie, and Anne Rosenthal provided student examples and classroom-based advice.

Special thanks go to my friends in industry who keep me abreast of current trends and needs. Among them are Carl Jones, Susan Constable, Norm Burchard, Marie Hofmeister Goodell, Jackie Turner, Judy Wilbur, Bev Lewis, Tom Dewan, J. B. Saint-Leger, Erich David, and Roger Connolly. The 20 professionals who answered my many questions about writing and their work have contributed valuable information and helped me personalize the writing process. In many ways, their responses form the core of this book. Thanks to Gordon Anderson, Suzanne Birdwell, Roger Connolly, Carol Gerich, Kris Lundquist Frank, Judy Hopkins, David Irwin, James Jahnke, Craig Johnson, Stacy Lippert, Michelle Marchant, Ann Petersen, Jamie Oliff, Pamela Patterson, Kate Picher, Anne Rosenthal, Mary Shah, Antonia Van Becker, Jeffrey Vargas, and Karen Zotz.

My students continually delight and enlighten me. In my writing and editing classes, my undergraduate students have helped me see the text through their eyes, so I could rewrite effectively. In the graduate seminars, students acted as my reviewers on portions of the new chapters and as resource persons for industry concerns. Special thanks to those students who contributed examples, who are acknowledged by name in the text. Additional thanks to the following: Maggie Black, Jane Bratun, Kathy Cline, Tony Cyphers, Daniel Franks, Rick Geimer, Anne Gmelin, Doug Mendoza, Jennifer Rhodes, and Gayle Van Briggle.

At St. Martin's Press, my thanks to Denise Quirk and Emily Berleth, who provided excellent editorial support.

And throughout the process, my husband, Bob, has been my exacting critic and chief supporter. His hours of computer work have helped make this book possible. Again, this book is for him with my love and thanks.

Brief Contents

Contents

Chapter 2   Considering Legal and Ethical Responsibilities   56

Chapter 6   Making Information Accessible   158

Chapter 7   Reviewing and Revising the Document   177

Chapter 8   Editing   186

Part Two	Techniques in Technical Writing

Chapter 9	Defining and Comparing 221

Chapter 10	Describing Objects and Places 234

Chapter 13 | Documenting Your Sources   298

Chapter 14 | Writing for International and Multicultural Audiences   313

Chapter 17   **Résumés and Job-application Letters   376**

Chapter 23   Research and Project Reports   546

Chapter 24   Speeches and Oral Presentations   567

Introduction to Technical Writing

Process and Practice

Second Edition

Introduction

Technical Writing and the Workplace

If you are reading this chapter, chances are that you are enrolled in a class called technical writing, engineering communication, scientific and technical writing, or some variation of that title. You may be studying technical writing because you are in a technical or scientific major, or because you see technical writing as a possible career choice. In either case, if you are like many of the students who appear in my technical writing class at the beginning of a term, you have some questions about technical writing, its importance in the workplace, and its value to you as a professional. Specifically, you may want to know:

- What is technical writing and what does a technical writer do?
- What is technical editing and what does a technical editor do?
- How much writing will I have to do on the job?
- Why should I care about my technical-writing skill?
- How is the working environment different from the college environment?
- How can this book help me?

All those questions are important, and you deserve honest answers. This chapter is designed to provide answers to those broad questions: to define the terms, examine the requirements of the workplace, explain corporate culture and workplace collaboration, and introduce you to 20 professionals from all over the United States who are working in a wide variety of occupations. In preparing this edition of *Introduction to Technical Writing*, I interviewed these 20 men and women, and their comments will open a window for you on the working world.

As you read and complete assignments, think of this book and your instructor as coaches. Our job is to help you build on the skills you already possess, introduce you to writing processes that will work for you, and give you practice in many types of technical writing. Practice, of course, is hard work—whether you are swimming lengths to build your stamina, singing scales to improve your accuracy, or writing and revising to communicate effectively. As you practice, remember that coaches are not judges but allies—people who want to help you write better and who will cheer you on to victory. We want you to succeed, and we know that you can.

To begin, focus on the meaning of the key terms and job responsibilities in the workplace.

---

1.   Key Terms and Job Responsibilities

Technical Writing

In the 1990s, the term *technical writing* may be too narrow to cover the kinds of communication it attempts to describe. Technical writing includes reports, proposals, manuals, procedures, and instructions. Documents like letters, résumés, memos, and sales brochures can also be called technical writing. Sometimes business writing and science writing are taught as separate subjects, but as you can see, there is considerable overlap in the content. Scientists must write letters and memos; engineers must report research; and business managers must prepare proposals. In fact, one expert in the technical-writing field claims that we need a new name for the kinds of communication professionals produce on the job. According to Professor Carolyn R. Miller (1992, 117), "There are no distinctive features of technical writing that set it clearly apart as a species from business writing—not purpose, techniques, rhetorical stance, forms, readers." While Miller does not propose a new term, she does suggest that what we are talking about is *workplace writing*.

In this book, I will continue to use the terms *technical writing* and *technical communication*, but you should understand that I am using those terms in the broadest sense. With that caution, here is a definition that clarifies technical writing.

> Technical writing is the communication of specific information to an identified reader so that the reader's understanding matches the writer's intention. The writer's responsibility is to make the communication accurate, clear, complete, concise, well organized, and correct.

If you break this definition into its parts, you can get a better idea of what it means.

1. The subject matter is clearly identified. It may be from the realm of science or technology, or it may be from law, education, or business.
2. The writing is directed to specific readers who seek information or advice, and who often will make decisions based on what is said.
3. The writing succeeds only if it is understood by the readers; the writer and the readers have, in effect, a contract. Therefore, what you write should be
   - *accurate:* supported and verifiable
   - *clear:* having only one meaning for the reader
   - *complete:* containing all the necessary information
   - *concise:* economical and direct
   - *well organized:* accessible through the logical relationship of ideas, short individual sections, and good headings and layout
   - *correct:* free from errors in form, grammar, usage, and punctuation

One useful distinction is that existing between technical writing and *academic writing*. As a student, you have already written essays for English classes and reports for classes such as history and sociology. In freshman composition, you may have generated several drafts of an essay in order to determine what you really wanted to say: an essay of self-discovery. In history, you may have written a report to prove you had researched the topic thoroughly or analyzed the problem and reached conclusions. In both types of writing, the intended reader was probably your instructor, who already knew more about the subject than you did and was reading your work, not to gain new knowledge or advice, but to evaluate your accomplishment.

Notice that what we call technical writing is different: In technical writing, the primary reader is not the instructor but a specific person in the workplace who needs your information. The purpose is not to prove you have done the work, but—very specifically—to define, describe, instruct, report, or recommend action. This is the kind of writing done everyday by professionals in the workplace.

The Job of a Technical Writer

Virtually all people who work in science, business, law, and technology—like civil engineers, research biochemists, operations managers, nurses, cattle breeders, and programmers—must write as a part of their jobs. In addition, a small group of people are full-time, or professional, technical writers. According to Keith Kreisher of the Society for Technical Communication (quoted in Kleiman 1984, PC 1), their primary job is "to bridge the gap between the technology of the product and the user, to make that technology understandable to the average person."

As computers and other complex technical tools move into the general market, writers are needed who can explain what the tools are and how to use them. Many technical writers have majored in English, liberal arts, or journalism; others come from technical fields but discover that they like to write and choose to do it full time. Technical writers need a feel for language plus excellent writing and editing skills. In addition, they need experience or coursework in the appropriate technical field and familiarity with word processing. According to Christine Browning, manager of a software manual writing group at Tandem Computers (quoted in Kleiman 1984, PC 1), "Technical writing is the best writing job you can have. It's the only one that gives you security. If you can write well and know the science you are writing for, you can get a good job."

The job market for technical writers is not only strong, but growing. In a 1991 article in *Money* magazine, author Diane Harris said, "The skills most in demand today are computer experience, technical writing, managerial prowess, and foreign languages, especially those spoken in Eastern Europe as well as Japanese and German" (133).

What's more, technical writers are well paid; for example, in most parts of the country, entry-level salaries are only slightly lower than those for engineers. The profession is demanding, however. According to technical writer Chris Klemmer (1984, 1), "You should be able to work under pressure and against deadlines. Most projects run late, and the writer is called in at the eleventh hour." In addition, Klemmer says, "You should be able to take criticism. The words you thought were so golden may not seem so to your superior. Be willing and gracious about rewriting."

Technical Editing

Technical writing is only one part of the process of producing a workplace document. Technical editing is another; it involves looking critically at the written document for accuracy, completeness, correctness, and effectiveness of presentation. You must always edit what you write, stepping back and reviewing the draft with a critical eye. Editing means correcting grammar and punctuation, but it also involves checking facts and assessing and revising organization. Above all, it means guaranteeing that the reader will understand the writing to mean only what the writer wants it to mean. Editing may even require selecting graphics, compiling a glossary, and designing each page in preparation for printing.

Like technical writing, then, technical editing includes many tasks. As a student writer or as a writer in a small company, you usually must be your own technical editor.

Be aware, though, that writing and editing tasks are different and that you should not try to do them at the same time. Writing (even technical writing) is creative, so when you write, you need to shut your critical eye for a little while. Don't worry about spelling, punctuation, or choosing the exact word. Let the words flow and get the main ideas down on paper. Then, with a completed draft in your hand, slowly open your critical eye to evaluate what you've written. When you edit, you pick apart and polish, check facts, rearrange sentences, correct spelling and punctuation. As one technical editor (quoted in Cavanaugh 1985, PC 12) put it: "The difference between a writer and an editor is that a writer creates something from nothing, while an editor analyzes and works with that creation." This book will teach you both writing and editing skills and tell you at what points you should become an editor.

The Job of a Technical Editor

Although you may edit your own material on the job, in many companies you will work with a professional technical editor. Often it's easier for a trained outsider to review a draft than it is for the writer. In addition, editing itself involves many specific skills, and not all scientists and other specialists have the training or the time to learn them. Many editors have majored in journalism, linguistics, or English and have excellent grammar and punctuation skills, an interest in language, concern for detail, and the ability to work well with people. In many companies, editing is an entry-level position that can lead in time to appointment as a writer. In others, editing is a high-level job because editors work with writers to shape the finished documents.

2. Technical Writing and Your Career

While some people are professional writers, it is safe to say that most people in the workplace are professionals who write. If you are preparing for a career in engineering, agriculture, law, consumer affairs, business management, scientific research, or any of the hundreds of other jobs that require a college degree, chances are that you will spend a significant portion of your workday speaking, reading what colleagues have written, or writing.

That writing begins as you apply for your first position. You already know that when you apply for a job, you must fill out an application form, prepare a résumé, and write a letter of application. All the employer knows about you is what's on those pages, nothing more. That means you will be judged by what you say and how you say it. One executive who is responsible for 100 engineers put it this way (McAlister 1984, 47): "We consider the ability to communicate to be as important as grade point average when we interview engineering graduates." The district manager for a large chemical company in Illinois said this about an application he rejected (Galm 1977): "This application is pretty pathetic in terms of language usage . . . The young man applying for a position with us will probably not be hired as a sales engineer because the ability to communicate is even more important than his technical skills and other fine qualities. This young man obviously has a lot going for him and would be just the kind of person we would like to hire—except that he obviously can't write an adult sentence."

When you write a résumé and job-application letter, you are using your persuasive, organizational, analytical, and even document-design skills. If these documents win you an interview, you can sometimes enhance your chances of being hired by bringing along samples of your writing—especially if those samples are neatly arranged in a portfolio that showcases your work.

Once you have secured a position, you will find that one of the ways you increase your visibility to managers is through your writing. As one supervisor in a research and development company noted (quoted in Barabas 1990, 8): "Even if our company did not require written reports, I would still insist that my people write them, for the sheer act of writing (and knowing that their reports will be read by management) forces them to reflect upon the significance of what they are doing, and that is really the main reason they are hired—to think and solve problems." In addition, since writing is a permanent record, it is often used at employee evaluation time. An often-cited survey of a research and development group in an oil company (MIT 1984, 27) revealed that "evaluations in over half the categories in the annual . . . performance reviews are based at least partly on an individual's writing."

One reason writing is viewed as important in evaluations is that professionals do so much of it. The U.S. Air Force, for example, estimates that it turns out 500 million pages of writing per year. "For the U.S. Navy," according to Nancy G. Wilds (1989, 189–90), "a vice admiral recently calculated that, on one of the smaller surface-warfare ships, the day-to-day paperwork and the file cabinets to hold it added 20 tons to the weight of the ship."

In the civilian sector, "the aerospace industry predicts that the planes of the 1990s will be documented by million-page manuals, a substantial increase from the 250,000 pages in the 1980s and the 1,000 pages of the 1940s" (Little 1990, 28).

Another way to calculate the importance of writing in the workplace is to look at the amount of time professionals spend doing it. The oil company survey (MIT 1984) indicated that all the research and development professionals spent a substantial portion of their time on technical communication, whether writing, reading and editing, or speaking. Managers spent 50 percent of the work week on communications, supervisors spent 40 percent, and staff members spent 35 percent. More recently, Notre Dame law professor Teresa Godwin Phelps (1989, 364) reported, "Most lawyers spend at least 50 percent of their time writing, and those in their first five years of practice can expect to spend 75 percent of their time writing. To be a lawyer means to be a professional writer, and lawyers write a variety of documents, each with its particular rhetorical situation."

My own interviews included 12 professionals who bear titles other than "writer." They work as engineers, consultants, analysts, planners, and project directors, and they reported spending an average of 40 percent of their time writing. The range, however, was very broad. Dave Irwin, a director of manufacturing operations, said he spent only 1 percent of his time writing (but 80 percent speaking to large and small groups), while environmental consultant Jim Jahnke and university extension specialist Karen Zotz reported their writing time as 80 and 85 percent of the work week. Three different kinds of engineers (software, civil, and design) averaged from 10 to 20 percent of the work week in writing tasks.

Fortunately, when you write in the workplace, you have access to a number of tools that make the writer's job easier. In nearly every occupation, computers have replaced typewriters, and sophisticated software and hardware have greatly simplified the production of text. But you must always remember that computers, laser printers, and graphics software are simply tools. By themselves they will not generate text, nor will they ensure that the words they print out (however beautifully) are clear, organized, and correct. That responsibility lies with the writer, and it is a serious responsibility. A wrong statement in a set of instructions, for example, could damage equipment or seriously injure the operator. In addition, that error could lead to an expensive lawsuit.

In other words, you are responsible for what you write. As the chairman of a large forest-products company said: "If your writing has even a minor error in it, your reader may question how much care and thought really went into it. And be sure to punctuate carefully. Even a single misplaced comma can change the whole meaning of a sentence. Grammar, spelling, and punctuation are the cornerstones of good writing and should be carefully considered at all times" (Hahn 1985, F 7).

What's more, you can't depend on the department secretary to type and correct rough drafts, for secretaries have nearly disappeared. It is not unusual for a department to have 100 professionals and only two secretaries. When you have your own computer terminal, you'll do your own typing, and you will be responsible for organization, accuracy, and correctness.

3. Corporate Culture and the Workplace

As a student preparing to enter the workplace, you also need to know that when you graduate, you will be moving from one so-called culture (the school setting) to a different culture (the organizational or corporate setting). In order to be an effective writer in that new setting, you need to understand what *corporate culture* is in general, and what the corporate culture is at the place where you work.

Corporate culture can be simply defined as "the way we do things around here" (Barabas 1990, 80). More specifically, corporate culture involves the "patterns of thought, belief, feelings, attitudes, behaviors, products, and publications that result from the experiences and common goals shared by people working in the same organization" (Lutz 1989, 114).

The thoughts, attitudes, and beliefs of the company itself and the people you work with are important, because to be happy and successful in that setting you need to be comfortable with those attitudes and beliefs. Corporate cultures can vary enormously. Fourteen of the 20 professionals I interviewed described their corporate cultures, with the descriptions ranging from "formal and bureaucratic" to "very relaxed" to "decentralized and benevolent." Some companies were viewed as having a "highly technical environment with tight deadlines" or "a concentration on doing things better, faster, and cheaper." Others were described as having corporate cultures that were "flexible, giving each person an area of responsibility with a degree of independence" or "extremely team oriented, with employees regarded as a family." One person said that the organization had "no single corporate culture, but 'cluster cultures,' where people with similar responsibilities informally define standards, congratulate or chastise each other, and have a jaundiced view of the other cultures."

Why is corporate culture so important to you as an employee? Researcher Christine Barabas (1990, 77) says that "problems can occur when there is a mismatch between an employee's character and that of the company, as for instance, when individuals who are by nature or experience rational, regimented, and mechanistic find themselves in an organization where inspiration, creativity, and innovation are the prevailing values—or vice versa." That statement warns that you should ascertain—as well as you can—the culture of an organization before you accept a job. Indeed you should, but for you as a writer, the kind of corporate culture has additional constraints.

Corporate culture can be conveyed internally and externally by employee dress, architecture and furnishings of buildings, social events like Friday afternoon "beer busts," and even the company logo. But communication—whether advertising, manuals, business letters, or reports—is the primary tool "through which members participate in an organization, through which nonmembers learn about an organization, and through which a corporation's climate and image are codified" (Lutz 1989, 114). As someone who will be writing within and for the company, you have to learn both acceptable formats (how documents are organized and designed) and company goals (how the company sees itself and wants others to see it).

Does this mean that you have to compromise your own views or ignore principles you have learned in school? Not compromise, but certainly clarify. Writing—the act of putting policies or decisions in words—also has the important function of crystallizing issues and forcing people to think about and take responsibility for what they are doing or ought to be doing. As Chapter 2 of this book points out, you have legal and ethical responsibilities to yourself.

However, you also must realize that for your writing to be effective, it must fit organizational designs and concepts. Barabas says, "Writers need to understand the interrelationship among individual, departmental, and corporate goals, and which of these, given the document's intended readers, most needs to be addressed" (1990, 54). As L. P. Driskill and J. R. Goldstein put it, "What is most crucial for the writer to supply is what the reader must know but does not know, must feel but does not feel, must believe but may not believe. In this gap lies the definition of the document's necessary content and persuasive challenge" (1986, 44).

Sometimes you can be guided in what you write by a company style guide; often, you will be given models on which to base your own writing. But you can also contribute to the shaping of the corporate culture by what *you* write—a task to which you bring your own background, education, and ideas. Thus corporate culture works both ways: you are influenced by the place where you work, and you can exert an influence on that culture by what you write.

4. Collaboration in the Workplace

As a student, for many years you have been a somewhat solitary player, mostly responsible for your own success by the work you did. You probably have participated in panel discussions or been a staff member of a publication; you may well have served on a classroom committee of some kind. But most of your communication activities—writing or speaking—have been done by you alone.

When you move into the workplace, however, chances are that you will find yourself engaged in collaborative activities—both in carrying out your professional activities and in writing about what you did. The word collaboration has obvious roots: it comes from the prefix *co* or *col*, meaning "together" and the Latin word *laborare* meaning "to work." Alternate terms for collaboration include *team writing, group writing,* and *project teams.* Whatever the size or orientation of the group, collaboration means that people are working together cooperatively.

In advanced writing classes, many instructors will ask you to work as part of a team because they know how prevalent collaboration is in the workplace, and they want to prepare you for teamwork. Two statistics are worth considering. Researchers Andrea Lunsford and Lisa Ede reported in 1986 that in a survey of 12,000 members of six professional organizations, 87 percent sometimes wrote as part of a team. In my own interviews with 20 professionals from across the United States, I found that most of them reported spending approximately 25 percent of their time in collaborative writing.

There are four main types of collaboration in writing, and—depending on your job—you may well engage in all four at one time or another. The four types are

- collaborative planning
- coauthoring or team writing
- division of duties by job specialty
- collaborative review and revision

Collaborative Planning

Before undertaking a large writing project, many organizations gather a team to decide on the purpose, intended reader, schedule, and planned organization of the document. Planning meetings like this take advantage of the collective creativity sparked by group discussion: A kind of synergy develops in the give-and-take of discussion, which means that many ideas are generated and many angles to solutions can be discovered. Team members can come from different areas of product expertise (for example, engineering, programming, quality control, marketing, publications) or can have different functions in production of the document (for example, subject-matter expert, writer, editor, manager, graphics designer).

Planning meetings can settle the broad issues of purpose and organization; decide how the document will be written and reviewed; and—perhaps most important—assign areas of responsibility and set the schedule.

Once the team has finished planning, the writing itself may be done by one person or several people. Judy Hopkins, for example, who works for a probation department, reported that "the last grant application I wrote was collaborative in the formation of the idea, though I was the sole writer." Civil engineer Gordon Anderson, by contrast, says that after planning, "Each person is responsible for a particular section of a report."

Coauthoring or Team Writing

A second kind of collaboration involves several people in the draft-writing process. This kind of collaboration works in one of two ways: (1) writers sit down together at a computer and literally compose together, generating text to which they all contribute, or one writer drafts text and sends it by E-mail or modem to other writers who modify, condense, expand, or comment on the text, and (2) writers divide the planned document into sections with different people responsible for different sections. For example, Karen Zotz, an extension specialist for a major university, reports that in preparing a draft, "responsibilities are divided according to a person's expertise. Then everyone gets together and shares their work for critique."

In coauthoring, members of the group need to decide in the beginning on the organization, tone, and general style they will use; otherwise, the finished document will have ill-matched pieces and will be confusing to the reader. One way to help solve problems of different styles is the technique called *storyboarding*, which is explained in Chapter 21. Storyboarding allows for different writers' contributions while still maintaining a consistent organization and style.

Division of Duties by Job Specialty

Another kind of collaborative work divides duties by job specialty. In a high-tech environment, for example, a "product team" may consist of research engineers, development engineers, support engineers, product marketing engineers, and the learning products engineer (the writer of the instruction manual). Each person on the team brings his or her expertise to the task: The research and the development engineers are the technical experts, support engineers provide problem-solving and troubleshooting information, marketing engineers determine market requirements, and the learning products engineer looks at readers' needs and writes the needed documentation. It would be difficult for one person to have expertise in all these areas. In addition, with collaboration, development of the product and the supporting documentation can proceed concurrently. Learning products engineer Jeffrey Vargas says, "While R&D engineers develop the software or hardware, the writer develops the 'learning products' for users, all the while providing input to make the hardware or software more usable."

In the field of air-pollution monitoring and control, consultant Jim Jahnke works with a different kind of team: He himself is the subject-matter expert and the writer; other team members include a technical editor, an instructional designer, and a graphic artist. Jahnke notes, "Technical editors can be very helpful because they can see if a sentence or paragraph holds together. A good editor can point out when an exposition is not complete, when more descriptive information is needed, or whether the material is padded."

Collaborative Review and Revision

Whether a document is written by one person or a team, collaboration is common throughout the workplace at the review and revision step. Sometimes this review and revision is done through formal document review—a scheduled meeting or meetings to which interested parties are invited. Copies of the draft are circulated before the meeting, and participants are expected to study the draft carefully, making notes about technical, organizational, or clarity problems. At the meeting, a discussion leader (usually not the writer or writers, who take notes) moves the team members systematically through the document, resolving issues as they go. Software engineer Mary Shah talks about this kind of collaboration: "Currently, we are implementing an inspection process. Representatives of different disciplines (testing, programming, documentation, and so on) get together to 'inspect' a document. This inspection is to identify any major flaws that need correcting. The aim is to create an accurate and clean document."

Collaborative review can also be done informally by circulating a draft to interested parties and asking for written corrections and comments. If the draft is on the computer, some software programs allow split-screen online conversations between the writers and the reviewers; others allow reviewers to add written comments that the writer can later incorporate (or not) in preparing the final document. Medical marketing writer Jamie Oliff uses this kind of review: "The writing is done individually, but is 'proofed' by many people who work on specific products. Then there is considerable rewriting to ensure scientific and medical accuracy."

Disadvantages of Collaboration

As you can see, collaborative writing is common in the workplace. Because many people are involved, collaboration can cause problems, but if you recognize the potential for problems, you can plan around them.

Two potential problems in collaboration are related directly to the production of a document. First, collaboration may take more time than solo writing. Team members must allot time in their individual work schedules to plan, write, review, and attend meetings. If one person falls behind, the whole project can be delayed. Second, when several writers contribute, a document may appear to be a patchwork of mismatched sections; different audiences may be addressed, different purposes may appear, and different styles of writing may clash. Solutions involve careful planning; frequent meetings to discuss goals, readers, and organization; and meticulous review and revision to ensure a seamless document.

Other problems can arise from personal relationships. Writers often feel that they "own" their writing; they are protective and sometimes defensive when faced with comment and criticism. Technical experts sometime stray beyond their subject-matter expertise and try to dictate construction of sentences and even punctuation. Team members may not fully meet their responsibilities, forcing others to do more than their share. Other team members may try to dominate discussion and decision-making, forcing their will on their collaborators. These problems must be resolved through conflict management and good communication. If the project is managed well, and if all team members are open and supportive, the common goal can be realized.

Advantages of Collaboration

Fortunately, the advantages of collaboration outweigh the potential problems. The biggest advantage is that each team member can use his or her abilities to their maximum potential. Bruce Boston, editor of the newsletter *The Editorial Eye*, commented this way on his collaborative experience, "Teamwork succeeded because each of us brought particular strengths to the writing process. One had the gift of seeing the whole document at all times, another was a detail expert" (1988). Another advantage is the increased pool of ideas: Collaboration builds ideas from the ground up, adding and enhancing the raw material during the planning, drafting, reviewing, and revising process. As conference speaker Marjorie Hermansen explained, "Collaboration is superior to compromise, which suggests a surrender of ideas. The notion of compromise starts with two fully formed ideas, and you take away pieces from each idea until they finally fit together—ending up with a lowest common denominator" (1990, WE-15). In collaboration, professionals work together, building—not taking away.

As a writing student, you may feel somewhat threatened by the idea that what other students do in a collaborative project will influence the grade you receive. You are accustomed to being competitive in the classroom rather than collaborative. On the other hand, you probably have previous experience in collaboration on athletic teams and group projects. If you can transfer to a collaborative writing project what you already know about making team members work together, you may find that working on a team is easier than doing it all yourself, and that the final document is actually better than one you could have produced on your own.

5.

Technical Writers and Their Work

Whether you write alone or in collaboration, your duties as a professional will probably include several types of writing. Your own situation may be unique, but you can learn from writers who are currently producing some of those millions of pages of letters, reports, proposals, manuals, and memos.

For my interviews, I deliberately chose professionals with a range of locations, job responsibilities, years in their occupation, and fields of interest. Of the 20 people I interviewed, 12 are professionals who write—that is, their job title is something other than writer. The other eight are professional writers—or editors or managers in publications groups.

Throughout this book, you will be hearing from these professionals as they comment on steps in the writing process, their own triumphs and frustrations, or one of the forms in which they write. At this time, let me introduce you to them. The "professionals who write" include the following:

- Gordon Anderson, a civil engineer in construction and construction management with about 10 years in the field.
- Mary Shah, a software engineer at a firm producing graphics for personal computers. Shah has five years' experience.
- Craig Johnson, a design engineer of electronic test and measurement equipment for 12 years.
- Jamie Oliff, a two-year employee in the marketing planning division of an international pharmaceutical company. Oliff's degree is in communication.
- Kris Lundquist Frank, the consumer affairs coordinator at a not-for-profit dairy council; Frank has 15 years experience and a degree in comprehensive home economics.
- David Irwin, director of manufacturing operations for a company producing telephone cable splicing and connecting accessories. Educated as a chemical engineer, Irwin has some 30 years in the field.
- Karen Zotz, extension specialist for the college of agriculture at a large midwestern university. Zotz has degrees in home economics education and 20 years experience.
- James Jahnke, a Ph.D. in chemistry and an independent consultant in air-pollution monitoring and control on the East Coast with 25 years of experience.
- Michelle Marchant, a corporate attorney for a computer and instruments manufacturer.
- Pamela Patterson, a policy analyst for a large aerospace company. Patterson has degrees in English and about 10 years experience.
- Judy Hopkins, project director in the probation department of a large urban county. Hopkins has worked within the criminal justice system for more than 20 years; she has degrees in psychology, sociology, and English.
- Suzanne Birdwell, regional manager of the human resources department for a company providing satellite-based telecommunications systems and services. Birdwell has a degree in business and about 12 years in the field.

The "professional writers" actually perform a wide variety of tasks. Three of them are managers:

- Roger Connolly, director of the publications center at an international computer company. Connolly has degrees in English and 13 years experience in the technical-writing field.
- Carol Gerich, a communications manager for a national laboratory conducting research in nuclear weapons design and energy resources. Gerich's degrees are in English; formerly, she was a senior scientific writer and editor for more than 10 years.

- Anne Rosenthal, technical communications manager for a manufacturer of scientific instruments in the biotechnology and biochemistry fields. Rosenthal has about six years experience and degrees in biology.

The other "professional writers" bear titles that are some company variation of technical writer. They include:

- Antonia Van Becker, senior technical writer at a semiconductor company. With six years of experience, Van Becker has degrees in English and music.
- Jeffrey Vargas, a learning products specialist at a manufacturer of measurement and computer equipment. Vargas has a degree in music with a minor in technical writing.
- Ann Petersen, a senior medical writer at a company making drug-testing equipment and assays. Petersen has seven years in the field and a degree in English with added pre-med classes.
- Kate Picher, a learning products developer at a company that builds machines to automate transactions for hardware stores and auto parts wholesalers. Picher has a degree in German and a certificate in training and human resources.
- Stacy Lippert, a senior editor at a mainframe computer manufacturer. Lippert's degree is in English; she has about six years of experience.

It is my hope that reading what these professionals have to say about their writing tasks will help you understand how you can apply the information in this book to your own career challenges.

6. The Effective Use of This Book

Introduction to Technical Writing will help you become a better technical writer and editor. The eight chapters in Part 1 guide you through processes you can follow in writing anything from a short letter to a long report. Those eight chapters cover

- planning
- considering legal and ethical responsibilities
- gathering information
- organizing
- writing the draft
- making information accessible
- reviewing and revising
- editing

For ease of discussion, each chapter isolates one step in the process. In practice, though, you'll probably find yourself dropping back to earlier steps from time to time. When you gather information, for example, you may want to return to your plans to refine them. And when you write the draft, you may have to gather more information to fill some holes. If you think of each chapter as a step on a staircase, you can see how this works. As Figure I-1 shows, the planning step *underlies* the whole staircase, so you can easily drop back to that step when you need to—and the same is true for each subsequent step in the process. Together, the steps will lead you where you want to go with your writing.

Part 2 discusses six specific techniques that you can use to develop and clarify your technical writing assignments. You may have learned some of these techniques in a more general way in freshman composition, but now you can apply them to your major field. You will learn how to

- define technical terms
- describe objects and places
- illustrate with graphics
- design the document
- document your sources
- write for an international or multicultural audience

Finally, each of the 10 chapters in Part 3 takes up one of the common forms in which technical writing appears. You will learn how to write each of those forms, and you can study both student and professional examples. Part 3 begins with the documents you'll write most often—memos and short reports—and moves to longer and more complex projects like proposals and comparative, feasibility, and research reports. Your instructor may ask you to read chapters in any order because the book is

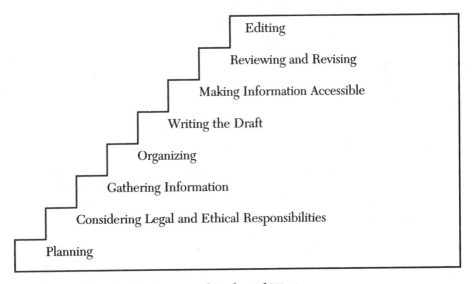

Figure I-1 Steps in the Process of Technical Writing

designed for flexible reading and writing arrangements. The ten chapters in Part 3 cover

- memorandums and informal reports
- business letters
- résumés and job-application letters
- instructions and manuals
- procedures
- abstracts and summaries
- proposals
- comparative, feasibility, and recommendation reports
- research and completion reports
- speeches and oral presentations

This book also includes many exercises to help you improve your skill in technical writing and editing. Some of those exercises are on the same topic—bicycles, for example—so you can learn how to develop a writing assignment from the first idea to a written draft. Every chapter defines the terms that are used, and the Handbook in the Appendix contains a guide and exercises for punctuation, numbers, capitalization, and spelling. Throughout the book, you'll find many cross-references to other sections of the text. In addition, you can always use the index to find specific topics.

I've talked about the importance of careful editing and of clear technical writing. Now let me be your coach, encouraging you to try your hand at it and do your best. No matter what your previous experience with writing has been, keep an open mind about the tasks you are assigned. As you read this book and learn more about technical writing and editing, and as you apply your new skills to your major field, I predict that you will learn to respect and even enjoy the technical writing process. Whatever your technical specialty may be, *you can become an effective technical writer*. With practice and a process that works for you, you can do it. Begin now by reading about *planning*, the first step in the writing process.

REFERENCES

Barabas, Christine. 1990. *Technical writing in a corporate culture: A study of the nature of information*. Norwood, NJ: Ablex.

Boston, Bruce O. 1988. *Language on a leash*. Alexandria, VA: Editorial Experts.

Cavanaugh, Wanda. 1985. Untwisting their words. *San Jose Mercury*, 3 March, PC 12.

Driskill, L. P. and J. R. Goldstein. 1986. Uncertainty: Theory and practice in organizational communication. *Journal of Business Communication* 23, 41–56.

Galm, John. 1977. Personal letter. San Jose State University. 11 July.

Hahn, T. Marshall, Jr. 1985. Don't let bad writing damage your career. *San Jose Mercury*, 8 May, F 7.

Harris, Diane. 1991. How you can live better. *Money*. October, 133.

Hermansen, Marjorie S. 1990. The writer/editor relationship: Collaboration—a step beyond compromise. *Proceedings*. Washington: 37th International Technical Communications Conference, WE-15.

Klemmer, Chris. 1984. *STC Connection*, March, 1.

Kleiman, Carol. 1984. Computer boom helps give lift to technical writers. *San Jose Mercury*, 24 June, PC 1.

Little, Sherry Burgess. 1990. Preparing the technical communicator of the future. *IEEE Transactions on Professional Communication*, 33 (March).

Lunsford, Andrea and Lisa Ede. 1986. Why write . . . together: A research update. *Rhetoric Review* 5:71–81.

Lutz, Jean Ann. 1989. Writers in organizations and how they learn the image. In Carolyn B. Matalene, ed. *Worlds of Writing*. New York: Random.

McAlister, James. 1984. Why engineers fail. *Machine Design*, 23 February, 47.

Miller, Carolyn R. 1992. Textbooks in focus: Technical writing. *College Composition and Communication*. 43. Feb.

MIT Industrial Liaison Program. 1984. *Communication skills: A top priority for engineers and scientists*. Report No. 32929.

Phelps, Teresa Godwin. 1989. In the law the text is king. In Carolyn B. Matalene, ed. *Worlds of Writing*. New York: Random.

Wilds, Nancy G. 1989. Writing in the military: A different mission. In Carolyn B. Matalene, ed. *Worlds of Writing*. New York: Random.

Part One

The Process of Technical Writing

Chapter 1

Planning

"My primary job is to make order out of chaos." Kate Picher is a technical writer at a computer manufacturing company, but she could be any writer faced with a writing project and a deadline. At the very beginning, all writing projects seem chaotic.

When *you* are the writer, your job will be to take an assignment and some abstract ideas about that assignment and transform them into a well-planned document that meets the reader's needs. The key to successfully completing a writing assignment is in the word *well-planned*. Any writing project takes careful and detailed planning; before you begin to write, you need to spend time thinking about your document's purpose and your reader's needs, and balancing those essential requirements with matters of length and coverage, tone, form, and available time.

Such planning abilities do not come naturally; you will need to learn planning skills and consciously practice them. "Writing is not an elementary skill, like riding a bike, divorced from thought and content," says probation officer Judy Hopkins. "It has an essential connection to learning and the expression of knowledge. It is not a ceremonious way of demonstrating facts, but a way of showing their value." Hopkins is a project director in a large county criminal justice agency, and she spends more than 50 percent of her time writing. Her writing assignments are very different from those for a computer manufacturer, but her approach to the planning process mirrors Picher's. "Always keep your overall objective in mind as you do each part of a report," Hopkins says. "Ask 'What am I to make of this?'" Picher's words of advice are an echo: "Take your time with the analysis."

This chapter will help you learn some planning techniques by breaking planning into six activities. Whether you are writing a half-page memo or a 50-page report, you will want to carry out each of these planning activities before you begin to write.

1.1 Determining the Purpose

The first question to ask in planning a technical-writing project is "What is my goal or purpose in this piece of writing?" As a student, determining a writing purpose may be new to you. Until now your primary goal has probably been to pass a writing class with a decent grade. To do that, you tried to please your instructor with each assignment, but the purpose of the document itself was not important to you. As a working professional, though, writing will be your way of communicating information and recommendations to other people. If you do not understand your purpose before you write, your reader will not understand what you are trying to accomplish.

For example, suppose your manager says: "Contact Laura and Bill over in the design lab and find out how that new control box they're working on is going to affect our power consumption. Then write it up for me. I need some solid information for my meeting on Monday."

If you analyze your manager's request, you can see that your assignment is to determine and then *explain in writing* the impact of the new control-box design on power consumption, an explanation that may involve your *evaluation, judgment,* or *recommendation*. If you write a description of the control box, if you tell how the control box operates, or if you compare the new control box to an existing one, you will have failed in your writing project because you did not achieve your assigned purpose.

What's more, your manager will not have the solid information needed for that Monday meeting.

How might an accurate purpose statement for this assignment look? Perhaps like this:

> This memo explains the electrical power requirements of the newly designed A26 controller and the consequent need for changes in our cabling.

A Definition of Purpose

The *purpose* in writing is the goal, the intended or desired result: in other words, the reason for doing the writing.

How do you discover that purpose from the writing situation? Your first task is to study the assignment carefully, looking for key words. In the assignment described above, the manager gives you these key words: *new control box, affect*, and *power consumption*. Of course you need more specifics than these, but you're also told the situation, or where to look for the information: *the design lab* and *Laura and Bill*. When you talk to Laura and Bill, you can refine your key words: *control box* becomes *A26 controller* and *power consumption* becomes *electrical power requirements*. The key word *affect* tells you that your purpose will be to explain. By analyzing the key words and asking "What is the result I want from this?" you can write a purpose statement based on a writing situation. Often you can explicitly state your purpose at the beginning of your document, which will help your readers know what to look for when they read. See Figure 1-1 for a flow chart showing this process.

You need to remember that in technical writing, *you* are the expert, the person who knows something. Your readers seek information, instructions—even recommendations—from you, and they depend on your accuracy and clear view of what you are

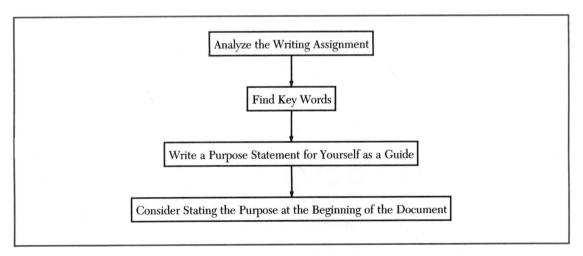

Figure 1-1 The Process of Determining a Writing Purpose

doing. That means you can't assume reader knowledge about your topic or reader understanding of your purpose. A good way to ensure that readers will understand your purpose is to state it in your opening section.

Possible Purposes for Technical Writing

Listed below are some major purposes for writing a technical document, each followed by an example of a purpose statement. Notice from the examples that a purpose statement can be written in many different ways. Sometimes the purpose is stated directly: "The purpose of this paper is . . ." or "This memo proposes . . ." At other times, the purpose is implied: "This article discusses . . ." or "What you read here should cover most of your questions . . ." However the purpose is stated, that purpose should be clear to the reader approaching a new subject.

To define what something is or means

This report defines and describes the Shamal, a seasonal wind that causes dust clouds and jumping sand particles in the Persian Gulf region. For comparison, it also defines other seasonal winds that occur around the globe, including the Santa Ana, Chinook, Foehn, Monsoon, Nor'easter, and Sirocco winds.

To describe what an object looks like or how a process works

Until very recently, most wood-burning stoves had overall energy efficiency ratings of 45 to 60 percent. Today a new generation of wood-burning stoves has overall energy efficiencies as high as 75 percent. The new technology that makes these high efficiences possible is the catalytic combustor. The purpose of this paper is to describe how a catalytic combustor works in a wood-burning stove.

To instruct how to do something

If you are called for jury duty, you will have many questions—from where you should report to what will happen during a trial if you are chosen to serve. Most of these steps are set by state law, a few by county rules. What you read here should cover most of your questions, although each county may be slightly different (California Judges Assn., n.d.).

To analyze as a basis for a conclusion

This paper discusses the sources of drinking water in the Lorimer Valley and the ways in which those water resources are managed to provide an adequate supply of safe drinking water to the valley's 400,000 residents. It discusses contamination problems that have affected the valley's drinking water or may do so in the future and the activities of various governmental agencies to respond to those problems.

To report status or progress

> To: J. L. Thompson
> From: LaVada Clayton
> Date: August 15, 19___
> Subject: Status Report on the technical evaluation of C.O.M. Recorders
> Reporting Period: July 22, 19___ to August 5, 19___

To inform or explain

> The process of creating hypertext documents and databases is called authoring. Since a hypertext database is an interactive program, the authoring process for hypertext is similar to creating any form of interactive instruction. On the other hand, it is a fundamentally different process than creating sequential media like print, audio, or video. This article discusses some of the major aspects of authoring including: structuring hyperknowledge, authoring principles, the root document, authoring tools, screen formats, graphics, collaboration, and cognitive processes (Kearsley 1988).

To interpret, evaluate, judge, or recommend

> Four Computer-Assisted Instruction (CAI) classrooms with a capacity of 25 each have been proposed for the new Humanities Building being built on campus at Central State University. Since software for these classrooms has not yet been decided upon, this report describes five different types of software that could be useful for a writing curriculum: word processing software, desktop publishing software, grammar checkers, spreadsheet programs, and virus protection software. Based on the present CAI classrooms' system, software packages for both the Macintosh computer and the IBM PC compatible computer were selected for evaluation in each type of software. The report evaluates these programs on the basis of five criteria: compatibility, capabilities, user friendliness, relevance to writing curriculum, and price. A recommendation is made for each type of software for both the Macintosh and IBM PC compatible computers. Each choice meets the needs of the students using the CAI classrooms at Central State University (Brooks 1990).

To propose, sell, or persuade

> In response to your Request for Proposal, this memo proposes a study of office telephone systems to determine the best system for Consolidated Life Insurance Company.

To request information

> I am interested in obtaining information about 6-inch to 10-inch reflecting telescopes and accessories produced by your company.

As you plan, remember that (1) you must understand your purpose before you begin to write, and (2) you should plan to let the reader know the purpose early in the document.

✔ **Checklist for Determining Purpose**

In determining the purpose of a document, ask yourself these questions:
1. Why am I asked to do this writing? What should the resulting document accomplish?
2. What are the key words that will help me?
3. Which of the following verbs describe my purpose?

    ___ define   ___ report status or progress
    ___ describe an object   ___ inform or explain
    ___ describe a process   ___ interpret, evaluate, judge, or recommend
    ___ instruct   ___ propose, sell, persuade
    ___ analyze   ___ other (name it)
    ___ request

≡ **EXERCISE**

Following are five situations that require a technical writing response. Read each situation carefully, and then choose a verb from the checklist that clearly describes the purpose of the response. Use that verb to construct a good purpose statement. Some situations may have more than one purpose.

1. TTT Plastics Company has designed a locking ski rack that will fit on the roof of most cars. The ski rack has passed all its functional and design tests, and the first racks will be built and ready to sell in two months. As one of the designers, you are asked to write the brochure that will be packed with each ski rack, telling customers the correct installation procedures.

2. Because of widespread concerns about environmental pollution, the Rock Springs Water Company has decided to send all its customers a booklet about the composition of the water that it supplies. Your job is to write that booklet, and you have access to the studies that have been done on water from the three wells that Rock Springs Water Company operates.

3. Ralston Engineering has the contract to build the new turbine on the Silver Bend Dam on the Green River. The contract specifies that reports must be submitted at monthly intervals to the Green River Watershed Advisory Committee. Those reports must include work done, problems, and schedules. You are to write the first report.

4. To ensure that mechanics understand potential problems with the AM/FM radios in its automobiles, Duvall Automobile Company wants a general information section in the front of the shop manual. This section will include information about how AM and FM reception works and will also tell what the terms *range, flutter,* and *capture effect* mean. You are to write this section of the manual.

5. Diamond Bicycle Manufacturing Company has successfully tested its new alloys for bicycle frames and wheels. The alloys will reduce the cost of its lightweight 10-speed bicycles and thus open the market to casual riders. To attract and instruct potential buyers, Diamond Manufacturing wants a well-written consumer booklet that will tell inexperienced cyclists all about the parts of a bicycle and how it works. Your job is to write the brochure.

1.2 Focusing on the Reader

The next two important planning questions are "Who will read this?" and "How can I meet that reader's needs?" All technical writing is read by real people who have real needs and wants, and considering that reading audience must be important to you. As you think about potential readers, remember what you have decided about your purpose, because audience and purpose often shape each other. You can evaluate your audience by looking at three factors: types of readers, the needs of those readers, and the number of potential readers.

Types of Readers

You can communicate a clear technical message if you have a good idea of the type of reader you are addressing, and if you *write to that reader*.

Have you ever tried to read one of the legal notices published in your local newspaper? Those notices sometimes concern class action suits and are published to alert citizens who might wish to join the suit. However, with their legal jargon and complicated sentences, many notices are almost incomprehensible, and probably only a few people understand them well enough to take action. Legal notices may be written *for* intended readers, but they are not written *to* those readers.

To help identify readers, you can classify them by *background* (what they know) and *function* (what they do).

Background

Suppose you were asked to write a discussion of photosynthesis—the forming of carbohydrates by plants from carbon dioxide and water. Because of their differing education, experience, and background, your discussion of photosynthesis would need to be different for each of the following readers:

- a high-school junior in a biology class
- a college junior in a botany class
- a college-graduate laboratory technician at a company that produces herbicides
- a scientist with a Ph.D. doing research on the properties of chloroplast membranes

Understanding your reader's background will help you determine matters like vocabulary (how many technical terms you can use), motivation (why the reader is reading this), and potential bias (what the reader already believes). Above all, though, you need to know reader background to determine *content* (what you will include in your writing), because what is necessary information for some readers is unnecessary information for others.

Function

On and off the job, readers fall into five major types: *generalists, managers, operators, technicians,* and *specialists*. As Figure 1-2 shows, you can arrange these five types along a continuum from the least technical to the most technical. Because their needs differ, each type of reader is described in the following paragraphs to help you know how to write to that reader.

Generalists   Generalists are consumers, casual readers, or laypersons. These readers have no special knowledge of the topic and are reading out of interest and for information and knowledge. All of us are generalists when we are reading outside our own specialty. An oil geologist reading about heart transplants is a generalist, as is a heart surgeon reading about coastal oil exploration. Much technical writing is aimed at generalists—instructions, descriptions, procedures, definitions, and reports are frequently aimed at first-time users or are intended for information purposes. In addition, many magazines are aimed at generalists, ranging all the way from the simplified *Popular Science* to the scholarly *Scientific American*.

When you write for generalists, you should make your writing interesting and easy to understand. To do so:

- Avoid specialized language (jargon).
- Use a clear organization and include examples and comparisons to what is familiar. If possible, avoid equations and formulas.
- Keep readers interested by providing anecdotes and background material.
- Present accurate information by checking facts and verifying conclusions.

In the following example, Albert Einstein writes for generalists about a Euclidean continuum:

> The surface of a marble table is spread out before me. I can get from any point on this table to any other point by passing continuously from one point to a neighboring one and repeating the process a large number of times, or, in other words, by going from point to point without executing jumps. We express this property on the surface by describing the latter as a continuum (quoted in O'Hayre 1966, 84).

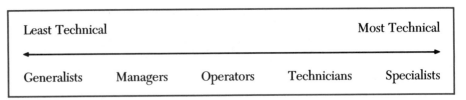

Figure 1-2 Types of Readers

Note Einstein's comparison to a marble table—a familiar image that is easy to visualize. Notice also the simple language and the definition—a way of adding to the generalist's knowledge.

Suppose you were writing to generalists about airplane controls and their effect on takeoffs. Figure 1-3 shows how your explanation might begin. Without "talking down" to such generalist readers, the explanation begins with something the reader knows, then progresses to simple explanations of needed terms.

Managers   Managers are only one step removed from generalists: they also read technical material for information, but they want that information to help them make decisions. Concerned with the day-to-day activities of a business or industry, managers must balance the varying demands of time, resources, people, and costs—but ultimately, they must produce a product or service. Managers are also called executives and administrators.

Some managers are technically trained in fields like engineering, agriculture, or science; others are generalists from fields like business, sociology, or psychology. Even those who are technically trained, though, often become generalists before very long. Management itself takes so much time that few managers can stay on the "cutting edge" in their technical fields.

If your reader is a manager:

- Avoid jargon and long explanations of theory.
- Organize the document to emphasize your conclusions and recommendations (maybe by putting them at the beginning).
- Check your facts and verify your conclusions. Because you are the authority on the subject, a manager relies on you for your technical expertise.
- Supply background information if necessary, but remember that a manager is busy and therefore unwilling to wade through unnecessary material.
- Be specific, providing the details that will help the manager make decisions.

If you had studied airplanes suitable for short-field takeoffs and landings in order to make a recommendation, you might write a paragraph to a manager like the one in

---

What makes an airplane climb? You may have watched the pilot in a flying movie pull back on the stick to lift the plane over a sudden obstacle like a bridge or a tall tree. In reality, though, it's not the stick that makes an airplane climb—it's power. It's true that pulling back on the stick controls the elevators and will move the nose up momentarily, but without increased power, the speed will soon decrease and the airplane will actually sink! Power from the throttle makes an airplane climb, and pulling back on the stick only controls the speed.

Figure 1-3 Writing for Generalists

Figure 1-4. Notice the emphasis on the recommendation, with only minimal technical detail. Notice too the specifics that are given—runway accessibility and payload. The manager is not expected to know what STOL means, so the acronym is explained.

Operators   Like generalists and managers, operators read for information, but they read a technical document because they need to know how to perform some operation. When you write for operators, make your material clear and easy to understand—usually in the form of instructions with a clear 1-2-3 method of organization. Be careful, though, not to talk down to operators; they may be well educated and very knowledgeable in their fields (for example, the operator of a nuclear energy plant or the operator of a flight simulator).

In writing for operators:

- Use reasonably short sentences and standard English word order, because operators may be reading and performing an operation at the same time, or they may need to remember the steps as they carry out the operation.
- Avoid equations, but include tables and graphs to make any math easily accessible.
- Ensure that all the information the operators need is at their fingertips in one document, so they will not have to go to a second source.
- Check the accuracy of the instructions, making sure that all steps are included and in the proper sequence.

When flying an airplane, a pilot is an operator, and a standard lesson for new pilots is how to make a takeoff from a short field over a 50-foot obstacle. Figure 1-5 shows how you might write to pilots on this subject. These instructions assume knowledge of some standard terms and activities (this is not the first flying lesson), but the sequence of actions is clear and simply stated.

Technicians   Technicians as readers share many characteristics with operators, but they also share characteristics with the last type of audience—specialists. Technicians are a bridge between operators and specialists: they usually have considerable practical, hands-on experience, and they are interested both in what to do and how

---

As a result of our study, we strongly recommend that Mountain Mining acquire a twin-engine STOL (short takeoff and landing) airplane for use by survey parties in Alaska and the Yukon Territory. STOL characteristics like full-span flaps and drooped ailerons will allow our crews to take off and land on short runways (less than 1000 feet) in the rugged mountain areas, while giving a payload capacity of two tons—enough for a crew of ten plus equipment. Details and specific recommendations appear in Section 2 of this report.

Figure 1-4 Writing for Managers

To make a short-field takeoff over a 50-foot obstacle:

1. Start the takeoff at the extreme end of the runway. Set 15 degrees of flaps. Hold the stick back and smoothly open the throttle.
2. As the engine develops full power and the plane is accelerating smoothly, apply right rudder to correct for torque. The plane will lift off when the minimum flying speed is reached.
3. Attain and maintain the airspeed for the recommended <u>maximum angle of climb</u>. Keep the throttle wide open.
4. At about 100 feet above the ground, assume a normal climb and use normal climb power. Ease the flaps off.

Figure 1-5 Writing for Operators

things work. Technicians frequently build and maintain complex equipment, translating the theory of specialists into machines and processes. Thus, they are interested in the practical application of theory, not theory itself.

As with any of the five types of readers, the educational level of technicians varies. Some will have college degrees, while others will have reached their status through years of practical experience. Like operators, technicians often need to perform operations, but the operations are more likely to be varied and individual rather than repetitive. For this reason, when writing for technicians:

- Provide background information and some theory, especially if it will help technicians understand *why* a particular application or function is necessary.
- Include explanations or definitions of technical jargon.
- Present tables, graphs, and other illustrations but keep complex mathematics to a minimum.
- Use comparisons and analogies to introduce new information.
- Verify the accuracy of the data you present.

A technician might modify or maintain an existing airplane for short-field takeoff and landing capability; he or she might also build a prototype of a new STOL model. To do that, the technician needs detailed technical information. Figure 1-6 shows how you might explain the function of different types of airplane flaps to a technician. Even without the figures, which are not included here, you can see how much more technical this information is than the paragraph written for generalists, yet how direct and specific the writing is.

Specialists   Specialists, or experts, usually have an advanced degree, years of experience in their fields, or both, and are, therefore, the most technically oriented of all reader types. When you write on a technical subject, readers usually assume that you are knowledgeable in that area; in other words, that you have done the research or design and are reporting your findings. Thus, when you write for specialists, you are

Plain trailing edge flaps operate by increasing wing camber, resulting in an increased maximum lift coefficient, decreased stall angle,
and no change in lift curve slope, as illustrated in Figure 7.3. Fowler
flaps provide an increase in area by aft translation, as well as a camber increase. They also make use of slots, or gaps, between the main
wing and flap, and between the various flap elements, to provide
natural boundary layer control to delay separation. Figure 7.4 shows
the lift characteristics of a single-slotted 30 percent chord Fowler flap
(Kohlman 1981, 124–25).

Figure 1-6 Writing for Technicians

writing to your peers—but you need to remember that they may not share all the areas
of your own expertise. For specialists:

- Assume a large shared technical vocabulary, but define any terms with
  unusual or restricted meanings.
- Follow accepted scientific methods in the work and in the writing.
- Provide theory, equations, and detailed supporting data.
- Refer readers to additional studies or sources.

As a student, you write to specialists when you submit papers to your instructors,
but in this case, the specialists probably know more about the subject than you do. As
you gain expertise in your field, though, specialists who are your peers will become
important readers.

A specialist in aeronautical engineering who is designing a wing would already
know the equation for lift ($L = C_L S \frac{1}{2} \rho V^2$), which contains the term C_L for the
coefficient of lift. Figure 1-7 shows how you might write for such a specialist. Notice
that the explanation relies on equations and technical terms, which the writer assumes
the specialist reader understands.

The maximum value of C_L, which occurs just prior to the stall, is
denoted by C_{Lmax}. . . . In comparison with the standard NACA airfoils
having the same thicknesses, these new Ls(1)-04xx airfoils all have

1. Approximately 30 percent higher C_{Lmax}
2. Approximately a 50 percent increase in the ratio of lift to drag
   (L/D) at a lift coefficient of 1.0. This value of $C_{Lmax} = 1.0$ is typical of the climb lift coefficient for general aviation aircraft,
   and a high value of L/D greatly improves the climb performance (Anderson 1984, 223).

Figure 1-7 Writing for Specialists

Needs of Readers

You will also want to determine your readers' needs: that is, *why* they want to read what you have written, and *how* they will use the information you provide. Readers' needs are heavily influenced by the *area* in which they work. In an industrial environment, for example, a reader might work in any of the following seven areas.

In Research

The emphasis in research is on finding and refining new materials and methods in a given technology. Researchers constantly strive to extend the outer limits of the current technology; frequently, instead of working on a specific product, they study things that seem to have no immediate practical value. Most readers in a research environment are *specialists* or *technicians*, and they will be looking for data that can be synthesized and applied to new situations. Researchers ask questions like these: What is it? Where does it come from? How does it change? Why? What else works like this? What if . . . ?

In Preliminary Design

In the design area, readers will be the planners who set standards for new products, determine materials, spell out requirements, and establish operating goals. Again, these readers tend to be *specialists*. Based on their previous experience with products, they plan months and years in advance. Designers ask questions like these: What should it be made of? What must it do? How fast should it perform? How reliable will it be?

In Development

People in development labs translate the criteria and standards of the designers into detailed specifications and then into working models. While they may be *specialists* or *technicians* in their field, they are less interested in theory than in practice because they must convert design theory into performance. Developers ask questions like these: How can we meet that standard? What changes are necessary in the design? How can we adapt or improve the current technology? How does what you are telling me mesh with what I already know? How can we stay within the cost projections?

In Production

Readers in production or manufacturing must take the processes developed in the laboratory on one or two prototypes and make those processes work efficiently in mass production. They can be *specialists, technicians, managers*, and/or *operators*. Their concerns are intensely practical, centering on these questions: How can we speed up that process? How can the reliability be tested? How can we cut costs? How can we ensure accuracy?

In Marketing

Readers in marketing are concerned with selling the product. They must focus on customer needs and from those needs forecast sales. They are concerned with costs, service, and customer relations. They may be *specialists, generalists,* or any type in between, but again, their concerns are practical. Their questions might be the following: How can we meet that specific market need? What will give us a price advantage? What will be the primary selling points?

In Installation and Maintenance

Operators and *technicians* are the readers in this area. They want "how to do it" instructions and descriptions. They ask questions like these: How do I install X? What do I do when Y happens? How do I fix Z when it fails?

In Administration

Administrators must coordinate the activities of all the other groups to ensure that products or services are delivered on time, with low production costs, and with reliability. Administrators must worry about people as well as products; *managers* and *generalists* make up this group. Their questions include these: Where is the most logical place to produce X? How can we increase production at that plant? Who is the best manager for that project?

While the needs of readers in other environments may be different from those in industry, they too will require answers to a wide variety of questions. In agriculture, for example, you might find researchers, planners, developers, farm owners, field workers, and administrators. For all your readers, you need to determine the kinds of key questions they will be asking. Knowing their potential questions will help you both in gathering information and in writing the draft.

Number of Readers

You cannot assume that only one person is going to read what you've written, so you should never dash off even a memo or a letter without careful planning. That memo or letter might go to a reader you hadn't anticipated, and it could have repercussions for both you and your employer. Therefore, additional questions about readers are "How many people will read this?" and "How can I meet the needs of various readers?" You'll want to think of both the primary reader and secondary readers.

The Primary Reader

The primary reader is the person who makes decisions or acts on the information you provide in a piece of technical writing. Often the primary reader is the person who assigned or requested the document. Sometimes, though, you'll be asked to write something by your supervisor, who will then transmit the information to the primary reader. You should characterize your primary reader by type and needs, and you should also, if possible, know the person's name. If you are writing instructions to consumers, you won't know a specific name, but it will help to visualize a real person you know and write to that person.

Secondary Readers

Frequently, what you write will also be read by secondary readers who may be reading for information or advice. These readers are usually one step removed from the decison or action, but want or need your input. If you can, you should also classify secondary readers by type and needs, and you should know their names.

What if you discover that your primary and secondary readers are of different types or have different needs? Then you should (1) design your document for the primary reader, and (2) modify it for any secondary readers.

For example, suppose you are writing a report "Improving the Efficiency of Greenfield Generating Station No. 4." Your primary reader is a manager who needs to make decisions based on your conclusions. You will focus on monetary implications, keep technical terms to a minimum, and eliminate theory and equations. You know, however, that technicians and specialists in fossil fuels will also read this report. For them, you might add an appendix to cover the theory and include the equations and other information they need and want.

Suppose that you are assigned to report on "The Feasibility of Providing 500 Student Bike Lockers near Campus Buildings." The campus facilities manager might be the primary reader, but your writing instructor (who assigned the report) would be a secondary reader. The primary reader has different needs—cost, specifications, demand for the project, alternatives—than does the instructor, who wants to see how well you can research, organize information, and write.

Different readers may also read the document differently. The primary reader, who assigned the document, may read it from beginning to end; a higher-level executive may read only the abstract and recommendations; still another reader may read the conclusions and then go to an appendix for calculations supporting those conclusions. One way to meet the needs of different readers is to organize the document in discrete sections. You will learn how to do this in Chapter 6.

As you can see, you must consider many factors to understand your potential reader. All these possibilities may seem overwhelming, but if you work with the checklist that follows and a planning sheet like the one at the end of this chapter, you can quickly characterize your readers. At the planning stage, it pays to take the time to understand your potential reader or readers. Then when you organize your material and begin to write, you can be assured that you will meet your readers' needs.

✔ Checklist for Focusing on the Reader

1. Who is my primary reader?
   - __ generalist
   - __ manager
   - __ operator
   - __ technician
   - __ specialist

   Name _____
   Background _____
   Needs _____
   Job title _____

2. Who are my secondary readers?
   - __ generalist
   - __ manager
   - __ operator
   - __ technician
   - __ specialist

   Name _____
   Background _____
   Needs _____
   Job title _____

3. How will I meet the needs of multiple readers? _____
_____

4. Which content factors do I need to stress for my particular readers?
___ avoiding technical jargon
___ using examples and comparisons
___ providing interest through examples and anecdotes
___ giving accurate facts and statistics
___ emphasizing conclusions and recommendations
___ providing background
___ organizing in a chronological sequence
___ using short sentences
___ avoiding equations and substituting tables and graphs
___ having all the information in one document
___ supplying theory
___ explaining technical jargon
___ assuming understanding of jargon
___ providing equations and detailed supporting data
___ referring readers to additional studies or data

EXERCISE

In a group of four to six students, discuss one of the situations below and choose from the checklist as many categories of reader as seem to apply. Elect a spokesperson from your group to defend your choices to the rest of the class. Be prepared to discuss why you have selected the specific potential readers and how those readers would influence the content of the writing.

1. Henry Chan will graduate from General University in June with a B.S. in chemistry. He has a B average, and a B+ average in his major. He hopes to work in the laboratory of a pharmaceutical firm; his primary interest is in chemical synthesis and purification. He has already written his résumé and now must write a letter to accompany that résumé as he applies to Optimum Drug Company. Who are his readers?

2. Paula Swink works for a paper manufacturing company that produces cartons of various types for containers. The company has designed a carton that can be used as a single file drawer; it is a double-sided box with a separate cover. Both box and cover are precut and scored for easy folding but are to be shipped flat and unassembled. Paula's current assignment is to write the instructions telling how to assemble the carton. Who are her readers?

3. Asa Tower works at Precision Manufacturing Company. Yesterday Precision's Plant 2 suffered damage from a flash fire. The fire started when sparks

from a carbide cutting tool used by workers repairing the roof ignited a bin of aluminum shavings. Asa's assignment is to describe the damage. Who might read his report?

4. Carol Rojanski works for Quality Metals Corporation, a small tool manufacturing company. The company wants to purchase a computer-aided design (CAD) system to increase its tool design productivity. This is a major purchase, with $200,000 allocated in the budget. Six different CAD systems are currently on the market that seem to meet some of the criteria of Quality Metals engineers. Carol's job is to evaluate the six systems based on five criteria that have been established and, in a formal report, to recommend one system for purchase. Who might read her report?

1.3 Limiting the Scope of the Topic

Still another question you'll want to ask as you plan is "What is the scope of my topic?"

A Definition of Topic and Scope

The *topic* is what is to be treated in a document or part of a document. In other words, the topic is the *content*, whereas the purpose is what you want to accomplish with that content. At the beginning, you need to plan both what will and what will not be treated and, by focusing on your topic, to set limits beyond which you will not go, thus setting the *scope* of the discussion.

The Scope of Professional Topics and Student Topics

The task of choosing a topic and limiting its scope may differ for professionals and students. A professional is usually given the topic, which is rather narrowly defined. For example, on the job you would probably not be asked to write about something as broad as "geothermal activity." If you worked in that field, your assigned topic might be as specific as "Potential Power Sources from Geothermal Activity in the Northwest Section of Wyoming." The scope of your discussion is already defined. As a technical-writing student, however, you are often writing in simulated conditions, and sometimes you will be able to choose your own topic. When you choose, you must also decide how much you are going to include—or the scope.

Even as a writing professional, you will need to set your scope, simply because you have a limited amount of time to write, and your readers have a limited amount of time to read. If you are asked for a "short" report on the causes of a switching failure, be aware that the request implies "done quickly" and "to be read quickly." On the other hand, if your assignment is to produce a "thorough" evaluation of methane gas migration from filled city-dump properties, you can expect more time for writing (maybe) and a potentially longer document.

The key word in setting scope is *limit*. You can decide on the limits of your topic by considering:

- *Time*. How much time do you have to write? How much time will the primary reader allot to reading?
- *Length*. How long is your written piece expected to be? For example, readers only spend 15 to 20 seconds reading a résumé for the first time. Those readers expect a résumé to be limited to one or two pages.
- *Useful life*. How long will this document be used? Will it be read and used only for a week? Or will it be filed as a permanent or five-year reference?

The primary considerations, though, are fulfilling your purpose and meeting your readers' needs. What you include in your document should help advance those goals.

Checklist for Choosing and Limiting a Topic

1. What is my purpose?
2. Who is my intended reader?
3. What is my topic (the content of my writing project)?
4. How much time do I have to write?
5. How long should this document be?
6. How long will this information be used (i.e., one week, six months, five years)?
7. How much shall I include?

1.4   Planning the Proper Tone

When you were a teenager, did an angry parent ever tell you, "I don't like your tone. I don't like your tone at all!" What did you say to provoke that outburst? What did your angry parent mean by *tone*? More than likely, the anger was provoked not so much by what you said (the content of your message) as by how you said it (the manner of your presentation). Sometimes, in fact, your angry parent may even have misinterpreted your attitude, influenced in gauging your tone by his or her own emotions.

If you transfer an experience like this to the realm of technical writing, you can understand how complex the idea of tone is. Just as the tone of your voice can alienate a listener, so the tone of your writing can alienate a reader. Knowing that, it makes sense to plan carefully the tone of anything you write.

A Definition of Tone

Tone is the attitude you project to readers based on the manner in which you write. Readers react to what you say, and their reaction is based on how you "come across" to them. Tone thus involves both writer and reader: it is projected through content (what you say), and writing style (how you say it), and it is influenced by your relationship to the reader.

Varieties of Tone

You can divide adjectives describing tone into three categories. Look at the three lists in Figure 1-8, and determine which list has positive words, which negative, and which neutral.

You probably labeled the first list positive, the second negative, and the third neutral. After reading the lists, you might think that to "sound technical" when you write, your tone should come from the neutral list. But neutrality won't always work.

Suppose you are writing a job-application letter. Do you really want the reader to think you are "cool" and "detached"? When you write a proposal (selling your services to undertake a project), do you want to be thought "indifferent" and "objective"? On the other hand, if you write instructions for installing brake shoes on an automobile, do you want the tone to be "pompous" or "condescending"? As you can see from these examples, you need to *choose* your intended tone; you can't just let it happen.

Factors That Influence Tone

When you are planning, you can choose a tone based on three factors:

- your position as writer in relation to the reader (also called your "stance")
- the content of the document
- the form the document will take

First, you must consider *writer's stance*: that is, where you are positioned relative to your reader, or what the reader's relationship is to you. In any technical writing situation, you can be writing *up* to superiors, *horizontally* to peers, or *down* to subordinates. You can also be writing inside or outside the company. Figure 1-9 shows how stance might look.

If you are writing to your superiors (inside the company) or to clients or customers (outside the company), you are writing to people in higher positions. If you are writing to subordinates in your company or to outside readers dependent on your

assertive	sarcastic	professional
informal	condescending	straightforward
kind	stuffy	firm
courteous	conceited	no-nonsense
cheerful	begging	cool
tactful	pompous	indifferent
friendly	hostile	objective
polite	rude	detached
confident	brusque	impersonal
knowledgeable	blunt	serious
lively	angry	formal

Figure 1-8 Varieties of Tone Possible in a Document

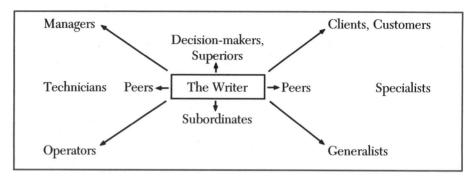

Figure 1-9 Writer's Stance in Planning Tone

good will (a supplier who depends on your business, for example), you are writing to people in lower positions. If your readers are peers (equivalent in rank and experience), your power positions are equal. What's important about stance is that the less power a writer has over a reader, the more important it is that the tone affect the reader positively. But for continued cooperation with peers and subordinates, you must also achieve a tone that will affect them positively. You can never afford to appear condescending or contemptuous.

Your relationship to readers also depends on their position. Before you decide on tone, you need to put yourself in your readers' shoes: where do they see themselves in relation to you? What is their attitude toward you? *Their* perception of the relationship will also influence your choice of tone.

Second, think about your topic or *content*. Is the content of your message positive (good news), negative (bad news), or neutral (informational)? When you write a status report to your manager reporting the solution to a vexing problem, the content of the message is positive. When you write a letter to a job applicant you're not going to hire, the content is negative. And when a report discusses the operation of the air scrubber for the laboratory cleanroom, the content is neutral.

If the content is positive, planning tone is not too difficult because the reader's primary response will be to the good news. But even here you could spoil good news with a tone that's grudging or condescending. If the content is neutral, you should be careful not to intrude on that neutrality. And if the content is bad news, you must plan tone very carefully. If you don't, your document may end up crumpled in the wastebasket—and only partly read. Perhaps even worse, you may provoke an angry response.

As you can see, in planning tone you must consider not only what the reader's attitude toward you and the subject is at present, but also what you *want* it to be after reading what you have written.

Finally, determine the *form* the document will take. The two forms most likely to be tone sensitive are letters and memos, because both are written personally *from* you and directly *to* a specific reader. A third tone-sensitive form is the proposal, in which you want to persuade the reader to accept your plans and ideas. However, even seemingly objective forms like instructions, descriptions, and reports can be tone sensitive if you have not thought about your intended reader.

Ways to Control Tone

Whether the reader finds your tone pushy or cordial, rude or polite, depends on

- how you use words
- how much "personalness" you include
- whether you stress the reader (*you*) or the writer (*I, we*)

Be aware that when you accuse ("you failed to," "you should have"), the reader will react emotionally. When you talk down ("you should have known," "everyone knows"), the reader will respond negatively. The words you choose directly influence the tone you project.

If you totally remove yourself and your personality from your writing, you will often produce a heap of lifeless words. Great scientific writers, like Thomas Huxley, Charles Darwin, and Albert Einstein, were not afraid to be present as persons in what they wrote. Later science and engineering writers (those of 30 to 40 years ago) tried to be "only scientists," and they wrote with third-person pronouns, in the passive voice, and in past tense, trying to reveal almost nothing about themselves in their writings. But such total self-obliteration is not necessary most of the time. In letters and memos, for example, readers will be more accepting and interested if you write to them with the same tone you would use in talking to them. That means using their name and their company name, and using the pronouns *I* and *you*. Reports tend to be more neutral and objective, and often do use third-person pronouns (*it, he, she, they*), especially in sections of description and explanation. In conclusions and recommendations sections, though, contemporary technical writers frequently use first person (*I, we*) to convey the sense of one real person writing to another real person.

Remember, however, that the reader should be at the center of the communication, not the writer.

Not We have reviewed your complaint about the missing parts in order #2116, and we will make the following adjustment. (Writer centered)

But You are correct; order #2116 should have included four sets of bearings, and we're sorry for the inconvenience. Your new invoice reflects the price adjustment and extends your discount period an additional 30 days. (Reader centered)

Reader-centered writing, sometimes called the "you-attitude," concentrates on information that is important to the reader and responds to the reader's needs. The writer is subordinated; *you* pronouns are more important than *I* or *we* pronouns.

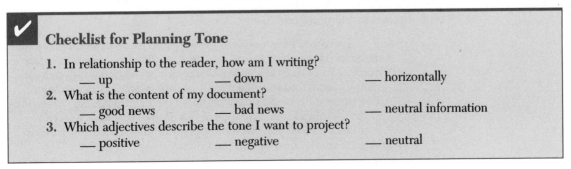

✔ Checklist for Planning Tone

1. In relationship to the reader, how am I writing?
   ___ up   ___ down   ___ horizontally
2. What is the content of my document?
   ___ good news   ___ bad news   ___ neutral information
3. Which adjectives describe the tone I want to project?
   ___ positive   ___ negative   ___ neutral

EXERCISE

The following excerpts come from a variety of technical-writing situations. After reading each, identify two to four adjectives from the lists in Figure 1-8 that describe the dominant tone. Then write a paragraph telling how you would react as a reader to what is written and why you would react that way. Submit your explanations to your instructor, and be prepared to discuss them in class.

1. Memo within a company:

> To: Supervisors on All Shifts
> From: Manager of Production
> We have several problems in our coating area in regard to substrate rejects that each of you must address. Define the cause and initiate corrective action on a sustained basis in regard to your responsible shift. . . . Effective immediately, the following will be done on all shifts without fail. . . . If you have a problem in the implementation of this directive in any way, please see me as compliance is mandatory, effective immediately.

2. From the package, and directed to the consumer:

> You hold in your hand the all-time indispensable kitchen tool—the vegetable peeler. You'll use it, use it, use it! It'll never let you down until the blade dulls, and eventually the blade will dull. We think you should take a good look at all your peelers. Chances are they're several years old, and don't do the job they used to. Modern technology hasn't worked out a way to keep the blade sharp forever, but we have managed to keep this tool inexpensive.

3. From a newsletter to fellow employees:

> I and my colleagues on the Part-Time Employee Council are appalled at the apathy and general laziness of the part-time employees regarding issues of concern to us. If you want to improve the conditions under which you work, you must work for it. You must work for it because no one, not even those of us who are committed to improving our situation, can do it alone or for you.

4. From a brochure for customers:

> Have you ever wondered what's in a glass of water? Most of us have. That's why we are sending you this first report on water quality. At Springview Water Company, we are committed to providing a safe and reliable supply of high-quality drinking water. The water delivered to our customers meets all existing safe drinking water standards. We hope you will take this opportunity to learn more about your water supply and the standards that measure its safety, now and in the future.

5. From an investment letter to stockholders:

> By far the most impressive thing Fifield Company has going for it is its management. There are thousands of public companies out there and this is the only legitimate one I have been able to find that is being successfully run by beautiful women! I know of few public companies that have a woman president. Three of the four board members are business women with extensive business experience. . . . The most impressive thing to me about Fifield Company's management (besides the beautiful women) is their philosophy about entrepreneuring. No salaries, low salaries, hard work, keep expenses down, and come up with fresh ideas. Not a bad formula for success. That's management! That's a formula I have never seen fail when applied. The problem is that it is not applied often enough. As you can see, I highly recommend this company.

Now revise each of the five paragraphs so the tone is appropriate for the indicated readers. Be prepared to discuss the specific changes you have made and why you made them.

1.5 Choosing a Suitable Form and Style

An important part of planning involves choosing the *form* and *style* in which your writing will appear. In Part 3, "The Forms of Technical Writing," you will find examples of each of the 10 major forms in which technical writing usually appears. When you begin any writing project, you can refer to those forms for models and examples. Now at the planning stage, you simply need to consider which of the 10 forms to choose for a particular project.

Definitions of Form and Format

The *form* is the way the written material looks based on organization and structure (for example, sections and paragraphs), whereas the *format* is the way the material looks based on typeface, type size, and page design. These two terms are often used interchangeably, and from the definitions, you can easily see why. Think of it this way: *format* is a way of showing the reader the *form* you have chosen as a writer.

Factors That Determine Form

You have already learned about the major factors that determine form: purpose, reader, and topic. Often on the job, the form of presentation will be dictated by your assignment: "Karen, write a letter to the transit authority detailing our objections to the projected overpass." "Bob, prepare a memo for my signature spelling out the new procedures for testing the modules before they are packaged." "Jamie, you'll be working on our proposal for the automatic ticket dispenser starting the first of the month." In each of these cases, someone else has determined the purpose, reader, and topic and has chosen the form. When you choose the form yourself, you should also be guided by purpose, reader, and topic. Technical writing forms have evolved out of practical experience, and each form meets a special need.

Definitions of Style

Whether a piece of writing should be formal or informal in style also depends on purpose, reader, and topic. While you are planning, think about the degree of formality required for the occasion. *Formal style* is writing that is highly structured, impersonal, and deliberate. *Informal style* is writing that is casual and conversational, like the ordinary, everyday language of speakers.

The key words that clarify the differences are *conversational* and *speakers*. If you write the way you speak, your writing will be conversational and informal. You will include contractions like *can't, won't,* and *I'll*. You will refer to people by their first names, and you'll talk directly to the reader, using the pronoun *you*. You'll use short sentences and the active voice. Because it projects a friendly tone, informal writing can be effective in one-to-one situations like memos and letters, especially if you already know the person to whom you are writing.

However, much technical writing is directed to readers personally unknown to the writer, who may be affronted by an informal approach. For example, many people dislike being called by their first names by a stranger. Also, if your document will go to more than one reader or be used over a long period of time, you may need to write more formally because those readers are less likely to know you. That means using third-person pronouns (*it, he, she, they*) instead of *I* and *you*, and eliminating contractions. It may mean longer and more complex sentences and increased use of the passive voice. Formal writing distances the reader from the writer.

A third possibility is to write in a *semiformal style*. Semiformal style uses the pronouns *I* and *you* extensively and may even use contractions occasionally. In this style, you do not address the reader by name, but you do write short and uncompli- cated sentences in the active voice. Even so, semiformal style is more carefully structured than informal style. This book is written in semiformal style. As you read, notice how often I directly address the reader as *you* and how many sentences use the active voice. If you count a sample of sentences, you'll find they are fairly short, averaging 18 words. Still, this book is designed as a written document, not a conversation, and I deliberately follow some conventions of formal writing.

The last sentence three paragraphs before this one illustrates what I mean:

> Because it projects a friendly tone, informal writing is effective in one-to-one situations like memos and letters, especially if you already know the person to whom you are writing.

In informal, conversational style, that sentence might end, "know the person you are writing to." In formal style, that sentence might end, "especially if the writer already knows the person to whom he or she is writing." I am more comfortable with the semiformal style that preserves the *you* pronoun of informality but follows the formal style of not ending the sentence with the preposition *to*. Table 1-1 summarizes the characteristics of each style.

Table 1-1 Characteristics of the Three Writing Styles

Type	Use of contractions	Use of first names	Person	Sentence length	Voice
informal	yes	yes	first and second	short	active
semiformal	sometimes	no	first, second, third	fairly short	primarily active
formal	no	no	third	medium	active or passive

Style is directly related to tone; the style you choose will influence the tone you project. But specific forms of technical writing are typically written either in formal, semiformal, or informal style. Thus, in planning you need to understand the forms that are available to you as a technical writer.

Ten Major Forms of Technical Writing

The 10 major forms in which technical writing appears are previewed for you in Table 1-2. Study them to see what your choices are. For detailed descriptions and examples of each form see the appropriate chapter in Part 3.

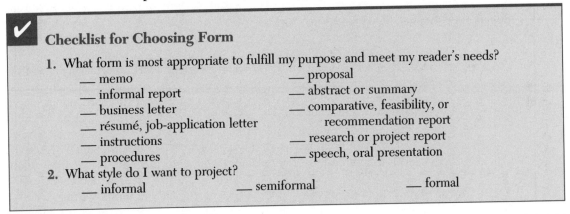

✔ **Checklist for Choosing Form**

1. What form is most appropriate to fulfill my purpose and meet my reader's needs?
   ___ memo
   ___ informal report
   ___ business letter
   ___ résumé, job-application letter
   ___ instructions
   ___ procedures
   ___ proposal
   ___ abstract or summary
   ___ comparative, feasibility, or recommendation report
   ___ research or project report
   ___ speech, oral presentation
2. What style do I want to project?
   ___ informal   ___ semiformal   ___ formal

≡ **EXERCISE**

For each of the following writing situations, you are given the purpose, reader, and topic. Refer to Table 1-2, and choose a suitable form and style. If more than one form or style is appropriate, note that. Be prepared to discuss your decisions in class.

1. purpose: to explain how to recharge it
   reader: operator
   topic: the air conditioning system in an automobile

Table 1-2 Ten Major Forms of Technical Writing

Form	Purpose	Reader	Topic	Style	Explained in chapter
1. memos and informal reports	to communicate *within* your organization, informing, requesting, proposing, confirming	any reader within an organization	no restrictions	semiformal or informal	15
2. business letters	to communicate *outside* your organization, sometimes internally, informing, explaining, confirming, proposing, requesting	any type	no restrictions	formal to informal	16
3. résumés and job application letters	to obtain an interview for a job	generalist, manager, specialist	your qualifications for a specific position	formal	17
4. instructions and manuals	to tell how to do some task	operator	no restriction	semiformal or informal	18
5. procedures	to explain how a natural or mechanical process works	generalist, manager, technician, specialist	no restriction	semiformal or formal	19
6. abstracts and summaries	to give an overview or condensation of a longer document	specialists and technicians, all types	no restriction, based on longer document	based on style of longer document	20
7. proposals	to suggest an idea or offer your services in performing a task	manager, specialist	the project that needs to be done	formal, semiformal, informal	21
8. comparative, feasibility, and recommendation reports	to inform, explain, evaluate, recommend	specialist, technician, manager, generalist	varies	semiformal, formal	15 22
9. research and project reports	to inform, explain, evaluate	specialist, technician, manager, generalist	no restriction	formal, semiformal	23
10. speeches and oral presentations	to inform, explain, recommend, orally rather than in writing	all types	no restriction	formal to informal	24

2. purpose: to present the main points, including the conclusions and recommendations, of a long report
   reader: your boss, a specialist in airport construction
   topic: Midland City's environmental impact report on extending the length of two runways at Midland Airport
3. purpose: to confirm in writing what was agreed to in a phone call
   reader: Susan Jasper, a mechanical engineer
   topic: the harmonic vibrations induced in the frame that holds the X-160 motor
4. purpose: to obtain information
   reader: Charles Hammel, Vice President of Sales at SASA Container Company
   topic: the most efficient way to ship bulk chemicals overseas
5. purpose: to report what you did during the month of January
   reader: your immediate supervisor
   topic: accomplishments, problems, plans to solve the problems

1.6 Scheduling Time and Dividing Duties

The final step in planning is to schedule your time and divide the duties of the writing task. A simple writing assignment can be tucked into available minutes in your work day, but a complex assignment needs a detailed work plan. Also, a complex assignment is likely to involve teamwork or collaboration at the planning, writing, and reviewing stages.

A Definition of Schedule

A *schedule* is a detailed work plan that extends from the beginning date to the completion date and shows intermediate deadlines.

For writing a short letter or memo, the schedule may be simply a mental note like this: "I have a status report due tomorrow afternoon. It'll take me about an hour to gather the material, organize it, and write a draft. Then I need somebody to check it over—Cheryl might do it—and that could take another hour. I need to allow 15 to 30 minutes to correct the draft and get it printed. Let's see, that's 2 $\frac{1}{2}$ hours. I have a meeting most of tomorrow morning. Guess I'd better start on it right after lunch to be sure I get it done."

For instructions, procedures, proposals, or reports of any length from 3 to 500 pages, your schedule should be more elaborate. It should be written down, so you can refer to it during the writing process to keep yourself on track. Long documents can take from two weeks to two years to write; obviously, the bigger and longer the project, the more carefully you must schedule the parts of the process. Long writing projects often include graphics like photographs, drawings, tables, and graphs. In scheduling, you must allow time for these to be completed, to be integrated into the text, and to be approved.

Also, long documents usually require several levels of review: for technical accuracy, for organization, for clarity at the sentence and paragraph level, and for details of punctuation and spelling. Since these reviewers also have other work to do, you need to schedule their participation and notify them well in advance of the time when you'll need their help.

A Definition of Collaboration

Collaboration literally means "working with" (*co + labor*), and the term has come to be applied primarily to the production of documents by teams or groups. Collaboration can also be called team writing or group writing. Collaboration may be between only two people (for example coauthors or a writer and an editor), but often it involves a group of people with differing expertise and job responsibilities, each of whom either contributes a portion of the document or undertakes a different task in the process of writing.

The advantages and problems of collaboration are discussed in detail in the Introduction, "Technical Writing and the Workplace." If you anticipate writing a document with a team, you might want to read that section now.

At the planning step of the writing process, you need to know that even though it has many advantages and often produces better documents, team writing can take from 20 to 30 percent more time than work done by one individual, and additional review may be needed.

In conclusion, even though a long writing project is more complicated than a short letter or memo, in *all* your writing—short or long—you should schedule time for each of these eight basic steps:

1. planning
2. considering legal and ethical responsibilities
3. gathering information
4. organizing information
5. writing the draft
6. making information accessible
7. reviewing and revising
8. editing

The eight steps of the writing process form the first eight chapters of this book, and you will learn the details to carry out any writing assignment as you continue to read. Remember that the steps often overlap, and you can drop back to a previous step whenever it is necessary.

The Writing Cycle in Industry

Schedulers in industry and business often plan time and resources with task-breakdown charts (sometimes called Gantt charts), which show how various tasks can overlap during the writing process. These charts are usually shown as a series of overlapping time lines. Figure 1-10 shows an example of the task-breakdown chart for a student's term project (Anderberg 1986).

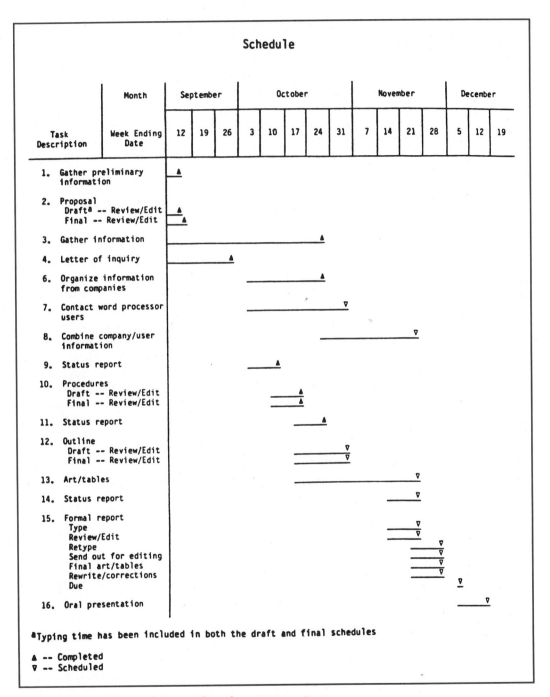

Figure 1-10 Task-Breakdown Chart for a Writing Project

Task-breakdown charts are also included in proposals because they show the careful planning that will ensure a project's completion by the deadline.

Industry planners schedule a total time of about 6.5 hours for each page of a document, including time for research, writing, editing, completing illustrations, and producing the final page. Even so, writers seldom have enough time to do all the research they would like. When you write, you should allow 2 to 4 hours to write each page, depending on the complexity of the task. Most writers spend 60 to 70 percent of that time on the first draft.

The Writing Cycle in the Classroom

The scheduling problems you face as a student are similar to those you will face as a professional. For example:

- You have to prepare assignments for other classes at the same time. Students sometimes say to me, "You don't understand. I have work to do for other classes too!" Of course I know that, but I must assume that students have allotted enough time in their schedules to do the work for each class. Likewise, on the job you will seldom have the luxury of doing only one project at a time. Phone calls, meetings, classes, and other assignments will all intervene, and you will be pressed for time.
- Someone else sets the deadlines. In the classroom, your instructor sets the basic schedule by giving due dates for your major assignments. Those dates are based on the college calendar and the instructor's experience. On the job, writing-project deadlines are also set by some outside force—perhaps the shipping date of a product or the due date of a proposal. Rarely will you set your own deadline.
- It's hard to estimate how long something will take if you've never done it before. This problem can only be cured by experience, and the only way to get experience is to begin scheduling. Whether in the classroom or on the job, you can begin by:
  - asking questions of those who have done it before
  - looking at schedules of similar projects
  - considering your own work habits
  - allowing extra time for emergencies

While you're a student, it's important to learn how to schedule a writing project and then how to work at it a little bit every day. Unfortunately, many students are like firefighters, rushing from one blaze to the next. That is, they only do those assignments each day that are due the next day, and they never are able to work ahead. Crisis scheduling like that won't work for a long-term writing project.

How to Schedule a Writing Project

You can successfully schedule any writing project by following these steps:

1. Determine the due date.

2. Complete a detailed planning sheet to clarify purpose, reader, topic, tone, style, and form. A sample planning sheet follows at the end of this section.
3. If your project is collaborative, meet with your team and establish the schedule, allowing more time for the collaborative process.
4. Start with the due date and work backward to the present, setting interim dates for the following:
   due date
   proofing
   assembling
   review and editing
   final page design and illustrations
   revision of the draft
   review
   first draft and preliminary illustrations
   outline revison
   review of outline
   writing outline
   gathering information
   considering legal and ethical implications
   planning and dividing duties
   assignment
5. Ask experienced writers or your instructor to help estimate the time needed for various phases, or estimate these yourself based on your work habits and your other assignments. When you're working, you'll need to add time in the schedule for management review.
6. Revise the schedule to fit the total time you have. As you work backward from the due date setting interim dates, you may arrive at today and be only part way through the list of tasks. Then you have to trim and slice—a few hours from this segment, a day or two from that. But be sure to allow extra time at the end—for emergencies like a broken copy machine or computer.

Then, as you follow your schedule, note places where you had to adjust the time. The next time you schedule a writing project, you'll have a better idea how long things take.

At this point, such elaborate scheduling may seem like more trouble than it's worth—especially for a short writing assignment. But remember that you are learning a new skill. If you spend five minutes now carefully planning a memo or letter, you are learning how to apply a useful set of questions to any writing task. When your writing tasks grow more complex, you'll remember that you need to consider purpose, reader, scope, tone, style, schedule, and division of responsibilities, and you'll build these considerations in as you go. In the long run, careful planning will save you time and make you a better writer.

The planning sheet that follows is designed to help you apply all the planning techniques explained in this chapter to your own writing. You can make several copies of it and then fill one out for each writing project, simply checking off the applicable categories.

PLANNING SHEET FOR _____ PROJECT   NAME:
                                              DUE DATE:

The PURPOSE of this project is to _____

_____ define	_____ report status or progress
_____ describe an object	_____ report or explain
_____ describe a process	_____ interpret, evaluate, judge, or
_____ instruct how to do	recommend
something	_____ propose, sell, or persuade
_____ analyze by	_____ request
_____ classifying	_____ other (name it here)
_____ dividing	
_____ comparing	

READER/S

Name of primary reader _____ Job title _____

    Type   _____ generalist
                _____ manager   Background _____
                _____ operator
                _____ technician   Needs _____
                _____ specialist

Name of secondary readers _____ Job title _____

    Type   _____ generalist
                _____ manager   Background _____
                _____ operator
                _____ technician   Needs _____
                _____ specialist

Ways to meet needs of multiple readers include _____

_____

TOPIC
   Scope _____

Time allotted for project   in calendar time _____
                             in work-hours time _____

Length of proejct _____

Useful life _____

TONE
My position relative to the primary reader, I am writing

    _____ up   _____ down   _____ horizontally

The content is _____ good news   _____ bad news
     _____ neutral information

I want the tone to be (choose 2–3 adjectives) _____

_____

My stance toward the reader(s) is _____

FORM

      _____ memo   _____ abstract or summary
      _____ informal report   _____ proposals
      _____ business letter   _____ recommendation report
      _____ résumé, job-   _____ research or completion
            application letter   report
      _____ instructions, manuals   _____ speech or oral presentation
      _____ procedures

STYLE

      _____ informal   _____ semiformal   _____ formal

SCHEDULE

    Due date _____
    Number of hours projected _____

BREAKDOWN OF SCHEDULE

Phase	Complete By	Time Allowed
Planning		
Considering Legal and Ethical Responsibilities		
Gathering Information		
Organizing (writing outline)		
Writing the Draft and Preliminary Illustrations		
Designing the Document		
Reviewing the Document		
Revising		
Final Copy and Illustrating		
Assembling		
Proofing		

WRITING ASSIGNMENTS

1. Follow your instructor's directions for choosing a topic for your term project. If you are not assigned a topic, consider these possibilities:

- a research project that reports the current thinking (within the last four years only) on a topic appropriate to your major
- an interview (supported by previous research) with someone working in the field you are preparing to enter
- a comparative study that researches a consumer product, evaluates competing brands of that product, and recommends one brand for purchase
- a collaborative proposal or feasibility study of a change or addition needed at your college campus in general or in a particular department. Campus possibilities might include a coffeehouse, ice-cream parlor, pub, record store, or travel agency—or even a parking garage or automated bank-teller stations. Department possibilities might include physical objects (adding or changing a laboratory or a major piece of equipment) or programs (adding or changing a major or minor). This will be a group project by four to six students, which takes early and detailed planning.

When you have chosen your project, write your instructor a memo. (See Chapter 15 for information about memos and for proper memo form.) Include the following information:

- purpose of the project
- anticipated reader or readers
- the specific topic and scope of your project
- reasons for your choice
- request for permission to pursue the project

2. Once you have received informal approval of the topic for your project, the next step is to write a formal proposal to your instructor (or another designated reader) setting forth detailed plans for your project. Before you write a proposal, you should read Chapter 21, "Proposals." Then, following your instructor's guidelines or Request for Proposal (RFP), plan the proposal using a copy of the planning sheet in this chapter. After planning, write the proposal, review it, and submit it to the intended reader.

REFERENCES

Anderberg, Nadine. 1986. *The comparison of four word processing systems*. Unpublished report. San Jose State University.

Anderson, John D. Jr. 1984. *Fundamentals of aerodynamics*. New York: McGraw-Hill, 223.

Brooks, R., S. Shen, S. Elahi, H. Lee, B. Mani. 1990. *A comparative study of software packages for a writing curriculum*. Unpublished report. San Jose State University.

California Judges Assn. n.d. *Jury duty: an honored service*. San Francisco.

Kearsley, Greg. 1988. Authoring considerations for hypertext. *Educational Technology*. November, 21-24.

Kohlman, David L. 1981. *Introduction to V/STOL airplanes*. Ames: Iowa State UP, 124–25.

O'Hayre, John. 1966. *Gobbledygook has gotta go*. U.S. Dept. of the Interior, Bureau of Land Management. Washington: GPO, 84.

Chapter 2

Considering Legal and Ethical Responsibilities

The process of technical writing is a series of steps, ongoing but interlocking, that lets you move steadily toward a goal of a finished document. So far in that process you have planned your purpose and analyzed your audience. Before you gather information and begin the writing itself, it's wise to pause and consider your work from a legal and ethical viewpoint. This will not be the only time you need to think about issues of law and ethics; such considerations should always be on your mind. Still, it's clear that whatever your profession will be—engineer, architect, scientist, writer, or something else—legal and ethical responsibilities will play a part.

What do we mean by legal and ethical responsibilities? In sections 2.1 and 2.2, I define these terms explicitly, but for now let's think more generally. When you think of the law, you think of society's rules: speed limits, construction codes, noise ordinances, minority rights. These are rules that we either follow or risk some kind of punishment if caught in violation. Laws, while sometimes complicated, usually are written down and have largely predictable consequences.

Ethics, on the other hand, are more like the golden rule: "Do unto others, as you would have them do unto you." Or ethics might be stated this way: "Do no harm." Ethics, then, are not so much rules as choices and behaviors: staying at work until 5:00 when there's nobody to check on you, paying for your coffee under the honor system, giving a colleague credit when nobody knows about the colleague's contribution. *Time* magazine, in a 1991 article called "Brushing Up on Right and Wrong," reported on a series of ethics seminars that proclaimed that "ethical values are more than a series of rules." In these seminars, people are encouraged to "look beyond the letter of the law to such principles of honor, fairness, honesty, and justice." We need to keep these principles in front of us as we work because, as ethicist and seminar leader Michael Josephson says, "We judge ourselves by our best intention, but we are judged by our last worst act" (Time 1991).

It is easy, of course, to make pronouncements on honor, fairness, honesty, and justice, but it is not so easy to apply them to specific work and school situations, especially under pressures of time, production, and someone else in authority. Students legitimately ask questions like "How can I have an effect on a product or a document if I am not the decision maker?" In this chapter, we will explore some of the responsibilities and goals that are yours as (1) a working professional or student in your field and (2) a writer or editor dealing with a written document.

Why are legal and ethical issues important in a writing text? The answer is simple: *It's when things are written down that they are taken seriously.* A second, related question is this: In this textbook, why are law and ethics incorporated into the writing process instead of being added somewhere at the end? This time the answer is not so simple, though it is important. Law and ethics are included here for two reasons: (1) You need to recognize legal and ethical constraints *before* you begin to write, keep them in mind *as* you write, and rethink them *as* you review and edit your document. (2) If you address the legal and ethical dimensions of a situation early in the process, it may be easier to take action than it would be later.

Understanding Legal Responsibilities in Writing

Let us look first at those areas that define and establish your legal responsibilities in writing technical documents.

Important Definitions

The word *law* came into the English language from a Scandinavian word meaning "that which is laid down." That source word is important, because *laws* are those regulations that have been established by government and are either written down (usually) or recognized by custom and judicial decision. *Legal*, then, means either permitted by law or established by law.

Applications

Issues of law surround you in all aspects of your student, professional, and personal life, and this chapter cannot discuss them all. But there are four areas of law pertaining to written documents that you need to know about at least in general terms: (1) copyright law, (2) trademark law, (3) liability law, and (4) contract law.

Copyright Law

Copyright law deals with ownership of written work, so-called intellectual property. The Copyright Act of 1976 includes as intellectual property computer software, photographs, music, and television shows as well as more conventional printed works like books and articles. This law has two important concepts: (1) intellectual property belongs to the person who created it, and that person has the right to profit from its creation and to prevent others from using or reproducing significant portions of the work without permission, but (2) such property also needs to be accessible to researchers and writers under the doctrine of "fair use."

Copyright law was further strengthened when the United States joined the international Berne Convention Implementation Act, effective March 1, 1989. Most industrialized countries belong to the Berne Convention, and a work first published in one country automatically receives copyright protection in all other Berne countries.

The very process of gathering information means that you will probably have in your data collection photocopies of articles, chapters, specifications, and direct quotations from authorities. How you can use that intellectual property both as a student and as a professional is determined by copyright law. Here are some general principles to follow:

- Identify and credit sources of graphics, quotations, and even ideas that you have summarized in papers you write as a student or articles you write for publication as a professional. Appropriate ways to document these borrowings are described in Chapter 13. You can use these graphics and quotations under the Copyright Act's provision of fair use, but you are limited in the amount of the source material you can use or quote. In general, publishers follow these guidelines to determine fair use: 10 percent or less of excerpts of prose works, and only one chart, graph, diagram, drawing, cartoon, or picture from any single book or periodical issue.

- Seek permission in advance for anything you intend to reproduce in a publication or use for commercial purposes. For example, using brief quoted material in a research paper or in-house newsletter requires acknowledgment but not permission, while using brief quoted material in a sales brochure requires both acknowledgment *and* permission. Likewise, quoting a source briefly in a document to be published requires acknowledgment only, but reproducing a substantial proportion of the original source in a document to be published requires both acknowledgment *and* permission. You may have to pay a fee to the creator for such use. In this book, for example, you will find a long list of acknowledgments and copyrights that identifies written material and graphics that come from other sources. You seek such permission with a letter to the copyright holder stating specifically what you want to use and identifying the document it appears in. Keep the official response as proof of the permission you have been granted.

- Use photocopying with restraint. The fair use provision of the Copyright Act was limited by a U.S. District Court decision rendered on March 28, 1991. This decision affected the reproduction of copyrighted material for educational purposes; the court held that no copyrighted materials could be used in course packets without permission of the copyright holder. You might expect that fair use provisions will be further limited.

If you break copyright law by using someone else's material without giving credit or by representing the work as your own, you will be accused of *plagiarism*. Interestingly, the word *plagiarism* comes from a Latin word that means "kidnapping"; in universities and colleges as well as in politics, science, and industry, plagiarism is considered a serious offense. The easiest way to avoid plagiarism is to keep careful records of the sources of your information, and then, when you write, to acknowledge the original source.

Trademark Law

Trademark law deals with ownership of names, symbols, letters, or marks by manufacturers to identify their products and distinguish them from others. Trademarks are usually registered with the government to ensure their exclusive use by the owner. Dictionaries identify words that are trademarks, and the trademark term is capitalized to differentiate it from the general term that describes the same item. So, *Ping Pong* is the trademark and *table tennis* is the general term. *Dacron* is a trademark, but *nylon* is a general term. (The 1982 *Random House College Dictionary* also notes that the word *nylons*—for stockings made of nylon—was formerly a trademark.)

As a student, you simply need to be aware that such trademarks exist and that you should be careful to acknowledge specific product names as such, in contrast with generic names. As a writer on the job, however, you will need more specific trademark information. For example, the Hewlett-Packard *Writing Style Guide* (1991, 151) notes that writers should "Use all trademarks as adjectives. Accompany each trademark with a lowercase generic descriptor." Thus, one writes "HP Laserjet printer"—in this case, *printer* is the generic descriptor and the capitalized company name is the trademark. Most large companies include in their official style manuals information on how to treat trademarks in company publications. Sometimes those trademarks must be acknowledged on the copyright page; sometimes they must be acknowledged the first time they are used in text; sometimes both.

The important thing about trademarks is that they can be lost through common use, so it is important to a company that writers use trademarked terms correctly. The Hewlett-Packard *Writing Style Guide* (1991, 151) notes, for example, that "the common nouns *cellophane* and *cube steak* were once registered trademarks. Loss of a trademark through common use is a significant loss to a company. It means that the resources that it took to develop, test, research, and promote the name have been wasted. More important, however, when a product name no longer represents a specific product, there is loss of product recognition in the marketplace." When you are working, avoid problems associated with the incorrect use of trademarks by consulting your company's legal department or style guide.

Liability Law

Liability law is a concern whenever you write instructions or procedures, because the Consumer Product Safety Act of 1972 makes manufacturers liable for injury to consumers due to defects in the product or to *instructions for its use*. Thus, important parts of both simple instruction sheets and complex manuals are (1) the adequacy and accuracy of the instructions themselves and (2) the warnings of possible hazards. These two aspects are intertwined, because in well-written instructions readers find the warnings incorporated into the instructions at the exact point where they need to know about them.

It will help you write adequate and accurate instructions if you plan carefully and follow the detailed guidelines in Chapter 18. A this point in the writing process, you simply need to be aware of your legal responsibilities and those of your company not only to ensure a safe product but also to teach people how to use it safely. The information in Figure 2-1 shows one aspect of this responsibility.

Contract Law

Contract law has to do with promises that are made by organizations to people who buy or use their products or services. Sometimes promises are carefully prepared and fully documented proposals, advertisements, or warranties; these documents are usually written by experts in the field, and they are reviewed before publication for their legal ramifications. But as a writer, you can also make promises casually, through a letter, a report, or an informal memo. Such promises can be housed in innocuous-sounding words like *free, new,* and *safe,* and you might be held responsible for what these words mean. The U.S. Food and Drug Administration, for example, recently forced manufacturers to abide by strict terminology when labeling foods *low fat, cholesterol-free, lite, lower salt,* and so on. You may not realize that the language you

In order to prevent the product from being unreasonably dangerous, the seller may be required to give directions or warnings, on the container, as to its use. Where warning is given, the seller may reasonably assume that it will be read and heeded; and a product bearing such a warning, which is safe for use if it is followed, is not in defective condition, nor is it unreasonably dangerous.

Restatement of the Law Second, Torts 2d
American Law Institute, 1965
Section 402A (comment: j, p. 353)

Figure 2-1 An Example of Liability Law

used in a document has legal implications; you may even have forgotten that you wrote such a document. Nevertheless, even if the promise is implied rather than explicit, you and your company could be held responsible for a consumer's *perception* of a promise.

You may also need to learn about writing a *disclaimer*, which is a limitation or restriction that avoids making a promise that can't be kept, and which can be a fair way to limit a customer's expectations. For example, you might say "This offer is valid in all states except where prohibited by law," or "The information in this document is subject to change without notice."

From this discussion of contract law, you should remember two important things: (1) you are responsible for every document you write, no matter how casual or seemingly unimportant it is, and (2) you should assume that anything you write could have long-term implications.

This general explanation of your legal responsibilities as a writer is intended to familiarize you with potential problems; it is by no means complete or exhaustive. If you have questions about legal issues, you should consult experts: professors in your field, the company style guide or legal guide, or the company legal department. (Notice that I have just written a disclaimer.)

Checklist for Legal Responsibilities

1. If I have collected material from published or unpublished sources, have I taken the following steps?
   — kept track of the original source so I can document the information when I use it
   — sought written permission for any material (other than a short quotation) that I intend to reproduce in a published or commercial document
   — stayed within the guidelines for "fair use" in using graphics and printed information
2. If I have any information that includes names, symbols, letters, or marks that might be company trademarks, have I checked the proper use of such items?
3. If I am going to write instructions and procedures, am I aware of possible hazards that I must warn against?
4. In preparing a proposal, report, or other written document, am I sensitive to the promises I might be making by my choice of words?

2.2 Developing Ethical Perspectives

Ethics can be both difficult to define and capable of more than one application. Nevertheless, as someone preparing for a professional career, you need to be aware of your responsibilities.

Definitions of Major Terms

In thinking about ethical perspectives, two key terms are *ethics* and *morals*. In contrast to laws—which are instituted by governments and either written down or supported by custom and judicial decisions—ethics are variously defined as "rules of conduct," "moral principles," or "professional standards." These phrases do not clarify whether ethics are simply customs or are written rules or codes; it appears they may be either. Often the terms *moral* and *ethical* are used to define one another, though one useful distinction may be that *morals* refers to generally accepted customs or personal behavior, while *ethics* implies standards and methods in professional and business organizations. Robert Garton, a state senator from Indiana, offers a useful definition of ethics when he says, "Ethical behavior flows from the ability to distinguish right from wrong and the commitment to do what is right" (1989, 25). Electrical engineering professor Joseph Wujek offers another when he calls ethics "the discipline governing *ideal* human behavior" (1990).

Applications

Ethical questions confront us daily in our academic, personal, and professional lives, and as writers we are faced with additional responsibilities when we commit ourselves to print. This section briefly examines three areas in which you may find yourself making ethical decisions whether you are a student, a professional in a technical or scientific field, or a writer. The three areas involve the actual choice of a job or profession, choices during the performance of your job, and choices directly related to writing and editing. Finally, we will look at some codes of conduct developed by companies and by professional organizations to help their members make ethical choices.

Choosing a Job

Career counselors always advise students to study carefully those companies and even those professions that offer potential careers. Often, however, in the heat of the job hunt, a person may accept a job without clearly understanding its ethical ramifications. For example, some people now believe that the risks of nuclear power development are so great to society that it is ethically wrong to work for such organizations. Other people are convinced that pollution caused by fossil fuels is equally bad, and that nuclear power can be so managed that it poses less risk than conventional fuels. Similar ethical debates rage over topics like logging, the alcohol industry, gun manufacturing, contraceptive research and development, and the use of animals in research. Before accepting a job, you will want to ask yourself how you will feel about working there and whether you can carry out the company's business in good conscience.

Daily Decisions in the Classroom and on the Job

Students working in the laboratory or writing up the results of lab experiments often face ethical choices when the experiment didn't go as expected or the data seem incomplete. We are so conditioned to "look good" and to "succeed" that it is tempting to pad the data slightly or write a discussion implying results that did not occur. With depressing regularity, we read about researchers who follow the same unethical practices on the job.

In gathering information and preparing to write, you may find yourself in a similar dilemma; it is very easy to adopt phrasing from an article that "says it just right" or to combine ideas from several sources into what appears to be your own structure of an argument. Sometimes this amounts to actual plagiarism, which can be a legal matter (see section 2.1), but there may be only a tenuous distinction between legitimate synthesis of ideas (making something new from the many ideas you have seen) and unethical appropriation of someone else's intellectual property.

Collaborative projects in the classroom or workplace involve teamwork, and collaboration is usually most effective when team members work closely and have excellent rapport. But the close ties of a team can also cause problems when the project goes awry and team loyalty takes precedence over honest reporting. Some authorities call this situation *groupthink* or *organizational peer pressure*.

For example, suppose that a small group of engineers is assigned a project to write a computer program that will automatically design a printed wire layout for a computer board. The group has a deadline to demonstrate their progress, but when the deadline approaches, the project is not ready for demonstration. No one on the team is willing to admit the problem. To meet the deadline and show significant progress, team members reverse the process: the layout is designed first and the program is written to accomplish that specific layout. The demonstration is successful, management is impressed, and the team leader is promoted. But now the whole team is in trouble because together they have misled their superiors. The demonstration gave the impression that they were solving the problem but they weren't. When the true story is revealed, the team is derisively dubbed the "Dummy Data Corporation," and the team members leave the company embarrassed and in disgrace.

For other engineers, an ethical choice may involve the actions described as "reverse engineering," where a company buys a competitor's product, takes it apart, analyzes the design, and copies that design in its own product—cutting costs by eliminating research and development. If you are the engineer assigned such a task, what will your response be? Do you have options? Are there conditions under which reverse engineering is ethical? If so, what are they?

For any employee, an ethical choice may occur when you discover evidence of falsified data, coverup of damaging evidence, or other unethical conduct within your department or the larger organization. Do you take your complaint to your manager or some other internal authority? What risks do you face by raising questions with management? Is there an employee "speak up" procedure or an ombudsman that would protect your anonymity? If not, at what point do you take your complaint to the press or to an outside authority—an action called *whistle-blowing*? What risks do you face by taking the problem outside the organization? What responsibility do you have to do so? As you can see, these questions have no easy answers, but at the minimum, you should be aware of possible courses of action before a situation occurs. Find out your organization's policies on ethics, employee grievance procedures, and job protection. Learn where you can go for help.

Ethical Responsibilities Directly Related to Writing and Editing

As the introduction to this chapter pointed out, ethics are important to writers because when things are written down, they are taken seriously. You make ethical decisions when you write up experimental results, when you recommend a course of action to your manager, and when you summarize what you have learned in your background reading. But ethical responsibilities can also appear in a wide range of other writing activities. As examples, look at the following three situations, all based on real incidents reported to me by former students:

1. You learn about an exciting job opportunity at a company in a nearby state. When you obtain a copy of the job qualifications, you realize that your education and experience don't quite match the rather detailed requirements. Course work or experience in statistics is listed, and while you once enrolled in a statistics course, you had to drop the course after three weeks when your work schedule changed. Likewise, the job description lists "minimum of one year's experience in reliability testing." You once held an assembly-line job during which you occasionally performed some reliability tests; that job lasted for nine months, then you shifted to another area. A friend tells you that your course work and experience are "close enough," and you can list the course (you *did* enroll) and the experience (it *was* almost a year). How will you craft your résumé to enhance your qualifications and still do it honestly and fairly?

2. The company for which you work asks you to plan, write, and design a publicity brochure that will serve as a recruiting tool at colleges and universities. This large company has an aging work force and is eager to hire graduates for a variety of engineering, technical, and support jobs. In the brochure, you describe the geographical location, the climate, the nearby cultural and educational attractions, the company facilities, types of jobs, and

other details that make your company sound like the great place to work that you believe it is. You decide to include three photographs of people at work to provide focus points, so you assemble ten colleagues for the photo sessions. These colleagues represent four different ethnic minorities, are split equally between men and women, and are—except for one manager—all under 35. The photography and writing go well, and you are pleased by your efforts.

When you circulate a mockup of the final version, you are stunned by charges from two reviewers that you have acted unethically: your photographs, they say, create an untrue picture of the real makeup of the company, which is 75 percent male, with an ethnic mix lower than that of the surrounding community, and mostly over 45 years of age. However, you know that the company is working hard to correct this imbalance, and the recruiting brochure is part of that effort. Should your brochure show the company as it is or as it would like to be? What is honest, honorable, and ideal human behavior?

3. You have been asked by your manager to edit the following memo intended for all the employees who work directly with customers of your organization.

> To: All Customer Service Employees
> From: K. Paul Jackle, Vice-President of Operations
> Subject: New Uniforms
>
> Thanks to your employee representatives, I have received a number of complaints about the new uniforms issued to all of you —at a substantial discount, I remind you—just six months ago.
>
> After many hours of designing and material selection, I came up with a uniform that was both practical and would give you people some pride. Instead of appreciation and thanks, all I have received is complaining about wrinkles, lint, and cleaning problems. I asked two secretaries in my office to clean these garments, and they reported no problems, so I am forced to believe that the complaints are greatly exaggerated.
>
> Since the company has made a substantial investment in these uniforms, and since they were designed with your comfort and style in mind, this office plans no changes in the requirements that uniforms be worn on the job. You people need to appreciate the positive image that a well-groomed work force presents to the public and the savings you realize by being able to wear our inexpensive uniforms instead of buying your own clothes.

When you read the memo, you realize that if it is issued in this form, it will enrage the recipients and seriously affect morale—perhaps even work performance. However, your job is to edit the memo for clarity, appropriate word choice, good sentence structure, and correct spelling and punctuation. What can you do about the offensive tone? What should you do? Is this an ethical problem?

2.3 Using Codes of Conduct

Companies and professional organizations often try to encourage ethical behavior by developing and publishing codes of conduct. Walter Manley, a Florida State University professor and attorney, says that such codes "pay off, creating a better workplace and fostering a sense of corporate community that translates into more good ethics and profit. Good codes drop to the bottom line" (1991).

However, according to business writer Claudia Deutsch of the *New York Times* (1990), companies have also learned that such codes need to have employee input, that "people who are involved in setting a code are invested in seeing that it is followed." Also, codes that are "worded in terms of acceptable, rather than proscribed, behavior" are more effective. Thus, instead of saying "You cannot be late for work," an effective code says "We expect you to be at work on time."

Not all companies and professional organizations have ethical codes, but many do. As a potential employee, you might want to ask about such a code when you interview for a job. Figures 2-2, 2-3, and 2-4 reproduce three codes from different areas so you can see how such codes are written.

Figure 2-2, the Triad Code of Service, is a company code written in 1989 and presented in a booklet given to all employees. The code is summarized here, but in the booklet, each of the 10 codes is explained and accompanied by a quotation from a customer or employee that shows how the code has been applied in a customer situation.

In Figure 2-3, the 1990 code for a professional organization, the Institute of

---

THE TRIAD CODE OF SERVICE

1. Remember that You Represent Triad
2. Communicate for Everyone's Sake
3. Create a Team Relationship with Other Triad Employees
4. Follow Up to Ensure Success
5. Exceed the Customer's Expectations
6. Really Listen to What the Customer is Saying
7. Treat Each Customer as a Unique Individual
8. Maintain Your Focus and Enthusiasm
9. See Triad From the Customer's Perspective
10. Customers Make Paydays Possible

Figure 2-2 The Triad Code of Service

Reprinted by permission. By Susan Constable. Copyright © 1989. Triad Systems Corporation. All rights reserved.

THE INSTITUTE OF ELECTRICAL
AND ELECTRONICS ENGINEERS, INC.
CODE OF ETHICS

We, the members of the IEEE, in recognition of the importance of our technologies in affecting the quality of life throughout the world, and in accepting a personal obligation to our profession, its members and the communities we serve, do hereby commit ourselves to the highest ethical and professional conduct and agree:

1. to accept responsibility in making engineering decisions consistent with the safety, health, and welfare of the public, and to disclose promptly factors that might endanger the public or the environment;
2. to avoid real or perceived conflicts of interest whenever possible, and to disclose them to affected parties when they exist;
3. to be honest and realistic in stating claims of estimates based on available data;
4. to reject bribery in all its forms;
5. to improve the understanding of technology, its appropriate application, and potential consequences;
6. to maintain and improve our technical competence and to undertake technological tasks for others only if qualified by training or experience, or after full disclosure of pertinent limitations;
7. to seek, accept, and offer honest criticism of technical work, to acknowledge and correct errors, and to credit properly the contributions of others;
8. to treat fairly all persons regardless of such factors as race, religion, gender, disability, age, or national origin;
9. to avoid injuring others, their property, reputation, or employment by false or malicious action;
10. to assist colleagues and co-workers in their professional development and to support them in following this code of ethics.

Figure 2-3 IEEE Code of Ethics

© 1990 IEEE. Reprinted, with permission, from The Institute of Electrical and Electronics Engineers.

Electrical and Electronics Engineers, Inc. (IEEE), is reproduced. Notice how this code differs both in tone and in approach from the company code in Figure 2-2.

Finally, since we are primarily interested in writing matters in this book, Figure 2-4 shows the Code for Communicators (1989) of the Society for Technical Communication (STC), the largest professional organization for technical writers and editors in the world. As you read the provisions of this code, notice that when you write, you may need to be aware of all three kinds of codes. On the job you will simultaneously be an employee, a member of a profession like engineering, biological science, or architecture, and a technical communicator.

SOCIETY FOR TECHNICAL COMMUNICATION
CODE FOR COMMUNICATORS

As a technical communicator, I am the bridge between those who create ideas and those who use them. Because I recognize that the quality of my services directly affects how well ideas are understood, I am committed to excellence in performance and the highest standards of ethical behavior.

I value the worth of the ideas I am transmitting and the cost of developing and communicating those ideas. I also value the time and effort spent by those who read or see or hear my communication.

I therefore recognize my responsibility to communicate technical information truthfully, clearly, and economically.

My commitment to professional excellence and ethical behavior means that I will

- Use language and visuals with precision.
- Prefer simple, direct expression of ideas.
- Satisfy the audience's need for information, not my own need for self-expression.
- Hold myself responsible for how well my audience understands my message.
- Respect the work of colleagues, knowing that a communication problem may have more than one solution.
- Strive continually to improve my professional competence.
- Promote a climate that encourages the exercise of professional judgment and that attracts talented individuals to careers in technical communication.

Figure 2-4 STC Code for Communicators

Reprinted with permission from the Society for Technical Communication, Arlington, VA.

Checklist for Ethical Perspectives

1. Have I considered my ethical responsibilities in choosing my job or profession?
2. In my daily responsibilities as a student and/or employee, do I act in the following ways?
   — state honest and realistic claims based on available data
   — address concerns about the safety, health, and welfare of the public
   — give credit for the ideas, recommendations, and work of others
3. When I prepare to write, am I aware of the following?
   — the need for precise language
   — the need to portray events and situations accurately yet fairly
   — the importance of tone
   — the ethical codes of conduct written by my company or profession

2.4

Discussing Legal and Ethical Scenarios

Legal and ethical responsibilities are seldom obvious; especially in the area of ethics, there may be more than one valid approach. Therefore, instead of a series of exercises, this chapter will conclude with three scenarios presented for you to read, consider seriously, and discuss with classmates and your instructor.

Your instructor may ask you to study one or all of the following scenarios, which merely outline the situation. In deciding on your response, you may want to consult the code of conduct published for your profession (ask a professor in your major field where to find such a code), or the sample codes printed in Figures 2-2, 2-3, or 2-4.

Scenario 1

You are an engineer charged with performing safety testing and obtaining appropriate regulatory agency and outside testing laboratory ("agency") approvals of your company's product. The Gee-Whiz Mark 2 (GWM2) has been tested and found compliant to both voluntary and mandatory safety standards in North America and Europe.

Because of a purchase order error and subsequent oversights in manufacture, 25,000 units of GWM2 ("bad units") were built that are not compliant to *any* of the North American or European safety standards. A user would be much more vulnerable to electric shock than from a compliant unit. Under some plausible combinations of events the bad-unit user could be electrocuted. Retrofitting these products to make them compliant is not feasible because the rework costs would exceed the profit margin. All agree that because of this defect the safety labels will not be attached to the bad units, per the requirements of the several agencies. Only two options exist:

a. Scrap the units and take the loss.
b. Sell the units.

An employee of the company notes that many countries have no safety standards of any kind for this type of product. It is suggested that the bad units be marketed in those countries. It is pointed out that many of these nations have no electrical wiring codes; or if codes exist, they are not enforced. The argument is thus advanced that the bad GWM2 units are no worse than the *modus operandi* of the electrical practice of these countries and their cultural values. Assuming that no treaties or export regulations would be violated:

1. What is your recommendation?
2. Suppose one of the countries under consideration was the country of origin for you or your recent ancestors. Would that affect your recommendation?
3. Now suppose you are not asked for a recommendation, only an opinion. What is your response?
4. Suppose it is suggested that the "bad units" be sold to a third party who would very likely sell the units to these countries. Your comment?
5. You are offered *gratis* one of the bad units for use at home, provided you sign

a release indicating your awareness of the condition of the unit, and that it is given to you as a "test unit." (Assume you can't retrofit it, and that the product could be very useful to you.) Would you accept the offer?

6. Suppose it is suggested that the offer above (question 5) be made to all employees of the company. Your comment? (Wujek 1990).

Scenario 2

You and a colleague, Pat, are working on a similar problem that will require extensive computation. Each of you owns a *Little Giant* (LG) personal computer. You are aware of a software package, *The Universal Wondrous Calculator* (UWC), which runs on the LG. The software has been well-received by the user community, and can perform all of the computations needed to solve your problem. The UWC package sells for between $150 and $300, dependent upon discounts and seller mark-ups.

You tell Pat of this software, and both of you agree that it is highly desirable to have access to it. Later that day Pat visits you with news of having the program, and offers to share it with you. When Pat hands over the program diskette, you observe that it has a crude, handwritten label rather than the usual printed kind. When you ask Pat about this, you are told, "I copied it from the original disk of a friend who bought the program."

You insert the diskette in your machine, and the title screen shows clearly that the program is *copyrighted*. Pat notes that it is reasonable to make a backup copy; indeed, many user manuals for software encourage this practice. Pat believes, "If they didn't want you to copy it, they would have made the program copy-protected!"

1. What is your comment on Pat's statement regarding copy protection?
2. Is it legal or ethical for you to *use* the program? (After all, *you* didn't make the copy.)
3. Is there any difference, legally and ethically, between the act of your using the UWC program and the act of Pat's copying it?
4. Now suppose that you and Pat are students, with all other conditions identical to those given above. Respond to questions 2 and 3 above.
5. Now suppose that you and Pat are *professors*, with all other conditions identical to those given above. Respond to questions 2 and 3 above (Wujek 1990).

Scenario 3

You have been working with a group of four students on a collaborative project in your technical-writing class. Your project so far has involved 12 weeks of a 15-week semester, and you are now writing the report: a feasibility study on the addition of a professional pilot training program to the existing aviation department at your university. When you planned the project, each student agreed to write one major section of the report and to be available to edit and review the other sections. They chose you as the project manager. Because of your conflicting class and work schedules, the group has not often met as a whole, but you have kept in contact with each member, and they have assured you they will meet their due dates. Student J.J., however, has been absent for a week; when he returns, you find he won't have his section written on time, and because he is behind in his other classes, he can't review anybody else's writing. He's sorry, but he's under pressure at work. You put together a rough draft of his section based on his notes, and the group agrees to cover his responsibilities for editing.

Now you are preparing the title page and the memo to your instructor that explains each person's part in the project. You estimate that J.J. has done only 25 percent of his share, and you are tempted not even to list him as a writer. You also know that he's a graduating senior and his company has been paying his tuition. This project counts 60 percent of the course grade.

1. As project manager, what is your best course of action?
2. Should you cover for him?
3. Should you tell the instructor about his lack of commitment?
4. Should you confront J.J. himself?
5. What is an ethical course of action?

EXERCISE

You have been asked to produce the next issue of your company's newsletter, *Right Angles*. Recently, you saw an article in *Some Professional Journal* that you'd like to reproduce. Visit the library and choose an article from a professional publication in your field, then write a letter requesting permission to reproduce the article in your newsletter. The request for permission should be addressed to the copyright owner, which may or may not be the journal in which the article appears. Find out who owns the copyright and direct your letter to that person or organization. See Chapter 16 for information on writing business letters.

REFERENCES

The Chicago manual of style. 1982. 13th ed. Chicago: Univ. of Chicago Press, 221.

Deutsch, Claudia. 1990. Codes of conduct: Involve the workers. *San Jose Mercury News*, 26 August, PC 2–3.

Garton, Robert. 1989. The business of being ethical. *Retail Control 57.5, 23*.

Hewlett-Packard writing style guide. 1991. Palo Alto: CA, 5091-0839E.

IEEE Board of Directors. 1990. *Code of Ethics*.

Manley, Walter. 1991. Qtd. in Ethics translates to the bottom line. *Modern Office Technology*. August, 10.

Society for Technical Communication. 1989. *Code for communicators*. Washington: STC.

Time. 1991. Brushing up on right and wrong. *Time* 137, 15, 15 April.

Triad Systems Corp. 1989. *Triad code of service*. Livermore: CA, 1002527-12m0012.

Wujek, Joseph. 1990. Some practical principles of engineering ethics. Conference notes. San Jose State University, February.

Chapter 3

Gathering Information

Information is the foundation of any piece of technical writing, whether the document is a one-page business letter, a 20-page proposal, or a manual telling readers how to replace automobile brake shoes. Therefore, whether you are a student assigned to write a feasibility study or a civil engineer preparing a completion report, an important part of preparing to write is gathering information.

Listen to the comments of people who write on the job. Kate Picher, who writes for a computer manufacturer, stresses the variety of sources: "I read product specifications and design specifications; I also interview marketing, programming, and customer support personnel. And yes, I talk to customers too!" Roger Connolly, director of a large technical writing department, talks about the value of information: "Obtaining appropriate and relevant data is often a problem," he says. But Stacy Lippert, a marketing editor and writer, is reassuring: "If you keep track of your resources (people as well as library and other references), gathering information should be the easiest part."

As a student, you may hesitate to plunge into gathering technical information, fearing that you won't understand it or won't know what to look for. But remember that information gathering is itself a learning process; as you read articles and data sheets and talk to experts in the field, your own thinking about the subject will become clearer.

If you are writing a short memo or letter, you may already know everything you need to say. In that case, information gathering simply means jotting down those ideas for later organizing and writing. For a short report or proposal, you may only need to call a fellow worker or two for their information and expertise. But a long report, proposal, or manual may require the use of many kinds of sources; for that you'll need careful preparation and a methodical search procedure. A thorough search procedure will (1) reassure you that you've checked all the reasonable sources, and (2) save you time and possible embarrassment in the long run.

Whether you are a college student or a writing professional, your potential information sources can be divided into three areas: recorded information, knowledgeable people, and personal experience. Recorded information includes books, articles in journals and magazines, company documents and reports, government reports and studies, sound recordings, laser and video disks, and computer data files. Knowledgeable people includes fellow workers, consultants, customers, professors, salespersons, and maintenance and repair persons. Personal experience is what you have been involved in yourself; it could include results of experiments, investigations, trips, calculations, or previous projects. Figure 3-1 shows the likely information sources in each area, all of which are discussed in this chapter.

3.1 Preparing for the Search Process

While you worked on planning, you probably thought in a general way about what you wanted to write, but now that you're about to start gathering information, you need to think more specifically about what you already know and what you need to know.

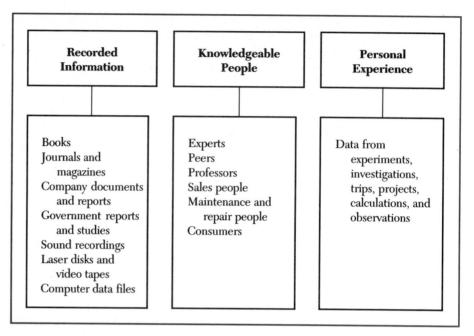

Figure 3-1 Potential Sources of Information

Determine What You Already Know by Brainstorming

Brainstorming is a good technique to determine what you already know (and don't know) about a subject. Begin with your purpose statement and your topic, and then quickly jot down all the things that come to mind related to purpose and topic. Ask yourself the key questions: who, what, why, how, where, and when. Write down ideas as fast as you can without worrying about order or relationship. Use short phrases, abbreviations, key words, sentences—whatever will remind you later of the points you want to include. Write down everything, even those things you're not sure are relevant.

Here are five different ways to brainstorm. Read through the list, and choose a way you feel comfortable with. You may want to try different methods until you find the one that works best for you.

1. Use a big sheet of paper, and write ideas down as you think of them, leaving some white space around each idea. Later you can cut this paper into sections for sorting, leaving one idea to each section.
2. Use index cards or small pieces of paper and write each idea on a *separate card or sheet.*
3. Use a big sheet of paper, and write in the center what you think is the key word. Circle it. Then jot your ideas down around it, working out from the center. In this clustering technique, as you generate ideas, circle the terms and join them with lines to group related ideas.
4. List your ideas—one after another—on a sheet of paper. Then study that list and group the ideas on another sheet of paper.

5. Talk your ideas into a tape recorder. When you're finished, listen to the tape and group related ideas on paper.

At this point don't try to write polished, complete sentences or paragraphs. You are generating ideas, and it's important now to capture those ideas before they float away like helium-filled balloons. Don't get bogged down in details either; you can always go back later to work out the details. Also keep in mind that the ideas you generate while brainstorming don't need to be correct or proven; being critical at this point can kill spontaneity, destroying potentially good ideas.

Occasionally, you'll find that you already know everything you want to say. When that is true, follow the suggestions in Chapter 4 to organize the items on your brainstormed list. Most of the time, though, you will need to do some more thinking. If possible, leave your brainstormed ideas for a while, turning your conscious mind to another task to let your subconscious mind continue working.

 EXERCISES

1. Suppose that you are going to write a short report on 10-speed bicycles suitable for commuting and recreational riding. As part of that report, you must describe a typical 10-speed bicycle. Information from your planning sheet appears below.

Purpose   To describe the whole and the parts of a 10-speed bicycle.

Readers   Fellow college students. (How do you describe these readers? Educated? Interested? Nontechnical? Why would they read your report? As potential consumers?)

Topic   Limited to general description of whole and parts. How the parts work and how they are built are not considered.

Tone   Objective, descriptive.

Choose one of the five ways of brainstorming and generate a list that will describe the parts of a 10-speed bicycle. Since you have not gathered information for this topic, work with what you already know about bicycles. Don't be concerned if you are unfamiliar with the names of some of the parts. Be prepared to discuss your list in class.

2. Think of an object in your major field with which you are familiar (for example: an oscilloscope, a microscope, a lathe, a word processor). Identify a typical purpose, reader, topic, and tone for a description, and then use one of the brainstorming methods to generate a list that describes the parts of that object.

Get an Overview

Often when you write something technical on the job, you will be writing out of your own experience. You will have inspected, planned, or designed something, and you will be reporting your results. Because you already have a basic idea of what you're doing, any information you gather will be to fill in details. As a student, though, (and sometimes as a professional), you may have to write about a subject that is new to you.

Then it's helpful to get an overview of the subject before you begin researching details or interviewing experts. You can find such overviews in several places:

- **ENCYCLOPEDIAS.** Most of us think first of the general encyclopedias like *World Book* and the *Encyclopedia Britannica.* For technical subjects, though, you may want to consult more specialized encyclopedias. These range from general scientific encyclopedias like the *McGraw-Hill Encyclopedia of Science and Technology* to those in specific fields like the *Encyclopedia of Chemical Technology* and the *U.S. Pharmacopoeia and National Formulary.* Check the reference section of your library and talk to a reference librarian to find the most appropriate general source for your particular project. Because of the long preparation time for encyclopedias, the information is usually not current, but these references will provide useful background.
- **DICTIONARIES.** Many good technical dictionaries are available to help you define terms. Ask a reference librarian for help.
- **HANDBOOKS, ALMANACS, YEARBOOKS, AND ATLASES.** Again, each scientific and technical field has many general sources in this category. As with encyclopedias, the information here will not be current, but it will help you understand the subject.
- **A GOOD GENERAL BOOK ON THE TOPIC.** For general books, consult the library catalog under the subject heading. Look for books that have words like *introduction to* or *basic* or *general* in the title. Look also for those books with the latest publication date. If the subject is totally new to you, you may want to look at a book like a high-school text, which can give you a simplified overview.
- **COMPANY RECORDS.** If you're working, you can often find earlier reports that have dealt with the same subject in the company files. At this stage, you are looking for design documents or those with background or overview statements. Ask colleagues, your company librarian, or the document controller.

While you read overview material, be on the lookout for the key terms writers use in talking about this subject. You'll need up to 10 key terms to help you in further research, so jot them down as you notice them. For example, some key terms for a study of assistive listening systems for the hearing-impaired might be *hearing-impaired, audio induction loop, infrared, AM or FM systems, TDD, captioning decoder.*

Determine the Kind of Documentation Required

Before gathering information, find out what kind of documentation (bibliographic citation) you will need to provide in your final paper and what system of documentation you should use. Ask your writing instructor or those working in your major field for advice, and see Chapter 13 for information on documentation systems. The more you know at the beginning about your future needs, the easier it will be to plan your search and strategy.

Whatever documentation system you use, you are sure to need basic bibliographic information, such as author, title, city, publisher, and date. You need to capture that information *while* you are searching because it is very difficult and time consuming to go back to your sources and retrieve it later. If you set up a plan before you start gathering information, you can proceed efficiently.

Set Up a Plan for Searching, Notetaking, and Documentation

The planning you need to do for efficient searching, notetaking, and documentation is a housekeeping chore—unglamorous but necessary. To understand the importance of this kind of planning, follow Joe P. Smith as he begins gathering information. Joe is a student about to write a recommendation report. Since he's also working, he plans to compare copy machines to determine which one to recommend to his boss, a commercial real-estate broker. He has determined his purpose, reader, and potential topic. He's brainstormed to see what he already knows, and now he's ready to begin gathering information—he thinks.

Joe begins his research at the campus library during a free hour between classes. Since he knows that computer searches can save him time, he goes first to a terminal that displays information in the library's online catalog. He searches under *machines* (too many choices), *copy* (nothing relevant), and *copiers* (four books, but the two newest ones are marked "checked out"). Disgusted, Joe decides to look for periodical articles at another terminal, this one accessing information in periodical indexes. Unfortunately for Joe, these terminals are very popular with students, and there is a line. Ten minutes pass before Joe is able to punch in his keyword *copiers* and begin to analyze the entries. The list is extensive, and it takes Joe several minutes to isolate and flag eight entries that look promising. Joe is in a hurry now because it's time for his next class, so he quickly prints those entries and stuffs the paper in his book bag.

That afternoon Joe calls two office-machine distributors to set up appointments with sales representatives. The next day when Joe arrives for the first interview, the sales rep demonstrates two machines and talks enthusiastically about the characteristics of each. Joe leaves that interview and goes immediately to his second appointment. There he learns about the advantages of still another copy machine. Next, he's introduced to the head of the repair shop, who talks about repair records and implies that a competitor's product is prone to breakdown.

When Joe gets home, his head is spinning with information and he's tired, so he watches television. Two days later he tries to reconstruct what he learned in his library research and his interviews, but all the information has jumbled in his mind like a tossed salad. He looks for his list of article titles, and he can't find it. He didn't take notes at his interviews, and the only written material he has is a sales brochure from his first interview. He remembers that the repair record is important, and he'd like to use that information. He can't quote the repair person, though, because he forgot to write down his name; now Joe is too embarrassed to call the distributor and ask for that information again. Joe actually used a good search strategy by relying on computers to save time and to get bibliographic information quickly. He did find potential leads and some solid information for his report, but because he didn't have enough key words and didn't plan, he didn't have time to finish his library search. Also, because he didn't take notes at the interviews, he couldn't use most of what he heard.

To avoid Joe's plight, here are some helpful guidelines:

In searching written sources

1. Use 3 × 5-inch index cards and note the complete bibliographic information of each source on a separate card as soon as you find it. At the top of the card write the author's name (last name first), title of the article or chapter, title of the book, place of publication, publisher, publication date, library call number, and—for articles, the title of the periodical, volume number, and date. Later you can use these cards to arrange your list of references or works cited, and you won't have to go back to find missing information. On the back of each card write down how useful this source is and what information you find from it. If you are required to annotate your references or works-cited entries, you use this information to write the annotations.

2. Before you start examining the sources themselves, supply yourself with plenty of change for a copy machine or buy a copycard that can be used instead of change. Rather than copying lengthy material onto index cards, photocopy relevant pages. *Be sure* that you write the author's name or a cross-reference to your bibliography card on the copied article, and be sure you record the page number.

3. Use 4 × 6-inch cards to take notes on the source material. You can list statistics and summarize factual information, but write down the author's exact words for anything that expresses opinion, judgment, or controversy. Enclose these direct quotations in quotation marks so you can differentiate when you write between paraphrase (restatement in your own words) and the author's original wording. See section 2.1 for help on avoiding plagiarism.

In contacting other people for information

1. Clearly identify each source on a separate 3 × 5-inch index card—name, title or position, place of business, phone number, and the date of the interview.

2. Prepare questions in advance (see section 3.3).

3. Consider bringing a tape recorder to any interviews, but ask for permission first. Whether you tape the conversation or not, write down all you can remember *immediately* after the interview, while the material is fresh in your mind.

In developing your own information

1. Carefully document all experiments or observations in a laboratory notebook.

2. Keep track of calculations and the methods used to obtain them.
3. Make notes of personal inspections. Otherwise, you'll find yourself wasting time backtracking.

Determine the Time Frame of Your Project

Knowing the time frame of your project will help determine the limits of your search. *Time frame* refers to how much time you want to cover and how far back you want to go. For example, if you want only the latest and most current data, you will not look in books because book publishing takes at least two years, with the information probably three to four years old. But if you want a broader view, the history or background of developments that have led up to current technology, books will be useful. For more current information, you should consult periodicals. To find relevant articles, check the most recent indexes to periodicals in your field, perhaps one regular index and one that includes abstracts. You might also look at government reports.

For the very latest information (newly announced products and technological advances), you may have to talk to people. Even as a student, you can often interview experts in the field if you are willing to call companies and tap into the network of those in the know. One way to begin is to contact professors in fields related to your subject either as experts themselves or as sources to recommend potential experts. You could also consult the *LC Science Tracer Bullet*, a short bibliographical guide to current "hot topics" in science and technology published by the Library of Congress. *Current Contents*, a weekly service from the Institute for Science Information (ISI), reproduces tables of contents from journals in specific fields and is another likely index source for very current information. A third source is *Science News*, a weekly that excerpts reports of the latest topics. Finally, electronic bulletin boards are becoming popular sources of current information. Such information may not have been checked for accuracy, so use it with caution. However, electronic bulletin boards do provide another way to access information.

Find the Key Words Used in Discussing Your Subject

Whether you are searching written sources, writing for information, or interviewing experts, much of your success will depend on your ability to ask the right questions. That means you must narrow your topic to a few *key words* and phrases. You can find these in the following ways:

- Review your purpose and topic for the key words in those statements.
- Go back to your brainstormed ideas, and from them cull up to 10 key words that describe your subject.
- Study the notes you took while reading overview or background material. Again, pick out the key words used by writers on that subject.
- Consult the book titled *Library of Congress Subject Headings*, usually found near the library reference desk. Additional terms listed for an entry (such as

Library of Congress
call numbers

Used for

Broader topic
Related topic

Narrower topic

Subdivisions

Bicycles   *(May Subd Geog)*
        ⌈*GV1040-GV1058 (Cycling)*⌉
        ⌈*HE5736-HE5739 (Transportation)*⌉
        ⌈*TL410-TL437 (Technology)*⌉
    UF Bicycles and tricycles
        Bikes
        Cycles (Bicycles)
    BT Vehicles
    RT Cycling
        Motorcycles
    NT All terrain bicycles
        Minibikes
        Minicycles
        Mopeds
        Pedicabs
    — **Law and legislation**   *(May Subd Geog)*
    — Parking
        USE Bicycle parking
    — Tariff
        USE Tariff on bicycles
    — **Tires**
    — — Tariff
            USE Tariff on bicycle tires
Bicycles, All terrain
    USE All terrain bicycles

Figure 3-2 Library of Congress Subject Headings

those listed for the sample entry in Figure 3-2) will give you more key search words. Be aware that sometimes the Library of Congress subject headings are outdated. Periodical indexes may include more current information.

Key words have always been important in finding information, but they become critical as more material becomes available for computer searching. Because all indexes do not use the same vocabulary, you also need to think of synonyms for your key words. Unfortunately, indexes often lag several years behind industry in redefining or using new terms, and phrases may appear in the literature long before they are used as index terms.

Once you have a consolidated list of about 10 words or phrases, you are ready to begin searching for information. If you plan to use the library, it's a good idea to discuss your search strategy with a reference librarian before you begin. New automated databases are becoming available all the time, and a reference librarian can help you take advantage of new information sources and new techniques for finding and using that information.

For example, you might be able to bring a disk to the library and download bibliographic information directly from the library's sources. This step will save you from tediously copying bibliographic information onto cards, and will ensure that the bibliographic information is complete and easily accessible. Whether you search in written sources, by interviews, or by personal investigation, if you proceed in a methodical way and write down information *as you find it*, your search will be efficient.

Checklist for Setting Up a Search Process Plan

1. What basic information comes from my planning sheet?
   Purpose
   Reader
   Topic
   Tone
   Form
   Style
2. How much time is available for information gathering?
   By what date must I be finished?
3. What words and ideas were generated by brainstorming?
4. What is a likely source to read for an overview?
   What words and ideas come from an overview?
5. What are my most likely sources of information?
   ___ recorded information
   ___ knowledgeable people
   ___ personal experience
6. Do I have the necessary research materials?
   ___ 3 × 5-inch and 4 × 6-inch index cards
   ___ computer disk for downloading information
   ___ change for a copy machine
   ___ tape recorder
7. What system of documentation (bibliographic citation) will I be using?
8. How will I set up and maintain good records of my information?
9. What is the best way to begin a search of written information?
10. What is the time frame of the project?
11. What are the 3 to 10 key words and phrases I can use for searching?
12. Am I ready to consult with a reference librarian about my research strategy? What questions do I have?

3.2

Searching Written Sources of Information

Gathering written material involves continual decision-making, frustrating dead ends, and occasional pleasant surprises. This section covers strategies for both automated and traditional searching; you can read the whole thing to help you decide how to proceed, or you can read those parts that will be most useful to you.

How you approach an information search will depend on (1) how far back you want to go and how many years you want to cover, (2) how much time you can spend looking, and (3) what the most promising sources are. Assuming you have answered these questions, read an overview, and picked out key words or phrases, you are ready to begin looking.

Library sources can be divided into four major categories: reference material, books, periodicals, and government documents. Each category has its own sources, as Figure 3-3 shows. You may decide to consult all four or only one, but don't be afraid to ask a librarian for help. Each category will allow you access to a different kind of information, and most categories can be accessed by computer.

Searching Automated Databases

Your approach to an automated search for information will depend on (1) the computer facilities and databases available to you, and (2) perhaps your budget. Before you begin an automated search, get an overview of your topic and pick out 10 or so key words or phrases. Do not try to search by computer if your topic requires information more than 20 years old, because most databases go back only to the 1970s.

The computer is the most efficient tool to use in finding information because it can use your key words to search through thousands of files in seconds—a process that might take you many hours in a traditional search. However, since the computer doesn't think, it will respond only to what you tell it, and it won't "notice" information indexed under synonyms or under broader or narrower terms. Therefore, it's important to have a broad range of key words to help in your search.

Usually, automated databases provide three information sources:

1. online library catalogs for books and reference material
2. CD-ROM databases for periodicals and government documents
3. "mediated" computer search databases for additional periodicals and government documents

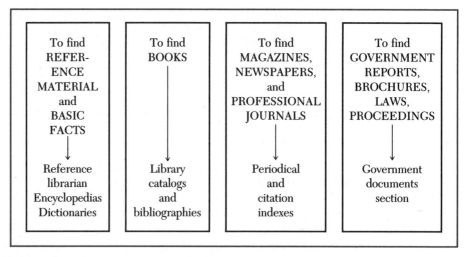

Figure 3-3 Four Major Library Sources

Online Library Catalogs

Many libraries have replaced or supplemented their traditional card catalogs for books with a computerized online catalog. Following the instructions, you type your request on a keyboard. You may be able to search by author, title, subject, key word, or call number. The information you seek will be displayed on the terminal screen. When you find a likely source, print the information, download it to your disk, or copy the complete bibliographic information on a 3 × 5-inch index card. Besides standard catalog information, the computer may tell you the location of the book in a multi-building library and whether the book is currently checked out. Catalogs may also include other databases, such as encyclopedias and periodical indexes.

One advantage of the online catalog is its accessibility; with a modem attached to your personal computer, you may be able to search the catalog from the comfort of your home, your office, or your dorm room. Also, libraries are able to mount additional databases on the primary catalog, so you may have access to even more information than the library's book collection. It pays to ask a librarian what is available.

CD-ROM Databases

One of the most popular ways to search for periodicals by computer is to use CD-ROM (compact disk—read only memory) databases. Here's how this search strategy works: The information in a heavily used database is loaded onto one or more compact disks, which are then kept at the individual library and updated frequently. You use a computer at the library to search the disk, and you search by entering key words or combinations of key words.

For example, the CD-ROM database called INFOTRAC™ indexes articles in

- general magazines like *Consumer Reports, Time,* and the *New Yorker*
- business journals like *Business Week, American Banker,* and the *Harvard Business Review*
- government publications like the *U.S. Tobacco Journal,* the *FDA Consumer,* and the U.S. Department of State *Dispatch*
- major newspapers like the *New York Times,* the *Wall Street Journal,* and the *Washington Post*

The articles in these periodicals can be located by key words, and the system gives cross-references to help you find what you need. When you find a reference you want to investigate, you simply press a print button, and the bibliographic information is printed out for you.

Another database indexes references from the Educational Resources Information Center, commonly called ERIC. In 1991, this index contained over 720,000 entries. Other indexes on CD-ROM are COMPENDEX PLUS for engineering and technology abstracts, MEDLINE® for medical citations, and DISCLOSURE® for extracts from documents on market share, profits, and sales filed with the Securities and Exchange Commission. A useful feature of some of these databases is their ability to print out the abstracts on command. Some databases even contain full text. For example, the newspaper index for the *San Jose Mercury News* gives not only the reference but also displays the article itself, which can then be printed or downloaded onto your disk.

Mediated Online Search Databases

A mediated online search means that—with a librarian's help—you can search databases that are not available for you to search by yourself in your own library. The key to finding what you want is to use the proper database and to select good key words and phrases. You should talk to a librarian about the databases available to you, the cost of a mediated search, and the procedures at your library for such searching.

Databases   In a mediated search, you might use an information retrieval service like DIALOG, which has more than 380 available databases and more than 260 million records in specialized areas. For example, DIALOG includes

- AGRICOLA (provided by the U.S. Department of Agriculture Technical Information Systems)
- GEOARCHIVE (provided by Geosystems of London, England)
- COMMERCE BUSINESS DAILY (provided by the U.S. Department of Commerce)
- WATER RESOURCE ABSTRACTS (provided by the U.S. Department of the Interior)
- MATHSCI (provided by the American Mathematical Society)
- TOXLINE (provided by the U.S. National Library of Medicine)
- SCISEARCH (provided by the Institute for Scientific Information)

Many of the DIALOG databases are available on Knowledge Index, a smaller and less expensive version of DIALOG that is available nights and weekends.

Costs   Costs vary according to the database you use and the time your search takes. Though search costs are high, the time you save may be worth the money, because you might locate information that cannot otherwise be found. The current cost varies by database, but runs from $10.00 to $20.00 for a 10-minute search. The cost for a typical student search is about $30.00; this cost includes connect time for access to data, the use of phone lines, and print charges. Ask if your library or major department will pay for some or all of the cost. In a company library, the cost will usually be covered by your employer.

Procedures   To conduct a mediated online search, you will need to work with a librarian. Ask at the reference desk about the procedure to follow; you may have to fill out a form to outline your topic, list your key words, and detail your time frame. With that information and your help, the librarian will conduct the mediated online search for you.

Searching by Traditional Means

Whether you search for information sources by computer or traditional means (that is, using a library card catalog and printed indexes), you have access to the same basic types of information: reference material, books, periodicals, government publications, and company publications.

Reference Material

Reference books include encyclopedias of all kinds, dictionaries, atlases, handbooks, and yearbooks. They are useful for background information and general factual data. Using your key words, consult three or four if you can. Examples of reference books are the *Chambers Dictionary of Science and Technology*, the *Encyclopedia of How It Works*, and the *Weather Almanac*.

Reference books will also be listed in the library catalog, but searching there may not be the most efficient way to find them. One good source is Sheehy's *Guide to Reference Books*, which lists and annotates some 14,000 reference books by subject, author, and title. Figure 3-4 shows sample entries from the *Guide to Reference Books*. Ask a reference librarian where to find it in your library.

Books

Many libraries still maintain card catalogs in addition to the online catalog. Each book has three cards—author, title, and subject. In most libraries, the author and title cards are in one file while the subject cards are in another. Use the author/title catalog only if you *know* the name of the title or author you want; otherwise, use the subject catalog and start looking under your key words.

Author/Title Publication date

Summary of contents

Second book by same author

Library of Congress call number

> **Huxley, Anthony Julian.** Standard encyclopedia of the world's mountains. N.Y., Putnam. [1962]. 383p. il., maps. **EE8**
>
> A popularly written work, arranged alphabetically by name of mountain; gives location, height, dates, and names of first climbers, etc., and a brief description and history. Illustrated with black-and-white and color photographs. Includes a gazetteer and an index. GB501.H8
>
> ——— Standard encyclopedia of the world's oceans and islands. N.Y., Putnam, [1962]. 383p. il., maps. **EE9**
>
> A popularly written work arranged alphabetically by name of ocean or island, giving location; dimensions, etc. (in case of oceans the maximum depth); and a brief description and history. Format is similar to that of the two preceding items. Includes a gazetteer and an index. GB471.H9

Figure 3-4 Typical Entries in Sheehy's *Guide to Reference Books*

Reprinted with permission of the American Library Association from *Guide to Reference Books*

Bibliographies will also direct you to books. Bibliographies are lists of books on a particular subject with author, title, and publication information. You can find bibliographies at the end of general articles or books on your subject, or you can look in the library's *Bibliographic Index* to find specific bibliographies on a subject. For current books, you can consult *Books in Print*, which lists books by subject, author, and title. You can also sometimes find bibliographies in the subject catalog under *Your Subject: Bibliographies* (for example, *Pollution: Bibliographies*).

When you find a likely source, copy all the bibliographic information on an index card. Continue searching with your key words until you've found a number of sources that sound promising. Now if your library's online catalog gives circulation information, you can find out which works will be on the shelves. If, after you search the shelves, you cannot find a book, ask for help. If you find a reference in some source to a book or article your library doesn't have, use the interlibrary loan service to get it from another library. However, plan ahead because it takes at least two weeks to get material through interlibrary loan. You also may have to pay for any copying costs.

Periodicals

Periodicals include magazines, newspapers, and journals of all kinds. Periodicals are especially useful for locating current information. A library at a large university may subscribe to anywhere from 6,000 to 25,000 periodicals and may have 350 or more books that index and abstract periodicals. Indexes provide easy access to the information you seek if you (1) find the appropriate index, and (2) use your key words to search for relevant data.

Many students are familiar with the *Reader's Guide to Periodical Literature*, which indexes articles in popular magazines. For most scientific and technical research, however, the *Reader's Guide* is of limited value because its material is aimed at general readers and is too broad. But many other indexes have the same format, so knowing how to use the *Reader's Guide* will help you use indexes like the *Applied Science and Technology Index*, the *Business Periodicals Index*, and the *General Science Index*. These indexes give you access to general magazine articles in technical and scientific areas and to articles in professional and trade journals. Remember that in many libraries these indexes are also available in CD-ROM databases. Figure 3-5 shows typical entries from the 1991 *Applied Science and Technology Index*.

Most index entries include the article title, author, and periodical title, date of publication, volume number, and inclusive pages. You'll find what you need in these indexes if your subject is indexed and if the article titles are complete enough to highlight any of your key words and phrases. Copy all the key information on your index cards. A few indexes also include abstracts (short summaries of the article contents). You may find abstracts more useful than simple title listings because you'll have a better idea of the article's contents before you search for it. In Figure 3-6 is a typical entry from the 1991 *Geographical Abstracts*.

Another type of periodical is the trade journal—a publication focused on specialist or technician readers, often sent to them free, and supported by advertising aimed at those readers. Trade journals frequently are easier to read than the official

Sand
 See also
 Aggregates
 Concrete—Aggregate
 Oil sands
 Sandstone
 Silica
 Soils, Sandy
Evaluating slow sand filtration. *Public Works* 122:104+
 Ap '91
Forties oil line replacement overcomes sandwave chal-
 lenge. W. J. M. Steel and R. Inglis. il map *Oil Gas
 J* 89:49-52+ My 6 '91
Local scour downstream of box-culvert outlets. H. Abida
 and R. D. Townsend. bibl diags *J Irrig Drain Eng*
 117:425-40 My/Je '91
Uplift behavior of screw anchors in sand. A. Ghaly
 and others. bibl diags *J Geotech Eng* 117:773-808
 My '91
Sand, Foundry
    Reclamation
Illinois examines the beneficial reuse of spent foundry
 sand. T. Burnley. diag *Mod Cast* 81:28-9 Mr '91
Sand and gravel industry
   Acquisitions and mergers
Who owns whom: sand and gravel. K. Zimmerman.
 Rock Prod 94:39-41 Mr '91
Sand and gravel plants
   Quality control
Morse zeroes in on quality. R. A. Carter. il diag *Rock
 Prod* 94:35-8 Mr '91

Figure 3-5 Entries from the *Applied Science and Technology Index*

Applied Science & Technology Index, August 1991, page 442. Copyright © 1991 by The H. W. Wilson Company. Material reproduced with permission of the publisher.

**91J/08801   The influence of forest decline and soil aci-
dification on water yield from forest catchments of the
northern Black Forest, FR Germany**
H. J. Caspary, in: *Hydrology in mountainous regions I*, ed
H. Lang & A. Musy, (IAHS; Publication, 193), 1990, pp
713-722.
Long-term monitoring of the Eyach catchment from
1973-1986 indicates a significant increase in water yield
and the runoff coefficient for the growing season, although
there has been no extensive cutting in the catchment. An
Ecohydrological Systems Model was developed by the
incorporation of field data and plant physiological
processes to describe the increase in water yield. The
model indicates that the observed increase in water yield is
likely to be caused by a reduction of forest transpiration.
This change in water yield is linked to forest decline and
soil acidification caused by anthropogenic sources of air
pollution.   -from Author

Figure 3-6 Entry from *Geographical Abstracts*

Reproduced from *Geographical Abstracts: Physical Geography*, 1991. Copyright Elsevier Science Publishers Ltd.

journals of scientific and technical organizations, and they often describe new products and technologies. Many trade journals are listed in indexes available at the library reference desk.

Don't neglect newspapers, many of which are indexed both in books and in CD-ROM databases. The *New York Times Index* is available in many libraries and provides access to nearly 150 years of *Times* articles. Many other newspapers are also indexed, including the *Christian Science Monitor*, the *Wall Street Journal*, and the *Los Angeles Times*. Use your key words to help you find current articles. If you find an article title in a newspaper index, you can probably find an article in your own area newspaper within a day or two of the listed date.

A citation index is another kind of index. If you know the name of an expert or authority in the field you are researching, you can look up that person's name in a citation index. Figure 3-7 shows entries from the *Social Sciences Citation Index* (1990). Under the authority's name are listed the names of other writers who have cited the authority and the articles they have written. These articles will help you find recent information related to your topic and comments on what authorities have said. Citation indexes also have subject and source indexes. They are useful both for finding information and for evaluating the reliability of information. If a writer is frequently cited by others, he or she is probably considered an authority. But be careful; sometimes writers are cited because they are wrong! The major citation indexes are the

Cited author

Cited reference, year, and period

S. E. Estroff's 1989 article in the Schizophrenia Bulletin *was cited by H. P. Lefley in the journal* Hospital and Community Psychiatry, *Vol. 41, page 277, 1990.*

```
ESTROFF SE─────────────────────
  81 J SOC ISSUES   37  117
  MCLEAN A      SOCIAL SC M     30  969 90
  81 MAKING CRAZY ETHNOGR
  COHEN LJ      SOC PSY PSY     25  108 90
  SHADISH WR    J SOC ISSUE     45    1 89
  81 MAKING IT CRAZY ETHN
  LEWIS DA      J SOC ISSUE     45  173 89
  LITTLEWO.R    BR J PSYCHI    156  308 90 R
  STEIN HF      SOCIAL SC M     30  987 90
  81 MAKING ITS CRAZY ETH
  LEFLEY HP     HOSP COMMU      41  277 90 R
  81 MAKING IT CRAZY    17
  BERLIN SB     SOCIAL SE R     64   46 90
  82 HOSPITAL COMMUNITY P   33  609
  BOND GR       SOCIAL W GR     13   21 90
  88 MED ANTHR Q    2  421
  PRESS I       SOCIAL SC M     30 1001 90
  89 SCHIZOPHRENIA BULL   15  189
  LEFLEY HP     HOSP COMMU      41  277 90 R
  SANDELOW.M    SOCIOL HEAL     12  195 90
  SHANER A      HOSP COMMU      41  332 90 L
  89 14TH INT ASS PSYCH R
  BLANKERT.LE   FAM SOC         71  387 90
```

Figure 3-7 Entries from *Social Sciences Citation Index*

Reprinted from the *Social Sciences Citation Index*®, Vol. 1 page 5068 with the permission of the Institute for Scientific Information® © copyright 1990.

Science Citation Index, the *Social Sciences Citation Index*, and the *Arts and Humanities Citation Index*. Citation indexes may also be available on CD-ROM.

In technical writing, it's often a good idea to look in indexes for the most recent titles; once you have about 10, it's time to find the articles themselves. You will save time if you gather information in blocks of about 10 items at a time. Ten sources give you enough possibilities so at least one or two should be useful. At the same time, the list of 10 is short enough so you won't spend all your research time listing references without ever getting to the sources themselves. In many libraries, these articles will be on microfilm or microfiche. Ask a librarian for help if you are unfamiliar with microfilm or microfiche or if you don't know how to use the machines for reading them.

Government Documents

More than 1,400 libraries in the United States have been designated as depositories for literature published by the U.S. government. These publications cover many topics, and intended readers range from children to highly trained scientists. However, government publications are sometimes separated from regular library holdings and cataloged in a different way. Therefore, you need to find out if your library has government publications and how you can locate specific documents.

In addition to your library's classifications of government holdings, you can consult special books like the *Monthly Catalog of U.S. Government Publications*, and the *Index to Publications of the United States Congress* (also called the *CIS/Index* after its publisher). Privately produced versions of both these indexes are now available on CD-ROM from various vendors. The three primary sources of technical and scientific documents are the National Aeronautics and Space Administration (NASA), the Department of Energy (DOE), and the Department of Defense (DOD). Documents from these agencies are available for sale through the National Technical Information Service (NTIS). Listings from NTIS can be found in regular indexes or in the *Government Reports Announcements and Index* (GRA&I), which is computerized in the DIALOG, ORBIT, and BRS databases. Some government indexes and documents are also available on CD-ROM. Indexes will give you a scale of document costs and the address for ordering them. Allow two to four weeks to receive documents.

Company Sources

Companies provide technical and scientific information in two ways: (1) through published (therefore external, or public) documents, and (2) through unpublished (internal) company reports, proposals, letters, and memos. Published materials include annual reports, sales brochures and specification sheets, and manuals and instructions of all kinds. You can find information on public companies in the database DISCLOSURE on CD-ROM, and information on private companies in the database called CIRR. Annual reports, especially those of local companies, are often housed in university and public libraries. For most other company information, you will need to contact the company itself with a letter of request. See Chapter 16 for help on writing request letters.

Internal company documents provide excellent sources of information on the job, but they may be unavailable to students. These documents will be in a company's files of previous work; they include memos, reports, specifications, flow charts, and even transparencies and videotapes. You can, of course, write a letter requesting specific information.

No matter what sources you use, you will probably find references to information that sounds perfect for your needs, only to discover that the information itself is not available from your library. Don't be discouraged. If you started searching early enough, you may be able to use interlibrary loan. If not, simply note on the back of your index card that the material isn't available and go on to another source. Remember that the word *search* is at the core of *research*, and searching means looking, and sometimes looking and looking.

When you seem to have exhausted all your potential leads, it's a good idea to consult a reference librarian once again. Tell the librarian what you have done, point out problem areas, and ask if you have missed potential sources.

Checklist of Potential Written Sources

Which of the following are likely sources of written information for my specific assignment?
___ reference books
___ online library catalogs for book titles
___ CD-ROM databases for article titles
___ mediated online databases for article titles
___ printed bibliographies
___ printed periodical indexes
___ government documents
___ printed company information

EXERCISE

Consult your planning sheet (see section 1.6) to refresh your memory about your topic, scope, time frame, and time constraints. Consult the checklist in section 3.1 to find your key words. Armed with that information and your source cards, visit your library to conduct a literature search. Ask your instructor how many sources you should consult or use and what documentation system you should follow. If this is a collaborative project, break up the tasks and assign one or two persons to each type of source: library, interview, survey, personal observation. Follow these general guidelines:

1. If your time frame goes back three years or more, consult the online library catalog by key words and subjects. Copy all the key information about potential books on your source cards, one book to a card. Find about 10 sources and then look for the books. If you have problems, ask for help.

2. If your time frame is very current (within the last two years), or after you have found the relevant books, look for periodical information. Search by computer, if possible, consulting the most relevant database for your topic. Use your key words and phrases. Get a printout or download the information or the articles themselves, or copy the information from the screen. If necessary, use traditional methods to consult the most relevant printed indexes for your particular topic. Ask a librarian for help in locating the best index to use. Use your key words and phrases to locate articles. Copy all the key information on your source cards, one source to a card. Find about 10 sources, and then look for the articles.

3. Consult government publications for information on topics like laws, statistics, regulations, scientific and technical studies. If you're unsure about potential government publications, talk to a librarian from that area.

4. Turn in your source cards if your instructor requests them.

Contacting Knowledgeable People

"People in the know" (specialists, technicians, and managers of various kinds) can often provide you with the most current information, and because technology changes so quickly, the most current may be the best. As a working professional, you will have relatively easy access to knowledgeable people either at your own company or outside, through vendors or customers. But even as a student you can talk to experts if you are creative and persistent. This section details three ways of contacting experts for information. You may choose to use one way or all three for a complex project.

Letters of Request

You can request information from a knowledgeable source with either a memo or a business letter. A memo is written to someone *within* the same organization, while a business letter usually is written to someone *outside* your organization. You should write a letter when

- you don't personally know your source
- you can't contact your source by telephone or in person
- you want to ensure either the accuracy or the formality of your request
- you can't get the information any other way

Note the last item in that list: "you can't get the information any other way." It's tempting to write for information as your first search strategy when, in fact, you should always do preliminary research to find out what is available in your college and local libraries and through brochures at local companies and dealerships. *Then* you write letters asking for very specific information that will fill the gaps.

Before you can write for information, you need to find the names and addresses of individuals, organizations, or companies that will be likely to answer your questions.

When you are a working writer, your company can help you locate these sources, but while you are a student, you will have to find the names and addresses on your own. Here are some places you can look in a library:

Industries and Businesses

- *AT&T Toll-Free 800 Directory (Business Edition)*. This volume lists toll-free 800 numbers. Call a company and ask for the name of a customer or sales representative to whom you can write for specific information. Be sure to get the complete address, including department, mail stop, and zip code. You can also call 800-555-1212 and ask for the toll-free number of a specific company.
- *Million Dollar Directory*. This three-volume reference lists companies in alphabetical order, by geographical location, and by the Standard Industrial Classification (SIC) number. A list of SIC numbers assigned to specific industries appears in the front. Company names, addresses, phone numbers, and division names are included. Similar information for larger companies appears in the *Billion Dollar Directory*.
- Standard & Poor's *Standard Corporation Descriptions*. This reference gives names and addresses as well as cross-references to subsidiaries.
- Standard & Poor's *Register of Corporations, Directors, and Executives*. This volume gives addresses, phone numbers, and products as well as listing key personnel.
- Thomas's *Register of American Manufacturers*. This reference gives company profiles with addresses, phone numbers, branch offices, asset ratings, and company officers. Use it to locate companies that produce specific products.
- Ads in technical journals or trade magazines. Often the ads will give a name and address to write to for information, or they will give an 800 number you can call.
- Local business directories or yellow pages.

Government Organizations and Studies

- *United States Government Manual*. This reference gives information on agencies of the legislative, judicial, and executive branches. It includes a Sources of Information section and gives mailing addresses. It is revised every two years.
- *Congressional Directory*. This reference has biographies of members of Congress and information about committees and departments, including addresses and phone numbers.
- State almanacs or government directories.

Government names and addresses change with each election and administration; therefore, you need to check carefully to be sure the information is current.

Once you have the name and address of a likely source of information, write a letter of request. Chapter 16 will help you write a clear, concise letter of request and show you accepted forms to follow.

Interviews

To most students, the word *interview* triggers thoughts of job hunting and that clammy-palm ordeal they both desire and dread. But not all interviews are for jobs; one important way to tap the knowledge of specialists and technicians is through an informational interview, a meeting between two or more people for the purpose of sharing information. Most people are gracious about sharing their knowledge with students in a well-planned interview, and they also might be valuable contacts in the future.

Professionals in the workplace, whether they are writers, engineers, or scientists, rely heavily on personal contacts for information. Many interviews are informally conducted by telephone, while others are formal face-to-face meetings. As a working professional, you will find it advantageous to set up formal interviews with key sources early in a project; then, later, knowing your source, you can phone with a question or problem. As a student, you can supplement data gathered from written sources by conducting informational interviews; often you can gain current information and opinions from specialists—information that is simply not available in written form.

However, be aware of the difference between an expert's oral opinions given "off the cuff" in an interview and the carefully structured, considered, and reviewed written opinion that appears in a professional journal. Because a published book or professional article has been reviewed by specialists and peer editors before publication, it is generally a more reliable source than an opinion expressed in an interview.

Choosing Someone to Interview

You need a clear idea of the purpose for your interview before you can decide whom to contact. Once you've decided on the purpose, write it down in one or two sentences. The interview purpose is related to the overall purpose of the report as listed on your planning sheet, but it is also more specific. In other words, you should do your background reading and literature search first; then you'll have good questions to ask, and you'll be able to fit the answers into a coherent whole.

For example, if you are studying 10-speed bicycles to recommend one for the touring enthusiast, whom should you interview? The answer depends on what you need to know. You might, for example, interview

- a bicycle salesperson for information about what options are available
- a repair person to determine repair records of competing brands
- a bicyclist to get the knowledge and opinion of an experienced rider
- a manufacturer to learn about advances in frame or gear construction
- an engineer or quality control expert to learn about a part's reliability or safety

You can find names by jotting down people mentioned in your reading, by calling manufacturers and likely sources, or by talking to bicycle enthusiasts.

Planning the Interview

Once you have chosen your "subject-matter expert," phone or write to arrange a specific appointment. Since your informant is doing you a favor, you should accommodate yourself to his or her schedule. Try to plan for 30 to 60 minutes of interviewing time.

To use that time most efficiently, write out questions in advance. My students like to put three or four questions on a page, leaving plenty of space between each question. Design the questions to call for a response beyond a simple yes or no. Thus, instead of saying, "Do you think . . .?" say "What do you think about . . .?" or "Why do you think that . . .?" Open-ended questions like those should get your source talking.

In your planning, consider tape recording the interview, but ask permission in advance, because some people will be uncomfortable if they are taped. A verbatim tape will let you fill in blanks in your notes and help you verify details of names and part numbers. If possible, use a battery-operated recorder so you don't have to depend on nearby electric outlets. But do plan to take notes, and bring a supply of paper and pens for that purpose. Also bring paper big enough for your source to use for sketching diagrams or flow charts. That way you can bring that valuable information away with you, and you won't have to copy it from a chalkboard or a scrap of paper.

Conducting the Interview

Review your planning sheet (especially the purpose section) before going to the interview. Arrive at your destination a few minutes early. Before you launch the interview itself, thank the person for his or her time and again ask permission to tape the conversation. Make your purpose for the interview clear at the outset. Opinion surveyor George Gallup (quoted in Brady 1977, 71) says, "When you start asking questions, the other person immediately wonders, 'Why does he want to know?' Unless your purpose is clear, he may be reluctant to talk."

Once you've established your purpose, begin with an appropriate question and listen carefully to the answer, writing down key words that will remind you of the points being made. Don't feel bound by the order of your questions; you may find that your informant skips from one topic to another. However, do use the questions to keep the flow of information coming and to keep the interview on track. Don't be afraid to admit that you don't understand something and to ask for examples and definitions of terms. Sometimes you'll find it useful to read back specific responses to confirm that you have clearly understood what was said. This technique assures the person being interviewed that you have recorded the conversation accurately.

As your allotted time nears an end, thank your source for his or her time and prepare to leave. A source who wants to prolong the interview will let you know, but usually an hour will be enough time for both of you.

Using Interview Material

As soon as possible after you leave the interview, fill in your notes. Your *short-term memory*, your mind's temporary storage, will be loaded to capacity, but if you immediately write down what you can remember, you'll be able to capture most of it. Within an hour or two, though, that information will be gone. If you taped the interview, you probably won't need to transcribe the whole thing. Experts tell us that only about 10 percent of what's said in an interview is important or relevant; therefore, the tape will be most helpful as a way of filling in your notes. When you write up the interview, be sure that you quote accurately.

As a courtesy to your source, send a brief thank-you letter along with a copy of the document you write based on the interview. Keep a record of the date and place of the interview as well as the name and title of your source. Include this information in the list of references or works cited of a formal report. See Chapter 13 for details.

EXERCISE

Use the following guidelines to plan and conduct an interview that will result in an informal report: Choose a person to interview who is working in the field you plan to enter. Phone or write a letter to request an interview. Following the directions in this chapter, complete your background reading. Write out a series of questions; if required, get your instructor's approval of the questions. Conduct the interview, recording it and taking careful notes. Go over your notes immediately after the interview to fill in blank spaces. See Chapter 15 for more information.

Surveys

If you are looking for information in breadth rather than depth, you may want to conduct a survey. Surveys, also called polls, are useful in learning opinions rather than facts; think of the publicity given to pre-election polls and the frequency of consumer surveys about brands of toothpaste or cereal. In technical writing, surveys may help you establish such things as

- the need for a product (market surveys)
- criteria for judging an item
- the relative merits of one proposal or product over another (comparative surveys)
- opinions about policies, procedures, or proposed changes

Kinds of Surveys

You can survey in two different ways: (1) with a written questionnaire mailed to potential sources, or (2) with telephone or personal interviews. Written questionnaires can be distributed over a wide geographical or social area and have the advantage of soliciting anonymous responses. Unfortunately, it's hard to get people to fill out questionnaires; response rates to mailed questionnaires typically are only 15 to 20 percent. In addition, you have the cost of printing and mailing. Personal surveys seem to work best for students, who usually have only a limited time to gather information.

My students have conducted successful surveys in many different areas. They have, for example, contacted

- aviation students to test interest in the establishment of a flight school
- hospital personnel to determine success in using various surgical lasers or EKG machines
- companies to check preferences in copy machines or surge suppressors
- parents to find the preferred child carseat or day-care options and the reasons for the choice
- repair shops to test the reliability of washing machines, touring bicycles, or videocassette recorders

If you intend to survey students, check with your college or university to see what the procedures are. You may need to have your survey approved.

Survey Questions

In designing survey questions, you want to ensure both validity and reliability. *Valid* questions elicit the desired information; in other words, they actually measure what you intend to measure. *Reliable* questions mean the same thing (1) to each person and (2) each time they are asked.

Effective surveys ask only a few questions, which can be answered quickly and easily. Answers fall into two groups: those that are quantitative (can be counted) and those that are qualitative (must be read and evaluated). Answers that are quantitative yield totals and percentages; they are good if you need to survey a large sample. You can also design questions so that the answers can be read and tabulated by computer, a process that can save hours of your time. The four kinds of quantitative answers are

1. yes/no and either/or types with only two choices
2. multiple choice
3. ranking, in which the reader arranges items in order by preference
4. rating, in which the reader rates questions across a scale such as "very important," "important," or "unimportant"

Two kinds of questions yield qualitative responses that must be read and evaluated: short answer (fill in the blank) and essay questions. Even though these answers can't be counted easily, you may find that the details of the responses are as illuminating as statistics would be. Figure 3-8 shows examples of the various kinds of quantitative and qualitative questions.

Whatever kinds of questions you use, design them carefully and pretest them on a small group to see if your questions are clearly understood.

Figure 3-9 shows a sample of a brief student survey testing student interest in a proposed new university program.

1. Questions with only two choices (yes/no, either/or)

    Would you use a copier that was stocked with recycled paper?
    Yes _____   No _____

2. Multiple choice questions

    How many copies do you make a month?
    0–10 _____   11–25 _____   26–40_____   over 41 _____

3. Ranking, in which the reader arranges items in order by preference

    Number in order, the location you most often use when copying documents.
    _____ copiers in the library
    _____ copiers in the student union
    _____ copiers in my major department
    _____ copiers where I work
    _____ copiers at a commercial copying vendor

4. Rating, in which the reader rates questions across a scale

    How often do you take things apart to find out how they work?

    | | Some- | | |
    | Often | times | Seldom | Never |

    +_____   +_____   +_____   +_____

5. Short answer or completion

    How many years have you worked as an engineer? _____

6. Essays

    How do you feel the organization of this guide could be improved?

    _____

    _____

    _____

Figure 3-8 Types of Survey Questions

AVIATION STUDENT SURVEY

1. Check your current class standing:

   Freshman _____   Junior _____
   Sophomore _____   Senior _____

2. After graduation, do your career plans include flying for hire (i.e., airline pilot, corporate pilot, charter pilot)?

   Yes _____   No _____

3. Do you currently possess any type of pilot's certificates?

   Yes _____   No _____

4. Would you be interested in a new aviation concentration entitled "Professional Pilot"? In this concentration the student would be trained from zero time up to certified flight instructor. It would be the student's responsibility to provide the funds necessary for the flight training.

   Yes _____   No _____

5. Please add any comments or recommendations you might have.

Figure 3-9 Student Survey

Keep in mind that without formal training in how to construct surveys and how to choose representative samples, you should be cautious in using your survey data for any major conclusions. Remember that even expert surveyors come up with wrong predictions. However, you can use survey data as support for information you gather in other ways. Be sure to credit your survey in your list of references or works cited, and perhaps include a copy of the survey questions in an appendix.

Checklist for Designing and Administering a Survey

1. What specific information do I want from this survey?
2. What target group shall I survey? How many people do I need? Do I need to seek permission for the survey?
3. What are my key questions? What kinds of responses am I seeking (i.e., statistics or comments)? Should I use quantitative or qualitative questions? Will these questions result in valid and reliable responses?

EXERCISES

1. Use the questions in the checklist to design and administer a survey that will provide information for a document you are writing. Tabulate or summarize the results, and incorporate the data into your document. Credit the survey in the list of references or works cited, and include a copy of the survey questions in the appendix.

2. Find a copy of an existing survey for analysis. Look in books and periodicals or ask professors in your major field. Determine from the questions what information the surveyor was seeking. Rate the effectiveness of the questions, and decide if they could be stated another way. Be prepared to discuss in class.

3.4 Developing Information by Personal Investigation

Most college students, whether in technical majors or not, have written laboratory reports for a science class. A lab report is a good example of writing up information developed on your own—that is, by personal investigation. In writing up a physics experiment, for example, you might be asked to

- state the object of the experiment
- describe the apparatus used
- describe how the experiment was performed
- state the method of deducing results from original data
- summarize the results
- give a physical interpretation of results

All this writing stems directly from the work you have performed in the laboratory.

Other ways of developing information on your own but outside the laboratory are trip reports, site reports, analyses of collected data, and progress or status reports. These kinds of reports present what is called *primary* information (you develop the information yourself), whereas research reading produces *secondary* information (you compile the information from other sources).

If you are majoring in a technical subject, much of the writing you will do as a professional will be primary reporting on your work; your main job will be to do the work: examine, investigate, experiment with different options, design, and so on. The writing will result from that work. That writing, though not your central concern, is very important; in a field like engineering it enables you to share what you have done, and in the sciences, publication is your way of officially announcing your findings. Unfortunately, many technical professionals forget to document what they are doing; for example, they build or test prototypes without writing down what they did and how they solved problems. This lack of written documentation causes serious problems when valuable information is lost or when co-workers pursue paths already investigated and repeat errors that have already been corrected.

Personal investigation can be divided into three major activities: observation, examination, and experimentation.

Observation

In observation, you simply report what you see. For example, you might report on a process you have watched taking place, like the manufacturing of cheese. You might describe a physical object like a transformer, or the results of some action like the damage caused to a warehouse roof by a violent storm. In a time-management study, you might observe the specific repetitive movements of workers in manufacturing plywood. If you are asked to report on potential sites for your company's office complex, you would observe and describe the sites, their size, physical features, and elevation. Observation usually results in some kind of descriptive or procedural writing. These are explained in Chapters 10 and 19.

Examination

Although the word *examination* is sometimes used as a synonym for *observation*, examination implies more active involvement by the writer. When you examine something, you might take it apart in order to see how the parts work in relation to one another. You might study a bearing surface under a microscope to see the wear pattern. You might also take raw data, break it down, and analyze it in order to develop conclusions or recommendations. For example, you might study the vandalism reports in city parks over a six-month period to determine the types of vandalism and patterns of occurrence, so you could recommend preventive measures. Examination and analysis are different from observation because they involve evaluation of relationships or cause and effect. The results of such examination and analysis may be written up in a short, informal report or as a section of a longer, formal report.

Experimentation

Experimentation requires the most active involvement on the part of the writer: evaluating, operating, or testing objects, processes, or equipment. A software computer manual, for example, is best written by someone who tests the instructions by hands-on operation. An analysis of protein exchange at the capillary-tissue level is best written by someone who has participated in the laboratory study.

My students have experimented on a wide variety of tasks to develop information for technical-writing projects. One student tested the strength and elasticity of various brands of monofilament fishing line. Another spent an afternoon in an electronics store testing various brands of stereo speakers within specific parameters, setting up distances and measuring responses. Still another student gathered a box full of the kinds of cartons, bags, cans, and bottles that he would keep in a refrigerator. Taking this box to an appliance store, he stacked these items into various models of refrigerators to test the usefulness of shelf and storage space. Results of experiments like these can be used in the body of a report to support conclusions and recommendations.

Personal investigation—whether by observation, examination, or experimentation—will vary depending on your field and your assignment; therefore, it's hard to be specific about how you should proceed. You should, however, keep these general principles in mind:

1. Have a purpose, topic, and one major point written down.
2. Prepare for any personal investigation by doing background reading.
3. Keep your key words in mind as you proceed in order to maintain your focus.
4. Take careful, extensive notes while you are observing, examining, or experimenting.
5. Complete the analysis or calculations as soon as possible.
6. Write up your conclusions or observations while the material is still fresh in your mind.
7. Be sure to document the details, including times, places, equipment, and supplies.

Checklist for Personal Investigation

1. What kind of personal investigation do I need for this document?
   ___ observation
   ___ examination
   ___ experimentation
2. How shall I proceed? Do I know my purpose? topic? one major point?
3. Is my background reading complete, and have I decided on key words?
4. Do I have good notes to support my investigation?
5. What have I concluded from this investigation? What shall I recommend?
6. Have I written down details of time, place, equipment, supplies?

EXERCISE

Meet with three or four fellow students from the same major, and compile a list of three to five kinds of personal investigation you have carried out in the classroom, field, or laboratory. For each example, discuss the problems that occurred and the things that went well. Be prepared to present this material to the rest of your writing class to help them understand how personal investigation is carried out in your major field.

WRITING ASSIGNMENTS

1. Once the formal proposal for your term project has been approved, you are ready to begin gathering information. Before you launch that information search, though, you should determine what you already know. Choose one of the five ways to brainstorm and generate a seies of key words and ideas to guide you in your information search. If you are working on a collaborative project, this brainstorming should be done by the whole group. You will find that together you can generate many key words, which you can then group for each member's individual research.

2. Using your library or company resources, find the names and addresses of 8 to 10 sources of information for your term project. Find the names of specific people whenever possible but substitute job titles (such as Consumer Information Representative) or departments (such as Service Department) if necessary. Follow the suggestions and forms in Chapter 16 to draft a sample request letter, and submit it to your instructor if requested. Then type and send a letter to each potential source. Consult your planning sheet to set cutoff dates for information.

3. Review the notes you took in your background reading and your library search to determine the specific information that would enhance your major writing project. Choose one or more likely sources of information and set up an interview. Following the suggestions given in this section, ask specific questions. Complete your notes as soon as possible. Credit your source in your list of references or works cited. See Chapter 13.

REFERENCES

Applied science and technology index. 1991. 75,2. August. Bronx, NY: H. W. Wilson.
Brady, John. 1977. *The craft of interviewing.* New York: Vintage, 71.
Geographical abstracts: Physical geography. 1991. J/08144-09306, 1020.
Library of Congress. 1991. *Subject headings.* 14th ed. Washington: Library of Congress.
Sheehy, Eugene P. 1986. *Guide to reference books.* 10th ed. Chicago: American Library Assn.
Social sciences citation index. 1990. Philadelphia: Institute for Scientific Information, 5068.

Chapter 4

Organizing Information

You have been working at your project for some time; now, where are you in the process of writing your technical document?

According to policy analyst and writer Pam Patterson, you are at the point of applying "critical thinking: the ability to see clearly through masses of information or technical data to see the 'figure in the carpet,' the underlying themes, the organizational structure that will best serve the needs of the project and audience." Patterson works in aerospace technology, but her advice holds true for any writer wrestling with details of information that must be put in order. Right now *you* are that writer.

You have determined your purpose and assessed your readers' identities and needs; you also have your information—or most of it. You have searched in the library, interviewed specialists, and written for specification sheets. The stack of information on your desk is impressive, and you have clarified legal and ethical issues. According to your task sheet, you are on schedule. Are you ready to begin writing? The answer must be "No"—unless you are willing to generate draft after draft of text while you figure out the major point you want to make and the best way to support that major point. While technical writers approach organizing information—the next step in the process—in many different ways, they all *do* organize before they write.

Any organizing task can be divided into two subtasks: (1) what you as writer must do to put the parts in a coherent order, and (2) what you must do to help your reader follow your organization. The first subtask is *organizing information*, which is the subject of this chapter. The second subtask is revealing that organization to the reader, which you do through content and format cues, a subject covered in Chapter 6.

When you organize, you do the thinking and arranging that will guide your writing. At this stage, you can (and should) experiment with different ways of organizing the same material, because almost all information can be arranged more than one way.

For example, Terri, a physics major, has been studying 8-inch reflecting and refracting telescopes. Her purpose is to find the one that will best meet the needs of an amateur (her reader) in the market for a deep-sky photographic and observational instrument. When she sorts her gathered information, this is what she finds:

The telescope models included the following:

- Celestron International model C8
- Criterion Instruments model 8000
- Meade Instruments model 2080

All the telescopes evaluated had the following common features:

- aperture
- f/ratio
- optical system
- stellar magnitude limit
- useful visual magnitude limit
- aberration

The telescopes differed in these features:

Important Features
- optical surface accuracy
- oculars (eyepieces)
- size and weight
- cost

Less Important Features
- gears
- tube construction

The major point: The Celestron International model C8 is the best choice.

How should Terri organize her material? She could organize this way:

Conclusion (Major Point)
Common Features of Telescopes
Differing Features of Telescopes
    A.   Important
    B.   Less Important

Or, she could do it this way:

Conclusion (Major Point)
Features (or Criteria) for Discrimination
    A.   Optical surface accuracy
    B.   Oculars
    C.   Size and weight
    D.   Cost
Common Features (or Criteria)
    (omitting in this outline gears and tube construction since they are not important)

Or, this way:

Important Features
Less Important Features
Common Features
Conclusion (Major Point)

Or, this way:

Introduction to Telescopes
Criteria for Discrimination
Analysis of Each Model by Criteria
Conclusion (Major Point)

These are only four possibilities; there are several other variations.

Terri might also want to analyze a document that was written to perform the same task she was assigned. This often happens on the job; you are given a document written by someone else and told, "This will give you an idea how to do it." You need to study that document carefully to see if its organization meets the needs of your assignment. If it does, you can preserve the organization, enhancing it with your own ideas. If it doesn't meet your needs, you should try a new organization pattern. At work you must, of course, follow company policy; nevertheless, you don't want to follow the organization of existing documents without asking questions.

This chapter will explain a process you can use to organize information. You begin by determining the major point you want to make.

4.1   Finding the Major Point and Sorting Information

Before you started gathering information, you may have spent a few minutes brainstorming to discover what you knew. Those words and phrases were randomly generated and may have appeared as a hodgepodge on the page. Nevertheless, from that brainstorming session, you probably found some key words and ideas that helped you focus your search for information. Using those key words, you searched in a variety of sources and gathered a mass of information—which may now appear just as much of a hodgepodge as a brainstormed list. But don't be discouraged, because now you are ready to sort the information and analyze it to determine your major point.

The first step is to dig out your planning sheet from Chapter 1 and reread your stated purpose and the description of your intended reader. Ask yourself if either one has changed. For example, you may find that you need a narrower purpose than you'd originally thought. Now look at the topic you originally set. Is the scope of that topic still the same? Do you have good information that will allow you to cover more territory? Have you discovered that information on some aspects of the topic is not available? Adjust your scope if necessary. When you're satisfied with the purpose, reader, and scope, spend some time simply reading through the notes you've taken from books and articles, and the information you've obtained through interviews, surveys, or examination.

As you read, you need to ask yourself four questions:

1. What is the major point I want to make?
2. Which information is useful and which is superfluous?
3. Is this information valid?
4. What am I missing?

If you have ever played any kind of card game, you have routinely asked yourself similar questions about the cards as you arranged your hand. Just as the rules of each card game determine the value and importance of each card, so the boundaries you have set for this assignment will determine the value and importance of each piece of information.

Determine the Major Point

Consult the purpose statement you wrote while you were planning, and use it and the information you have gathered to help you frame your major point. If possible, write that one major point in a single sentence or phrase. Here are some examples of major points:

> Araxo Corporation's portable generator is the one most suitable for Jones Co. to use on construction sites in northern Utah.

> At the planning and budget review sessions scheduled for Nov. 14–18, 19__, managers should provide details on assigned staffing, department budget, and program requirements.

> In hand developing X-ray film, six steps are followed in darkroom conditions.

> When you read my qualifications, you will want to interview me for the position of junior statistician.

This major point may eventually shape the title of your report, form the conclusion of a memo, or become the overview statement that begins your written document. It may even be an *unstated* point. (In a job-application letter you wouldn't, for example, state your desire for an interview as boldly as in the preceding example; nevertheless, that is precisely your major point.) At this step in the writing process, you should think of your major point as the goal of your writing (what you want the reader to understand).

What if you have more than one major point? Suppose that your purpose is to describe a new digital blood-pressure gauge and tell how to use it. What will be your major point? You could (1) divide your assignment into two sections with one major point for each section (how it looks; how to use it), (2) make how it looks the focus and explain the use of each part as you describe that part, or (3) make its use the focus and add description as you spell out the instructions. But notice that you make choices; you don't just let organization happen.

Checklist for Choosing the Major Point

1. Before sorting the information, review the planning sheet and ask
   - Has my purpose changed? If so, how?
   - Has my intended reader (or readers) changed?
   - Should the scope of my topic be enlarged? diminished?
2. After studying my purpose statement and my stack of information, what is the major point I want to make to my reader? Can I write it in one sentence?
3. Do I seem to have more than one major point? If so, what are the points?
4. Do I want to make one point more important than the others?

☰ EXERCISE

Examine the list you brainstormed for a description of a 10-speed bicycle in exercise 1 in section 3.1. Determine what your one major point would be if you were going to write that description. In one sentence, write down the major point. Be sure you have considered purpose, reader, content, and tone—all items on the planning sheet in Chapter 1.

Judge the Usefulness of Information

In many card games, you must judge the relative value of the cards you are given. Some cards will please you, and you'll begin planning ways to use them to your best advantage. Other cards will go in the discard pile.

In sorting information, you can follow the same principle. At this point, you can sort into two piles: information that will contribute to your major point, and information that may be interesting but is extra and unneeded. Librarians who monitor student research say that about 30 percent of the information most students gather is not useful for the final document, but don't be surprised if an even greater percentage never appears in what you write. Information gathering is a cumulative process, and you may go astray in the early stages as you follow leads that appear interesting. As more sources are covered, though, your mind begins to fit the pieces together, and your focus becomes more concentrated.

For example, if you were comparing different brands of answering machines, you might consider the manufacturer's warranty an important factor until you discover that all warranties are virtually identical. Then you must shift your focus to factors that do vary. Sometimes a source (a person you interviewed, perhaps) will open an avenue that you hadn't thought about and will shift your focus slightly or even radically. You also may have started a project with a preconceived idea of the "best" procedure or the "best" product. Students tell me they frequently are surprised by what they learn when they gather and analyze information. Be flexible in your thinking until all the facts are in, so you don't skew your information to fit your prejudgments.

Don't throw away notes that you put in the discard pile. You may want to include some of this material in an appendix to a long report or keep it in your file for later reference. You may even find as you begin to write that a discarded item can be used after all.

Evaluate the Validity of Information

If you are dealt a joker in a card game that does not use jokers, you will quickly ask for a redeal. In evaluating information, the same principle operates. You need to examine each bit of information by asking:

- Are examples or evidence available to support this?
- Where did it come from?
- Is the source respected and unbiased?

How can you determine the validity of any piece of information? One way is to be sure that generalizations are supported by facts, statistics, examples, or expert opinions. You may have to drop back to the information-gathering step to find additional support. Another way is to examine the source. If the source of information is a person, you can look at that person's credentials—that is, education, experience, position, and acceptance as an expert by others. You might check expertise by looking in a citation index (see Chapter 3) to see how often the person has been quoted by others. If the source is a journal article, you can estimate its validity by the kind of journal. Professional journals in all fields have panels of specialists who must approve submissions; in this case, publication itself may imply validity. If the article is in a trade journal, however, you may need to determine its validity by checking the writer's credentials, since trade journals do not generally use panels of specialists for approval. Also be aware that trade journals are supported by advertising; thus, they may highlight the services and products of their biggest advertisers. For this reason, articles in trade journals may be biased, so you should examine information from such articles with care. Many journals give short biographical sketches of each author, so you can evaluate education and experience that way. If you are in doubt, ask professors in those fields.

Find Any Missing Information

As you must discard in some card games, so too you can replenish your hand by drawing new cards. You always hope that those new cards will help fill the gaps in your hand. The information sorting and evaluating process will also show you what's missing and where your information is thin or unreliable. If you sort and evaluate soon enough, you'll still have a chance to find the missing information before your writing deadline approaches. Fill the information gaps as soon as possible.

4.2 Choosing Supporting Subpoints

Now that you have chosen the major point and sorted your information, you are ready to choose subpoints to support that major point. To do that, you need to group your pieces of information—always remembering what you wrote on your planning sheet about your purpose, readers, and topic. When you group or cluster related ideas, you must keep your reader's *short-term memory* in mind. Short-term memory is the mind's temporary storage. You use short-term memory when you hold a new telephone number in mind long enough to dial it, when you listen to a speech, or when you read a sentence. Harvard psychologist George A. Miller (1967) says that the capacity of short-term memory is seven plus or minus two items. For the reader to easily understand and remember, three to five points is excellent, seven is acceptable, and nine is the maximum. Personally, I believe short-term memory is related to the number of fingers we have. I tell my students that they understand best when they can tick off on one hand the main points they're hearing or reading: it's harder when they have to use both hands; it becomes impossible when they have to take off their shoes.

In order to make communication easier, then, remember the limitations of your reader's short-term memory. Choose from two to nine subpoints and subsume your other points under them.

Besides the *number* of subpoints, you need to consider their *relationship*. Your two (or four, five, or seven) subpoints need to be equal in importance. They also need to be logically related and grammatically parallel. In this chapter, I will talk about equal importance and logical relationship. In Chapter 6, I will talk about grammatical parallelism.

Think of your items of information as a pile of bones of different shapes and sizes. Some bones are long and sturdy, while others are small and delicate. Your job is to sort through the pile of bones and construct a skeleton that will fulfill your writing purpose. Remember too that there are all kinds of skeletons: you can't assume that you'll be building a human skeleton. Maybe your skeleton will be four footed; maybe it will have a tail and no feet at all.

In a human skeleton, you would probably consider the skull and sternum to be more important than toe bones, with arm and leg bones perhaps equally important. Think about your items of information in the same way. Which individual items are equal in importance, rank, or degree? Once you have determined that, consider how the points are related. Both arm bones and leg bones, for example, provide movement (one kind of relationship you might want to develop). But if you were looking at supporting structures, then leg bones might be more logically related to the vertebrae that make up the backbone. However you approach your subject, logical relationship means that the main subpoints are tied to one another in a recognizable way.

For example, suppose that in brainstorming for a description of a running shoe, your generated list of parts and their definitions looked something like this:

tongue	a cushioned piece of fabric or leather that lies on the top of the foot and below the laces
heel tab	a padded lip at the top of the collar that protects the Achilles tendon
outsole	the layer of rubber on the shoe that makes contact with the ground
upper	layers of fabric or leather that enclose the top and sides of the foot
eyelets	holes in the upper through which laces can be inserted
insole	the pad on which the foot rests inside the shoe
last	the means of securing the upper to the sole
lace	a fabric or leather strip used to fasten the two sides of the upper together
heel counter support	a firm material that buttresses the heel counter
midsole	a layered foam section between the outsole and insole
heel counter	a firm cup that holds the heel and restrains excess motion
collar	the top edge of the shoe that envelops the foot around the ankle
sole	the bottom of the shoe

You need to determine some way of handling all those pieces of information that will make sense to your reader. If you discuss them in the order they are listed here, the reader will not see the relationships or understand what your major point is. So, you study the list and, remembering that your purpose is to write a description, decide that your one major point is "A running shoe has three major parts." You decide on three major parts somewhat arbitrarily, remembering short-term memory. Now you can test your major point by choosing subpoints.

When you choose your main subpoints, suppose you list these:

1. outsole
2. laces
3. upper

If you choose these three, your subpoints are not equally important. Laces are needed to fasten the shoe together, but they are not equal in rank or degree to the upper and outsole. To complicate matters, neither are the outsole and upper equal in rank.

You look again at your list and choose new main subpoints. You need to find terms of larger scope that can gather the various parts under them in a meaningful way. This time you choose

1. sole
2. upper

Now you need a third term that can include all of the interior parts of the shoe, but there is none on the list. Think about your reader and ask, "How will the reader easily understand these parts and how they work together?" You try again and come up with two new terms: *back* and *front*. Now you can discuss all the parts of the shoe in a meaningful way:

1. sole (for all the parts on the bottom of the shoe).
2. back (for all the supporting and cushioning structures of the heel)
3. front (for the parts that cover the top of the foot)

In your choice of supporting subpoints, you need to show how the parts relate to one another. A different organization might use outside (or visible) parts and interior (or hidden) parts.

Remember that as the writer you can choose your organization. But since you want to communicate that organization to a reader, you must think about equal importance and logical relationship, and beyond that, what the reader knows, does not know, and will do with the information.

EXERCISE

Look back at your list of parts for the description of a 10-speed bicycle in section 3.1 and your choice of the one major point in section 4.1. From your list, choose the most important subpoints that support your major point. Remember to keep the number of subpoints under nine so you don't overload your reader's short-term memory. Be prepared to discuss your choices in class.

Arranging Subpoints Logically

Now that you have chosen a manageable number of subpoints that are of equal importance, you need to arrange them in some kind of meaningful order. To do that, you need to know how information can be organized.

How Information Is Organized

Theorists tell us that information can be presented in three different ways: at *random*, in *sequence*, and by *hierarchy*. You use all three ways every day.

Random Organization

You "draw straws" to determine who gets the extra piece of chocolate cake: it's a random choice. You enter a sweepstakes contest to win a trip to Hawaii; the prize is drawn at random. You shuffle the cards and deal a hand of poker: the cards are randomly distributed. You brainstorm ideas for a writing assignment: your ideas are randomly generated. In random organization, you treat each item as equally important, and you want the arrangement to be by chance.

Sequential Organization

To look up a phone number for the Cox Tire Service, you used the phone book, arranged alphabetically. Later, to write a check for a new tire, you used your checkbook, arranged numerically. Tomorrow you will write a set of instructions for adding a modem to a personal computer. You will tell the reader what to do first, second, and third. In all the kinds of sequential ordering, you treat each item as equally important, but you choose *one* identifying characteristic (the first letter of the word, an assigned number, the location in a series), and you organize by that characteristic.

Hierarchical Organization

This morning a revised company organizational chart was posted on the bulletin board where you work. When you studied it, you noted that a new management level had been added to your department and that your boss had moved up in the hierarchy. At the beginning of this term, you opened a new economics textbook to the table of contents to see how it was organized. You found the book divided into six key sections. Within each section, four to nine chapters covered divisions of the key points, and within each chapter three to eight subdivisions discussed details. You were looking at a hierarchical organization. In hierarchies, you no longer consider items equally important. Instead, you assign a value to each item, and by that weighted value you determine the rank, level, or tier in which you place it.

Combinations

Often an arrangement uses both a sequence and a hierarchy. To understand how a combination works, think of a typical family tree. A simple family tree, like the one in Figure 4-1, is primarily hierarchical because it shows a father and a mother in key

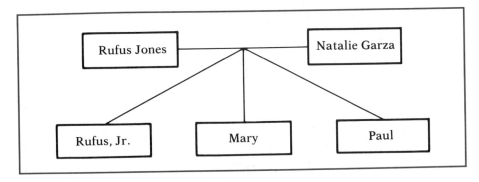

Figure 4-1 Hierarchical Organization

positions, with their children at a lower rank. However, in time, the family tree grows and extends; the lower-ranked children marry and reproduce, and the tree also shows progression in chronological time. Figure 4-2 shows how the extended family tree is a combination of a hierarchical and a sequential arrangement.

Because you are trying to help the reader see relationships, random arrangement doesn't work in technical writing. Instead, you use *either* sequential *or* hierarchical organization, or you combine sequential *and* hierarchical organization. Whatever arrangement you choose, the main points you select and the order in which you present them will provide the skeleton organization or framework for your writing. Skeletons and frameworks are supporting structures; with that interior framework in place, the writing task is easy because you can simply flesh out the skeleton.

The Relationship between Organizing and Outlining

What is the relationship between organizing and outlining? *Organizing* means arranging the information into some kind of structure, and—for most technical writers—*outlining* is the way of showing that structure. An outline is not, however, the only way to show structure. For example, suppose you choose three points to support your one major idea. You might show the same structure through a flow chart, an idea tree, or an outline, as shown in Figure 4-3.

An outline is merely one way of making your structure visible; at the organizing step, if you are more comfortable with idea trees or flow charts, use those visual means to keep track of the structure you decide to use. Whatever the structure, keep it simple.

Various *methods of organization* have become standards in technical writing because they help readers understand the relationships among pieces of information. If you study these methods of organization, you will be able to organize your information clearly.

When I talked about the ways of organizing information, I said that technical writers organize points in two major ways: sequentially and hierarchically. All the *methods of organization* described in this chapter can be classified in one or the other

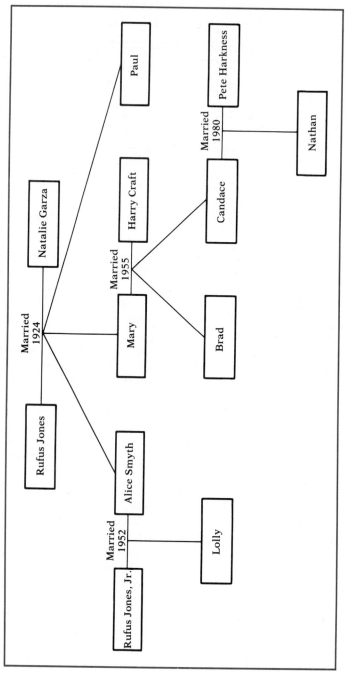

Figure 4-2 Hierarchical and Sequential Organization

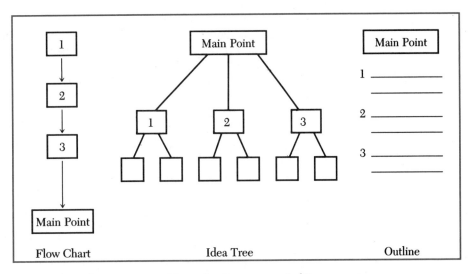

Figure 4-3 Three Ways to Show the Structure of a Document

of these two categories or a combination of both. Figure 4-4 gives you an overview of these methods of organization.

You have probably already used many of these in writing without thinking of them as methods of organizing. Some methods discussed in the explanations that follow may be new to you. Read through each explanation to see how the material is organized and what the potential uses are. Study the example. Later, you can come back to this section for help in choosing a method of organization that will fulfill your purpose and meet a specific reader's needs. When you organize, you need to be flexible, because often the same information can be handled in two or three different

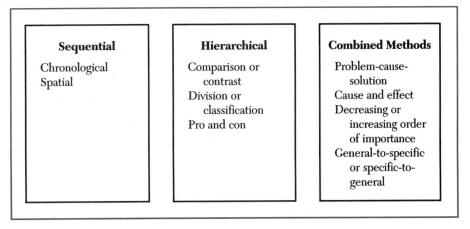

Figure 4-4 Overview of the Methods of Organization

ways, each way emphasizing significant information differently. Your job is to choose the most effective method of organization based on purpose, reader, and topic.

In any method of organization, you can number the main points in several different ways: with Arabic numbers (1, 2, 3), with Roman numerals (I, II, III), with decimals (1.0, 2.0, 3.0), or with letters (A, B, C). The type of numbering varies among organizations; what is important is that the reader understands the values of the numbers and the relationship among items.

Methods of Sequential Organization

Sequence means "following one another in order"; thus, a flow chart illustrates sequence visually. The two main types of sequential organization are chronological and spatial.

Chronological

In a chronological arrangement, points are arranged on the basis of time, from first to last. Chronological sequence is useful for

- work schedules
- instructions
- test procedures
- process descriptions
- the methods section of experiments
- historical reviews

The process description in Figure 4-5 is organized chronologically. This chronological arrangement works well because it follows the process: the microwaves must first be produced, then they are controlled and distributed, and finally, they are absorbed by the food. Subpoints under each of the three main points can explain each step.

Spatial

In a spatial outline, points are arranged as they appear physically, in relationship either to one another or the surroundings. For example, spatial arrangement may be

PURPOSE:	to explain how microwaves cook food
READERS:	generalists
ONE MAJOR POINT:	The microwave cooking process consists of three steps.

   I. Production of Microwaves
  II. Control and Distribution of Microwaves
III. Absorption of Microwaves

Figure 4-5 Chronological Organization

top to bottom, left to right, outside to inside, or front to rear. This organization is useful for describing

- objects
- places
- test procedures
- processes

The outline in Figure 4-6 is arranged spatially; that is, the skin is viewed in cross section, and the discussion is oriented to the layers of the skin. Viewing the skin spatially from the top layer down helps achieve the purpose of warning of the dangers of tanning salons. The reader can see that the deepest skin layer sustains the most damage.

Methods of Hierarchical Organization

Hierarchy implies choosing the most important points and assigning areas of responsibility or coverage for each of them. The idea tree and the outline both show hierarchy visually. Three types of hierarchical organization are commonly used: comparison or contrast, division or classification, and pro and con.

Comparison or Contrast

Points are chosen to show similarities or differences between elements of two or more subjects, or to explain a difficult or unfamiliar subject by relating or contrasting it

---

PURPOSE: to warn of the dangers of indoor tanning salons
READERS: generalists
ONE MAJOR POINT: The ultraviolet A rays produced in tanning salons
                    cause damage to the deepest layer of the skin.

1.0 Epidermis
   1.1 Penetrated by UVC rays
   1.2 Minimal damage because UVC rays are screened out of natural
      sunlight
2.0 Dermis
   2.1 Penetrated by the UVB rays in natural sunlight
   2.2 Damage
      2.2.1 Reddening of skin
      2.2.2 Thickening of skin
3.0 Subcutaneous Tissue
   3.1 Penetrated by the UVA rays produced in tanning salons
   3.2 Damage
      3.2.1 Harmful to blood vessels
      3.2.2 Can cause cataracts
      3.2.3 Causes skin swelling and injuries when mixed with chemicals
         or perfume on skin

Figure 4-6 Spatial Organization

to one simpler or already known. The most important ways to explain the comparison become the main points. The comparison or contrast method is useful for

- sales pitches
- recommendations
- analyses
- extended definitions

The sample outline in Figure 4-7 analyzes two kinds of woodworkers' carbide-tipped saw blades, those manufactured for the industrial market and those manufactured for homeowners. Figure 4-8 shows an alternate method of organization using comparison or contrast.

In both methods of comparison or contrast, the two kinds of blades are measured against each other. The first method stresses the criteria or points of comparison; when you want to measure something against judgment points, this is a good choice. The second method discusses first one type and then the other and requires a concluding

---

PURPOSE: to analyze various carbide-tipped saw blades
READERS: operators
ONE MAJOR POINT: While industrial-quality blades cost more than
                   homeowner blades, a comparison of materials
                   used, manufacturing processes, and the sharp-
                   ening process shows them to be a better buy for
                   fine carpentry.

  I. Materials
     A. Saw blade blanks
        1. Industrial blades
        2. Homeowner blades
     B. Carbide teeth
        1. Industrial blades
        2. Homeowner blades
 II. Manufacturing Processes
     A. Machining
        1. Industrial blades
        2. Homeowner blades
        (1 and 2 will be the same for each following subpoint)
     B. Heat treating
     C. Surface grinding
     D. Tensioning
III. Sharpening Process
     A. First grinding
     B. Polishing—number of faces

Figure 4-7 Comparison or Contrast Organization Stressing Points of Comparison

```
    I.  Industrial Blades
        A.  Materials
        B.  Manufacturing Processes
        C.  Sharpening Process
   II.  Homeowner Blades
        A.  Materials
        B.  Manufacturing Processes
        C.  Sharpening Process
  III.  Conclusion
```

Figure 4-8 Comparison or Contrast Stressing Types Being Compared

section to make direct comparisons. This method is useful when you want to present all the information about one type in one place.

Division or Classification

In division, points are determined by breaking a topic into its most important parts or functions. In classification, points are determined by grouping related items and assigning an umbrella term to each group. Division or classification is useful for

- analyses
- descriptions
- design discussions
- procedures

In the example in Figure 4-9, the writer wants to describe the design of a large computer information system workstation. Dividing the material into the component parts of the workstation makes it easier for readers to understand.

```
PURPOSE:            to describe the CompuCorp 1000 workstation
READERS:            managers (potential purchasers), operators
ONE MAJOR POINT:    The four components of the CompuCorp 1000
                    workstation work together to provide easy solo
                    operation and the ability to connect to both
                    local and distant networks.

1.0  Introduction giving overview of the workstation
2.0  The Four Major Components
    2.1  Processor
    2.2  Display
    2.3  Keyboard
    2.4  Screen Pointing Device
3.0  Network Connections
```

Figure 4-9 Division Method of Organization

Because the workstation is a complex product, the writer gives a broad view first and then breaks the material into small, manageable segments.

The discussion of running shoes in section 4.2 deals with organization by classification. You started with a list of parts, and—in order to show how those parts were related—you needed to find three main terms under which the parts could be grouped.

Pro and Con

In the pro-and-con pattern of organization, points are determined to be either advantages or disadvantages. The presentation is one category at a time followed by conclusions and recommendations. Pro-and-con organization is useful for

- evaluations
- proposals
- justification
- persuasion (sales and technical advertising)
- recommendations

The partial outline in Figure 4-10 comes from a report that analyzes grand pianos in order to recommend one for purchase. Part of the analysis looks at the "action" of the piano.

PURPOSE OF THIS SECTION: to evaluate chemically treated hammer heads

READERS: generalists, managers, operators (potential consumers)

ONE MAJOR POINT: The disadvantages of chemically treated hammer heads outweigh the advantages.

IV. Hammer heads
   A. Description
   B. Chemical treatment
      1. Advantages
         a. Shock resistance
         b. Resilience
         c. Shape retention
      2. Disadvantages
         a. Excessive hardening over time
         b. Production of harmonics
         c. Long-term metallic or tinny tone

Figure 4-10 Pro-and-Con Method of Organization

Giving both the pro and con sides of an issue provides balance and lets the reader participate in the evaluation or analysis. In this case, the advantages are listed first, and the writer closes with the point she wants to make—the disadvantages. In another situation, disadvantages might be listed first. Do not assume that you should always put your major point at the end, however, for in most technical writing the opposite is true: readers want to know your major point or your conclusion at the beginning. Then, if they wish, they can read on to see how you arrived at that major point.

Combined Methods

Sometimes sequential and hierarchical organization are combined to produce the following four methods of organization:

1. problem-cause-solution
2. cause-and-effect (or effect-and-cause)
3. decreasing (or increasing) order of importance
4. general-to-specific (or specific-to-general)

Problem-Cause-Solution

When you develop an outline using the problem-cause-solution method, points are grouped by the category into which they fall. This kind of classification is useful for

- recommendations
- problem-solving
- analyses

The outline in Figure 4-11 is for a "recall" letter from a major automobile manufacturer. The method of organization explains the problem, discusses the cause, and then recommends the solution (bring the car in for a no-cost refit). Stating the problem first gets the reader's immediate attention. Once the reader recognizes what can happen to the car if the engine rotates, he or she will pay attention to the causes and will be more likely to follow the solution—to bring the car in for a refit. This organization is primarily hierarchical, but cause also precedes solution in the sequence.

Cause-and-Effect (or Effect-and-Cause)

In a cause-and-effect organization, points are chosen hierarchically: that is, on the basis of which is more important, the cause or the effect. Once the supporting points are chosen, they are presented sequentially. This method of organization is useful for

- accident reports
- predictions or plans
- analyses

In Figure 4-12, the writer wants to use an airplane maintenance problem to push for better maintenance procedures. This part of the outline describes an "incident" that

PURPOSE: to recommend bringing the car in for a refit
READERS: generalists
ONE MAJOR POINT: The possible safety hazard is separation of an
                    engine mount caused by fatigue of the rubber
                    cushion; the hazard can be eliminated by the
                    installation of restraints.

1.0 Introduction explaining affected models
2.0 Explanation of the problem
  2.1 Description of engine mounts
  2.2 Engine mount separation allows engine rotation
    2.2.1 Can hold the throttle open
    2.2.2 Can disconnect the power brake vacuum hose
    2.2.3 Can interfere with the shift linkage
3.0 Cause of the problem
  3.1 Fatigue of the rubber cushion in the mount
  3.2 Separation of the rubber cushion from the mount
4.0 Solution
  4.1 Installation of restraints
  4.2 Procedure for arranging installation

Figure 4-11 Problem-Cause-Solution Organization

could have led to an accident. Effects are described first, then causes. Notice in this outline that the main divisions are the effects and the causes, a hierarchical organization. But within the effects section, the order is sequential: first flight problems and subsequent maintenance, second flight problems, and so on. Arranging the material this way stresses the time lapses and helps the writer make his point about inadequate maintenance procedures.

Decreasing (or Increasing) Order of Importance
    When organizing by decreasing order of importance, you first choose points by their importance (hierarchy) and then present them in sequence from the most to the least important. Decreasing order of importance gets the reader's attention with the key point and makes a strong initial impression. In increasing order of importance, you again choose the most important points (hierarchy), but you present them in order from least to most important. Increasing order of importance works well with a potentially hostile reader because you can meet the reader on common ground and then use the pattern to move to the points that are most important to you. Order of importance arrangements are useful for

- conclusions and recommendations
- scheduling reports
- purchasing reports
- explanations

PURPOSE: to describe a near accident and discover why it occurred

READERS: operators (maintenance), technicians, managers

ONE MAJOR POINT: The plane's problems in flight were caused by a tiny piece of potting compound not discovered for four months due to inadequate maintenance procedures.

2.0 Effects in flight and actions taken
 2.1 First flight
  2.1.1 Plane is unstable—pitches up.
  2.1.2 Autopilot system is checked; some parts replaced.
 2.2 Second Flight
  2.2.1 Plane is sloppy in pitch.
  2.2.2 Autopilot system is checked; other parts replaced
 2.3 Third flight
  2.3.1 Pitch problems continue. Trim problems noted.
  2.3.2 Pressure check of system. Small hole found.
 2.4 Fourth flight
  2.4.1 Pitch problems continue. Other trim problems noted.
  2.4.2 Entire autopilot system disassembled for visual inspection and testing.
3.0 Cause of problem
 3.1 Piece of potting compound 1 cc in diameter found in venturi.
 3.2 Obstruction causes decreased ram air pressure to trim and autopilot.

Figure 4-12 Effect-and-Cause Organization

- status reports
- persuasion and argumentation
- problem-solving

In Figure 4-13, the writer uses decreasing order of importance, starting with the most critical feature and moving down to those less critical. Notice in the Safety Features section that the first two points are critical; they prevent unauthorized use and permit the system to be shut off in an emergency. Although the other points help the surgeon observe safety, they are less critical.

General-to-Specific (or Specific-to-General)

When you organize from general-to-specific (the funnel method) or specific-to-general (the horn method), you classify points by degree of detail and then present them sequentially from broadest to narrowest or narrowest to broadest. In general-to-specific organization, the broad picture comes before the details. In specific-to-

---

PURPOSE OF THIS SECTION: to explain the importance of safety in choosing one surgical laser over another

READERS: managers, technicians, operators

ONE MAJOR POINT: Surgical laser safety depends on specific features and procedures.

3.0 Criteria for Evaluation
  3.1 Safety
    3.1.1 Safety Features
      • Keylock master switch
      • Emergency shut-off control
      • Simple meters and scales
      • Test controls
      • Audible and visual warning signals
      • Shields and shutter systems
      • Short focal lens

    3.1.2 Safety Procedures
      • Recalibration before each use
      • Use of focusing guides with short focal length and easy focal spot control
      • Use of wet towels near incisions
      • Avoidance of combustible anesthetic gases

  3.2 Versatility

Figure 4-13 Organization by Decreasing Order of Importance

general organization, the opposite is true. This method is also called simple-to-complex (complex-to-simple) or deductive (inductive) reasoning. It is useful for

- teaching or tutorials
- reviews
- recommendations
- evaluations
- introductions

In Figure 4-14, the writer plans an introduction that will move from the specific example to the general point. The same material could be arranged from the general point to the specific example. The details here catch the reader's attention; as the introduction proceeds, the points become more general—moving from Tucson to individual states to the United States as a whole. This organization is especially good for introductions when readers are generalists.

Remember that these *methods of organization* can be combined in any one writing project. An individual paragraph usually follows a single method of organization, but a long report or proposal might combine several methods.

PURPOSE: to introduce the food-waste problem and attract the reader's attention
READERS: generalists
ONE MAJOR POINT: Discarded food in Tucson is typical of what happens in the United States.

I. Introduction
   A. Tucson residents discard 9,500 tons of food annually.
      1. Plate scrapings account for one-third.
      2. Spoiled lettuce, partially eaten apples, rancid cheese, leftover macaroni, and other forms of waste account for two-thirds.
   B. Tucson residents discard 15 percent of the food they buy.
   C. Studies show similar patterns in Wisconsin and California.
   D. U.S. residents discard enough food every year to feed all of Canada.
II. Report on Consumer Food Waste

Figure 4-14 Specific-to-General Organization

Checklist for Organizing Information

1. What is the one major point?
2. What are the main supporting points?
3. Have I considered short-term memory by using fewer than nine points?
4. What method of organization best supports my purpose and meets the need of my targeted readers?

Sequential	*Hierarchical*	*Combinations*
chronological	comparison or contrast	problem-cause-solution
spatial	division or classification	cause-and-effect (or effect-and-cause)
	pro and con	decreasing (or increasing) order of importance
		general-to-specific (or specific-to-general)

EXERCISES

1. Listed below are purposes for specific writing assignments. For each one, choose an appropriate method of organization from the checklist. If you think more than one method is appropriate, also indicate your alternate choice.

   1. to *evaluate* the potential closing of ZBD's Denver manufacturing plant
   2. to *explain* the function of the autopilot in an airplane
   3. to *explain* why the thermocouple in the toaster failed
   4. to *teach* the reader about lasers
   5. to *describe* the organizational structure of ZBD Mfg. Co.

6. to *recommend* which leasing service should be selected to provide fleet automobiles for ZBD branch offices
7. to *report* on an automobile accident
8. to *explain* the responsibilities of the vice president of sales at ZBD Mfg. Co.
9. to *instruct* the reader in how to set the switches for operating the warehouse burglar alarm
10. to *explain* the priorities in scheduling the shutdown of the generator

2. Refer to the subpoints you chose to describe a 10-speed bicycle in section 4.2. Review the writing purpose and arrange the subpoints according to the most appropriate method of organization explained in this chapter.

3. The following table presents information that might be contained in a report on the storage conditions of paper stock at a printing company. It does not provide the details, only the main points. Remembering that information can be organized in more than one way, use this information to write three simple outlines, showing three different ways to organize these facts. Be prepared to discuss how each method of organization would be better for particular purposes or groups of readers.

CONDITION OF PAPER STOCK AFTER STORAGE

Kind of Paper	Location			
	Warehouse A	Warehouse B	Warehouse C	Warehouse D
coated	poor	good	good	poor
book	good	good	poor	fair
tag	poor	fair	poor	fair
newsprint	fair	poor	poor	poor

Adding Details to the Subpoints

Now that you have chosen your main points and arranged them in an appropriate *method of organization*, you are ready to add the details to each main point. You need to do two things:

1. Make sure that the details support their particular point.
2. Arrange each subset of details in the method of organization that will work for that section.

Details do not need to follow the same organizational method as main points. For example, the eight chapters in Part 1 of this book follow a chronological method of organization; in the process of writing, planning comes before gathering information. But the sections that make up Chapter 3 follow a division-and-classification method of organization. "Developing Information by Personal Investigation" is one way of gathering information, and "Contacting Knowledgeable People" is another way, but there is no reason that one of these steps must precede the other.

As you add details to your main points, you will also discover where you may need more information, and you can drop back to that step. It's easier to find additional information now than it will be when you begin the actual drafting.

Choosing an Outline System

Now is also a good time to choose an appropriate outline system. So far all your organization efforts have been on rough outlines, idea trees, or flow charts. Rough outlines will work for short documents, but before review of a long document, you may want to set up a formal numbering system. The two systems in Figure 4-15 are commonly used in technical writing.

Your company or instructor may favor one system over the other; increasingly, companies are using the decimal system, and military documents use it exclusively. In either system, if your instructor or company style guide permits, you might want to use "bullets" instead of numbers after the third or fourth level of subordination. Bullets are usually either filled-in circles or squares. Most word processors have a bullet; on the typewriter you can make bullets by typing a lowercase *o* and filling it in. Using bullets will simplify the document. Compare:

4.2.1.1	4.2.1.1
4.2.1.1.1	• _____
4.2.1.1.2	• _____
4.2.1.2	4.2.1.2

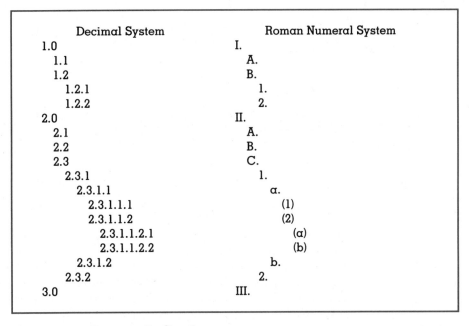

Figure 4-15 Common Outline Systems

Bullets are especially useful when you do not want to indicate any particular order in a list or when the list will not need to be referenced later in the text.

This textbook uses a modified form of the decimal system. Chapters are numbered, and the main divisions are denoted by decimals, but subdivisions are indicated by type size and style. The book also includes many examples of bulleted—as well as numbered—lists.

You need to remember four important things when you add a formal numbering system to your chosen organization:

1. Numbering systems are useless unless they are combined with clear headings.
2. Because you can't divide something into one, you should always have at least two subpoints under each main point. If you have only one subpoint, study your organization. Perhaps it needs revision, or your subpoint needs renaming. Perhaps the subpoint is an explanation of the main point, not a division of it.
3. The numbering system in an outline is less important than the actual relationship of ideas. The numbers are primarily an aid to the reader, but they also can help you see the relationships among your ideas as you write the draft.
4. You can sometimes avoid a numbering system by using type size and style (such as bold face) along with indented lines to show the hierarchical relationship. See Chapter 12 for more information about document design.

Checklist for Completing an Outline

1. As I add details to each point, have I
   __ chosen those that really do support that point?
   __ arranged them in a method of organization that is appropriate for that section?
2. If I have any gaps in my data, do I know where to find the missing information?
3. Have I chosen an outline system that is appropriate for my purpose and reader? How many levels of detail does it include? Do I want to use bullets at the third or fourth level? Do I want to use type size and indentation instead of numbers?

4.4   Reviewing and Revising the Organization

After you finish organizing, you will want to take a second or third look at what you have done. This review should be done after you have been away from your outline for some time. Ideally, you should let the outline rest a week or so before you review, but realistically, you may have to settle for letting it rest for a day or several hours. This section explains why you need to review at this stage and describes four possible kinds of review.

Understanding Why Review Is Necessary

When you're ready to review, think about what the word *review* really means—"to look again." While you were assembling and shaping your information into a structure, the arrangement was open to change. Once you finished your outline, though, you may have felt that the organization you chose was the only possible one. Now, in review, you can take a second look when the ideas are cold and you have achieved a certain distance.

In reviewing your organization, then, you want to ensure that whatever arrangement you chose is the best one. Remember that you've created a working outline—one that will guide you when you begin to write. Cooled though the ideas may be, your outline is not embedded in cement. You can change it now by looping back to the organizing process, and you can change it still more when you begin to write. What's important is the relationship of the ideas, the achievement of your purpose, and the clarity with which you give the readers the information they need.

Four kinds of review are possible at this stage: self-review, collaborative review, peer review, and instructor or management review. How elaborate your organizational review must be depends on the complexity of your assignment: for a simple letter or memo, self-review or a quick look by one other person may be enough. Collaborative review is necessary when more than one person is developing the organization. For a long document, most students and working writers appreciate feedback from three or four peers in a review workshop. Ironically, very short documents like résumés—which have multiple sections—also benefit from group review, because short and packed documents are very hard to write. The more sections the writing has, the more elaborate the review should be, because it is difficult to keep the relationship of several sections in mind. The fourth kind of review, by your instructor or manager, can be extremely helpful because this person can view your organization in light of considerable experience. Instructors or managers can also verify that you are doing the right thing—that is, fulfilling the assignment.

Self-Review

Many times you will do the reviewing yourself. Check your planning sheet and compare it with your assignment and then with your outline. Answer each question below:

1. Have I fulfilled my writing purpose?
2. Do I have a single major point?
3. Is the major point I started out with the major point I still want to make? (Sometimes as you work through your material, you find that it adds up in ways you hadn't anticipated. That means you have literally "written to learn"; you have discovered something you didn't know before you began, and that's good. But if so, now is the time to shape the new major point and reshape your working organization. You'll find it easier to change your organization now than when you begin writing.)

4. Do the subpoints support the major point? (Here you can ask other questions too: Are significant points emphasized? Are there two to nine subpoints—no more than nine? Have I chosen the right organizational pattern for my purpose? Are the subpoints arranged in the best sequence? Are the details logically organized under each subpoint?)

5. Do I have a numbering system that helps the reader understand the organization? Does every division have at least two subpoints?

6. How does my outline compare to existing documents? If your company has a format or outline for this type of report, it may help you to look at it. Also see Part 3 of this book for sample formats for particular documents.

Collaborative Review

When you work on a large and complex project, you may be only one of several authors. Usually each author is assigned a piece of the writing pie and is responsible for organizing that piece. The lead writer may assign the sections and also rework the various methods of organization so they are compatible. At the review stage, you may be meeting with the other authors to work out an overall organization, or you may be meeting with the lead writer to see how your contribution meshes with the rest. This collaborative review is a variation of self-review.

Just as you can't buy a single pie that contains one slice of cherry, one of strawberry, one of chocolate, one of pecan, and one of coconut cream (interesting as that might be), so you can't have a single document that contains six radically different methods of organization. With multiple authors, everybody has to compromise, and this review stage is a good time to alter your organization as needed.

Peer Review

Whether the outline was written individually or collaboratively, you should have it reviewed by outside reviewers (fellow workers or fellow students). They can see what you have written from another point of view and help you clarify your organization. The questions listed under self-review can also be used for peer review.

Small groups of three to four people are best for peer review. If possible, submit your outline to them to read in advance. During the review session, listen carefully to suggestions you receive and take notes, resisting the urge to interrupt the reviewers to defend yourself. Later you can adapt your organization to include those suggestions that improve what you have written.

Instructor or Management Review

Instructors or managers will approach your outline from a slightly different perspective than will your peers. With their depth of experience and their vision of a broader picture (such as how your contribution fits into the whole project), instructors or managers can offer sound advice about organization before you begin to write. And remember that the more help you can get at this point, the easier the writing will be, and the more likely it is to be successful.

While you may find the idea of review by other people intimidating at first, you'll soon see how helpful it can be. Because fellow students or colleagues at work will be unfamiliar with the material, they can tell you if your presented organization makes sense. They can spot holes in your information and sometimes suggest alternate methods of organization. After the review, you can adapt those ideas that will work and discard those that don't apply because, as the writer, you are still the final decision-maker. Students often tell me that their review sessions were among the most valuable parts of the class. After all, review and reorganization of your outline can save you many hours of time in the actual writing of your draft—which is the next step in the process.

Checklist for Reviewing the Organization

1. Which type of review should I use?
   ___ self-review
   ___ collaborative review
   ___ peer review
   ___ instructor review
2. What is the writing purpose?
3. What is the one major point?
4. Do the subpoints support that major point? Are any subpoints missing?
5. How many subpoints are there? Are they equal in importance? Should some subpoints be combined and others separated?
6. What is the primary method of organization? What is the method of organization of the subpoints? Is this the right organization to choose? If not, how should the order be rearranged?
7. Do the details support each subpoint?
8. Is the numbering system sound? Does each division have at least two subpoints?

EXERCISES

1. Add a formal numbering system to the organization you chose in the bicycle exercise in section 4.3. Then review it yourself. Answer each question in the checklist in writing, commenting on problems. Then review your outline as needed.

2. Pair up with another student and exchange your outline from the bicycle exercise in section 4.3. Review the outline using the checklist, and add questions you would like the writer to answer. Write your comments on your partner's outline and submit it to your instructor or to the writer, as directed.

☰ WRITING ASSIGNMENTS

1. Review your proposal, planning sheet, and the collection of data for your term project. Then choose the major point you will want to make when you write up the results of your study. Write that major point in one sentence and be prepared to defend it. If this is a collaborative project, choosing the major point should be a group decision.

Next, examine the list of brainstormed ideas for your project, the proposal you wrote for writing assignment 2 in Chapter 1, and the notes and other data you gathered in your information search. Read through that information, evaluating and sorting it into four categories by asking

- Is this useful?
- Is this relevant?
- Is this valid?
- What am I missing?

Then go back to information gathering to find any information you may be missing.

Third, review the sentence you wrote as the major point for your term project. Keeping in mind your purpose, reader, and topic, list from two to nine subpoints that will provide the framework for your paper when it is written.

You now have the main points of the outline for your term project. Add the necessary details and choose a formal outline system. This outline will be the basis for your paper. Your instructor may ask you to prepare several copies for a workshop review by fellow students.

For a workshop review, join a group of three to four fellow students who worked on a different project. Supply each member of the group with a copy of the outline for your project. As you review each group's outline, answer the questions on the checklist. Write your comments on the outline. Submit your review pages to the instructor or to the writer, as directed.

Finally, submit the revised outline for your term project to your instructor for review. Attach a page with three or four questions about your organization that you'd like the instructor to address.

2. If you are engaged in a term project, you should keep your instructor appraised of your progress with biweekly or weekly status reports. The completion of the organizing step is a good time to begin reporting the status of your project. Following the form in Chapter 15 or one suggested by your instructor, write a status report in memo form.

REFERENCES

Miller, George A. 1967. *The psychology of communication*. New York: Basic Books.

Chapter 5

Writing the Draft

Once you have reviewed and revised your outline, you are ready to plunge into writing the first draft. For many writers, staring at that first blank page is as frightening as standing atop the high dive and staring at the blue water far below. What's more, starting the actual writing can be as difficult for professionals as it is for students. "My biggest problem in writing a draft is getting started, particularly after coming off another writing project," says Jim Jahnke, an independent consultant in air-pollution monitoring and control. "I need deadlines to be forced to write."

Writers *do* have deadlines, but fortunately, writing does not require the same kind of all-or-nothing commitment as does stepping off the high dive. Writers usually think of the main part of a document as containing an introduction, a body, and a conclusion, and almost every piece of technical writing is composed of separate sections within those larger parts. These parts need not always be written in the order they will appear in the final document. This means you can ease into the writing process much as you might ease into chilly water with your toes first. For example, you might want to begin with the easiest sections of the body first: those for which you have the most information or those with a sequential method of organization. Often, the definitions and descriptions, because they are so specific, can be written first. Then you can go on to more difficult sections because your outline has given you the total picture and will help hold the separate pieces together.

If you are writing a short memo or letter, you might want to write the introduction first. For a long or complicated document, though, you may find it easier to construct the body first, write your conclusions next, and then drop back to write the introduction when you already know where you are going and how you are getting there. Even instructions and manuals—which may not have conclusions—can be written in sections.

5.1 Preparing to Write

Before you start writing, spend some time reviewing your original plan and revising it as needed. Then schedule your time.

Revise and Follow Planning Sheets and Schedules

In Chapter 1, you learned the details of planning a writing project, and you were encouraged to fill out a planning sheet like the one in section 1.6 for each writing project. As you complete short writing assignments, you can see how much time they took and how accurate your estimates on the planning sheet are. If you are involved in a long-term writing project, you need to review your original planning sheet before you write the draft to see if you are on schedule. If you are, congratulate yourself for good planning. If you have fallen behind, analyze the reasons. Remember that your original time estimates were just that—guesses—and maybe you needed more time than you thought. On the other hand, perhaps other assignments intruded on your schedule. In any case, think through the current status of your project and revise your schedule as needed. This revision will help you in two ways: (1) it will give you up-to-date information for any status or progress reports you have to write, and (2) it will prepare you psychologically to begin writing.

Now is also a good time to review the other items on your planning sheet and to revise those that may have changed. Pay special attention to the purpose you checked on the planning sheet, your written purpose statement, and the major point you wrote as you organized. Make sure these are still your purpose and major point because you'll need them to write a thesis statement.

Keep a Time Log and a Document Log

Planning practical time allotments is difficult for writers, but you can make that planning easier on the next project if you keep track of your researching and writing time now. The best way is to log your time as you work. Many professionals keep a log on a desk calendar—marking at the end of each day the number of hours or minutes spent on a particular project or blocking off time segments. Other writers keep logs, like the one in Figure 5-1, in an engineering or laboratory notebook or on a computer. Logs like these can help you charge for your time as well as help with future planning.

You also need to log your information sources for later documentation. When I talk to students at midpoints in the writing process, I often see them carrying notebooks stuffed with odd-sized scraps of paper in no discernible arrangement. Such students usually can't find what they are looking for, and they often forget to write down their exact sources. Unfortunately, at the draft-writing stage these students must backtrack to figure out where specific points of information came from.

You can avoid this common problem by planning now how you will keep track of source material. You can do this most easily by using your stack of bibliography cards (see Chapter 3). One way to keep track of information as you write is to alphabetize those cards by the author's last name, then number the cards consecutively. Then when you are writing and want to document information, you can simply note in your draft either the page number of the source and the card number, or the author's name, page number, and the year of publication. These two methods work well if you plan to use the *author-date* or *author-page* documentation system (see Chapter 13). Another

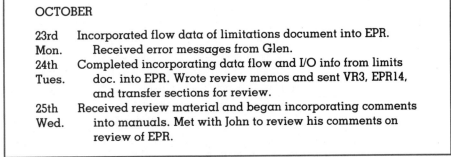

OCTOBER

23rd	Incorporated flow data of limitations document into EPR.
Mon.	Received error messages from Glen.
24th	Completed incorporating data flow and I/O info from limits
Tues.	doc. into EPR. Wrote review memos and sent VR3, EPR14, and transfer sections for review.
25th	Received review material and began incorporating comments
Wed.	into manuals. Met with John to review his comments on review of EPR.

Figure 5-1 Samples from a Writer's Computer Time Log

method is to number the index cards in the order you use that information in writing your draft. Then when you want to document information, you write down those consecutive numbers. This method works well with the *number* system of documentation explained in Chapter 13. Of course, if you change the order of documentation when you revise, you will also need to change the numbering. Your instructor may specify a documentation method or you may choose one now.

Learn to Write in Segments

A segment is one of the parts into which something naturally separates or is divided. Once you've written your thesis statement (see section 5.2), you can take advantage of the segmental nature of technical writing by beginning to write with the segment that comes easiest to you. That segment may or may not be the introduction. The introduction should provide an overview, and it's hard to have the perspective for an overview until you've written and assembled all the segments. But no law says that just because the introduction is *read* first, it must be *written* first.

Think about the segments of an orange. Each orange segment contains what is essential to making up the orange, yet a whole orange consists of eight to ten segments put together and then covered with skin. Any piece of technical writing is something like an orange because it has independent yet related parts. Those parts may be paragraphs in a letter or memo, but often they are divisions of a longer piece, each with its own subheading. Put together and unified with a thesis statement and introduction, those individual parts make up the larger whole. Together, the thesis statement and the outline are like the skin of the orange; they hold the segments together.

Even when you write in segments, you may find that different kinds of writing require different strategies:

- In a *development project* in engineering, your writing reports what you learned during the project. These drafts are best written incrementally (that is, in small segments while the project is still going on). You can write each section quickly at the end of a project phase, and by the project's end, you will have the paper already written in draft form. See section 15.3 on informal reports.
- In a *research paper* or *lab report*, the paper's structure is more rigid: introduction, materials and methods, results, and discussion. Here you might want to begin with the introduction in order to state the problem first, and then continue with the easiest section, which is probably materials and methods. The discussion section interprets the results, so that section is similar to conclusions. Chapter 23 explains the format and organization of research and laboratory reports.
- In a set of *instructions* or *procedures*, the method of organization is sequential, so you can begin with the first step and then proceed in order. However, you can put off writing any overview statements or introductions until after you have written all the steps. See Chapter 18 for instructions and Chapter 19 for procedures.

- In a *feasibility* or *comparative* project that leads to a recommendation, the writing itself may be a learning process—leading you to new ideas and to insights about the data. In this case, once you have a purpose statement and an outline, you can fairly easily launch into the body of the paper; then write conclusions and recommendations; and then write the introduction and abstract or summary. Chapter 22 covers feasibility and comparative reports.

To make the writing task easier, consider the following strategies:

1. Choose a comfortable place and time of day to begin writing. The place should be free of distraction and have a surface big enough so you can spread out your notes and papers. The time should give you about two uninterrupted hours. Many people write most easily in the morning when they are fresh; others are more alert in the evening. Chose the time that's best for you.

2. Whether you write in longhand, on a typewriter, or on a computer, begin with the segment you know best and are most comfortable with. Write fairly quickly, concentrating on ideas and content. Don't worry about spelling, punctuation, or perfect expression; just leave plenty of empty space on the page. Remember that this is a draft; later you can revise, add, or subtract, but now you want to get ideas on paper. If you use a word processor, additions and changes to the draft are easy to make.

3. Write everything you have to say about that particular topic. You may feel that you're writing too much, but you can always trim later. Remember how much more quickly a reader can read than you can write. An average person can read more than 250 words a minute; a hard-working writer may only write between 250 and 500 words in 30 minutes or more.

4. At the end of about two hours or when you finish a section, stop writing. Most people concentrate productively for two hours or so and then begin to tire. When you stop, spend a few minutes thinking about what you'll write next, and jot down a word or two to jog your memory when you begin again. Rereading what you have written is also a good way to get back into the writing mood.

 Checklist for Preparing to Write the Draft

1. Have I reviewed the planning sheet and revised it if necessary?
2. Have I established a time log for a long report?
3. What procedure shall I follow to keep track of documentation?
4. What segment will be easiest for me to begin writing?

5.2 Constructing the Body

The main part of a piece of technical writing—whether it is a one-page letter or a 45-page proposal—has three sections: the introduction, the body, and the conclusions. But these parts are not always written in that order. In fact, in a long report or proposal, they may not even appear in that order: the conclusions may come first, then the introduction, and then the body.

Thus, when you begin to write, you may want to begin with a section which contains specific details: the body. The controlling factor in the body is the thesis sentence, which you'll want to write first.

Combine the Purpose and Major Point into a Thesis

Early in your planning you determined the purpose of your specific writing assignment. That purpose may have been to explain how something works, to evaluate the feasibility of a new procedure, or to propose a specific course of action. Whatever the purpose, you were encouraged to write it down so it was clear to you and would be clear to your readers. When you began to organize, you decided on the one major point you wanted to make. Once you wrote down that single major point, it became a reference for organizing the rest of your material. Each subpoint in your outline contributed to moving you and your reader toward recognition and acceptance of the one major point.

For example, suppose that as part of a recommendation report on running shoes, you decide that your main purpose is *to describe* a running shoe. Then, as you organize, you decide that your one major point is "A running shoe has three major parts: the front, the back, and the sole." Now, as you begin to write, you need to combine the purpose and major point into a thesis statement. What will it say? Before you can answer that question, you need to think about what a thesis is.

Definition of a Thesis

A *thesis* is the main idea you want your reader to reach because it provides the answer to the problem or question being investigated. In technical writing, you can have one of two different kinds of thesis:

1. If you are simply presenting information or explaining a complex idea, the thesis is a *statement of content*. The running shoe description requires this first kind of thesis. Since your primary purpose is to describe, you are presenting information; thus, your one major point can also be your thesis: "A running shoe has three major parts: the front, the back, and the sole."

2. If you are analyzing data and drawing conclusions, the thesis is a *proposition to be defended*. This kind of thesis statement is more common. Here the writer adopts a specific view, defends that view by logical argument, and offers a valid solution.

The following section describes these two kinds of thesis statements more fully and gives advice about how to write them. Ideally, both kinds of thesis

- tell the purpose
- announce the topic and limit its scope
- express the main point
- map the order of supporting statements (or tell how many there are)

In addition, the thesis that is a proposition to be defended indicates the writer's point of view—often, in fact, taking a strong and persuasive stand.

You should always write a thesis statement for yourself because it will guide and focus your writing. Also, most of the time you'll want to state the thesis within your document as an aid to the reader's understanding. Occasionally, however, the thesis will be implied but not directly stated.

How to Write a Thesis Statement

Typically, you will write a thesis that is a *statement of content* when your purpose is to define, describe an object or process, instruct, or present narrative or objective information. As the writer, you do not take a strong stand, and your tone is objective. For example:

Purpose	to define a cave
One Major Point	A cave is a natural opening in the ground that extends beyond the zone of light and is large enough to permit human entry.
Thesis Sentence	The same as the one major point, with—perhaps—an elaboration to indicate the order of supporting explanations.
Purpose	to describe how a laser works
One Major Point	The basic process of laser action involves four steps: excitation, spontaneous emission, stimulated emission, amplification.
Thesis Sentence	The same as the one major point, since the four steps map the order of the supporting statements.
Purpose	to report how suspect Jones was arrested (a narration of events)
One Major Point	Suspect Jones was arrested by Reporting Officer Smith on May 14, 19__, in response to observed driving behavior, physical indications, and results of a field sobriety test.
Thesis Sentence	The same as the one major point, which outlines the sequence of events.

You write a thesis that is a *proposition to be defended* when your purpose is to interpret, evaluate, judge, recommend, propose, sell, or persuade. As the writer, you do take a stand, and you construct your thesis to spell out both your point of view and the support for your stand. For example:

Purpose	to propose a remote sensing survey of Portsmouth Harbor, NJ
One Major Point	My company should win the survey contract.
Thesis Sentence	Allied Marine Survey is best equipped to conduct a remote sensing survey of Portsmouth Harbor for sunken vessels because of its previous success with magnetometer and sidescan sonar, its recent acquisition of improved sensing equipment, and its experienced staff.

(Note that the thesis implies the main point, clearly states the writer's point of view, and supports that point of view with details. This kind of thesis could appear in an introduction and/or in conclusions.)

Purpose   to evaluate the effects of stabilizer buildup on operating costs of 20-30,000 gallon swimming pools

One Major Point   The stabilized form of chlorine in trichloro-s-triazinetrione is more costly to use over the long run.

Thesis Sentence   The stabilized form of chlorine in trichloro-s-triazinetrione, while initially appealing because the chlorine remains in the water longer, is costlier in the long run because the pool water must be partially drained at intervals to eliminate the stabilizer concentrations.

(Note that the thesis in this case *assumes* the purpose, clearly states the major point and the writer's point of view, and then outlines the support for that point of view. Such a thesis might go in the conclusions section, though the conclusions will probably be presented at the beginning of the report.)

Purpose   to recommend purchase of a four-track cassette recorder for a home recording studio

One Major Point   The Tascam 244 is the best recorder of the four evaluated.

Thesis Sentence   The Tascam 244 four-track cassette recorder is recommended for use in a home recording studio because (1) it best satisfies the six evaluation criteria, and (2) it combines all the features necessary for recording, processing, and mixing on a high-quality four-track cassette tape.

(Note that the thesis goes far beyond the major point by clarifying the topic and by setting out main supporting areas: the evaluation criteria and the features. Such a thesis would probably appear in the recommendations section, which might well appear at the beginning of the report.)

In summary, then, the first kind of thesis, the *statement of content*, often appears in the introduction where it can map out the main points or sequence of events. The second kind of thesis, *the proposition to be defended,* often appears in the conclusions or recommendations where it summarizes support for those conclusions or recommendations. For the second kind of thesis, you might also include a separate purpose statement in the introduction.

What is most important as you begin to write is that you know where you're going. Writing a clear thesis statement will help you construct meaningful paragraphs.

Checklist for Writing a Thesis

1. What is my purpose? What is my major point?
2. What should my thesis be?
    ___ a statement of content
    ___ a proposition to be defended

3. When written down, does my thesis accomplish the following?
   ___ tell the purpose
   ___ announce the topic and limit its scope
   ___ make the major point
   ___ map the order of supporting statements
4. If it is a proposition to be defended, does my thesis indicate my point of view as the writer?

EXERCISE

To sharpen your understanding of thesis statements, look in technical and scientific reports, instructions or manuals, or chapters in textbooks to find two examples of thesis statements. Identify their location—that is, are they in the introduction, conclusions, or recommendations section. Then determine if they are statements of content or propositions to be defended. Be prepared to discuss your choices in class.

Construct Meaningful Paragraphs

With a good thesis sentence, you have a controlling focus for your writing and are ready to write the paragraphs that will support that thesis. Fortunately, you also have a well-organized outline to help you construct those paragraphs. Paragraphs are the building blocks making up any major point, and most paragraphs contain a topic sentence and support sentences. Before you begin writing, refresh your memory about those terms.

Definition of a Paragraph

A *paragraph* is a group of related sentences, complete in itself but also usually part of a larger whole. Paragraphs are set off as independent items with white space: an extra line of white space in block form or an indented first line. You can write good paragraphs if you remember three things:

1. A paragraph should contain only one central idea, which will be stated in a topic sentence.
2. The rest of the sentences should support or explain the topic sentence.
3. Each paragraph should be logically related to the paragraphs before and after it.

Definition of a Topic Sentence

A *topic sentence* contains the main point of a paragraph. For example:

Computer Output Microfilm (COM) recorders can use either of two methods of film development: wet processing or dry processing.

With this topic sentence, what do you expect the paragraph to say? You expect it to discuss the two methods of film development—either explaining each one or comparing advantages and disadvantages of each one. Just as the thesis statement controls the whole document, so the topic sentence controls the paragraph. Technical readers usually must read quickly, and they depend on the topic sentence to tell them the main point. This in turn tells them whether or not they need to read the rest of the paragraph carefully. For this reason, most paragraphs in technical material place the topic sentence at the beginning—as the first or second sentence.

Support Sentences

Once you have stated the main point in the topic sentence, you should support that idea in the rest of the paragraph. Support can come through giving examples or statistics, through defining or describing, through comparing or contrasting, or through logical argument. Just as a document or section of a document should have a method of organization, so should each paragraph. Methods of organization are discussed in detail in section 4.3; you might want to review that material before you begin to write.

The important thing to remember is that support sentences should be specific, related to one another and the topic sentence, and sensibly organized. Look at the way this writer (Clayton 1985) supported the topic sentence on film development:

Topic sentence

Wet processing

Dry processing

Advantages and disadvantages of each

Computer Output Microfilm (COM) recorders can use either of two methods of film development: wet processing or dry processing. Wet processing involves printing the character images onto a silver-based film, which is then passed through five to nine chemical tanks. These tanks develop the film and wash off the chemical residues. Dry processing involves printing the character images onto a heat-sensitive film that needs no development and is immediately available for viewing. The advantage of wet processing is that the film can be stored for a long period of time in dark or light areas, whereas dry-processed film can be stored for a long time only if kept in a dark place. This can be inconvenient. On the other hand, dry processing requires no additional supplies—only the film. Wet processing requires chemicals, filters, and developing supplies in addition to the film; therefore, dry processing is less expensive in terms of cost of supplies.

Note that the paragraph first explains wet and dry processing, then gives advantages and disadvantages of each process—and both explanations support the topic sentence.

Paragraph Length

How long should a paragraph be? To answer that question, you need to keep your reader's short-term memory in mind. Short-term memory is the number of things you can hold in your mind at one time in order to process the information. The magic number is seven plus or minus two items. Thus, if your topic sentence is one point, and the reader also has to keep in memory the general point of the previous paragraph, you are left with three to seven items of information you can add. Typically, each item of information is in a separate sentence, so the number of information items will dictate the paragraph length.

The paragraph in Figure 5-2 is from the summary of a 350-page document written by the U.S. Office of Technology Assessment in 1991 about greenhouse gases. Note how the *topic* sentence (sentence one) is supported by six sentences explaining *how* the United States is the leading contributor of greenhouse gases. This paragraph is packed with details, but after you read it, ask yourself if the paragraph is clear.

Topic sentence

Support 1

Support 2

Support 3

Support 4

Support 5

Support 6

Among individual countries, the United States is the leading contributor of greenhouse gases. With 5 percent of the world's population, the United States accounts for about 20 percent of the world's warming commitment. U.S. CO_2 emissions (20 percent of the global total) originate almost exclusively from fossil fuel combustion. Anthropogenic sources of methane in the United States account for about 6 percent of global emissions from all sources; among the anthropogenic sources, landfills, coal mining and domestic animals account for most of the U.S. total. The United States also consumes between 20 and 30 percent of the world's CFC-11 and CFC-12, the two most damaging chlorofluorocarbons in terms of global warming. Roughly 60 to 70 percent of these CFCs are used in air-conditioning or in the production of thermal insulation; these gases are scheduled to be phased out by the year 2000 under the revised Montreal Protocol. U.S. nitrous oxide emissions (roughly 15 to 20 percent of the man-made global total) originate primarily from fertilizer breakdown and high temperature fossil-fuel combustion. Greenhouse gas emissions are closely entwined in the United States with energy use; currently, America uses about 15 times more energy per person than does the typical developing country.

Figure 5-2 Paragraph Illustrating Topic and Support Sentences

In shorter documents like letters and memos, paragraphs can also be shorter. But don't make a paragraph too short; a series of very short three-line paragraphs is hard to read because the topic becomes fragmented and the continuity is lost. With many very short paragraphs, the reader will try to keep them all in short-term memory in order to see the relationships. Remember too that most of your paragraphs will naturally set their own limits; the sense of the paragraph, the complexity of the document, and the intended reader's background all control how long a paragraph should be.

Paragraph indentations, or the white space setting them off, give the reader's eyes and brain a chance to relax by subtly saying, "We're changing subjects here." As a writer, you can also supply eye and brain relief by using headings, lists, and graphics to break up the solid blocks of print on a page. Chapter 6 tells more about headings; Chapter 11 gives information on graphics; and Chapter 12 explains how to design effective documents.

Types of Paragraphs

The paragraphs discussed in this section are *content* paragraphs, the type that present information, analyses, or recommendations to the reader. Content paragraphs make up the largest part of most documents.

A second type of paragraph is called *transitional*. Transitional paragraphs function as glue to hold a series of discrete ideas together. They do this by moving the reader from one idea to the next. Because transitional paragraphs are one of the ways you can show your reader how you have organized a document, they are discussed in detail in section 6.5.

Write Straightforward Sentences

When you are writing a draft, it's more important to get your ideas on paper than it is to perfect each sentence. Later you can revise for clarity. If you need help in revising or practice in writing good sentences, you'll find many exercises in Chapter 8 and in the Punctuation section of the Handbook. At the draft-writing stage, simply keep these things in mind:

- Help the reader understand in every way you can. As you write, ask yourself what the reader is supposed to do or think after reading what you have written.
- Keep your purpose and your thesis clearly in mind. If you aim at the purpose and thesis as you write, your writing will be focused and clear.
- Write as simply and clearly as you can. Don't use fancy words just to impress the reader, and define any terms that your reader might not know.
- Get at the writing; don't put it off. You will learn something from the act of writing itself.

Checklist for Writing Paragraphs and Sentences

1. Are my paragraphs self-contained units making one major point?
2. Do my content paragraphs each have a topic sentence?
3. Do the rest of the sentences support that topic sentence?
4. Are the support sentences organized in a sensible way?

EXERCISES

1. Search textbooks, technical reports, and magazine or journal articles to find two examples of well-written paragraphs on technical subjects. Photocopy the examples or clip and attach them to a sheet of paper. After each paragraph, write a short evaluation, telling what's good about the paragraph. If any improvements are needed, indicate them. Be prepared to discuss your findings in class.

2. Choose two content paragraphs at random from a paper you are working on. Bring the paragraphs to class and exchange with another student. Analyze your fellow student's paragraphs to see if each one has a topic sentence and clear supporting sentences. Check to see if those sentences are organized to make the writer's point. Write comments on the paper and submit it to your instructor if requested.

5.3 Writing Conclusions

Once you have written the central portion—or body—of a document, you have prepared yourself and your reader to reach some conclusions based on that information. You may think of technical writing as totally nonjudgmental and objective, but remember that usually you are writing from your knowledge and experience to readers who know less about the subject than you do. Therefore, readers look to you for (1) factual knowledge (in the body), (2) analysis and interpretation of that knowledge (in a conclusions section), and (3) recommendations based on your interpretation (either in the conclusions or in a separate recommendations section). In other words, you often must not only take a stand, but recommend action based on that stand.

Be aware that technical writers use the word *conclusion* in two ways: (1) as the final section of the body of a document, and (2) as the section of a report that includes logical interpretations and inferences drawn from the data in the body. In short documents like letters, memos, and informal reports, the logical conclusions may go in the final section. But in long reports and proposals, they are more often placed at the beginning.

Definition of Conclusions

When you write the body, you provide facts and even analysis, and the reader is free to interpret those facts, form judgments, and reach conclusions. But most readers want that job done for them; they expect the writer (who after all has been working on this) to weigh and evaluate the facts and the results of any analysis. Readers want to know what is important, what the relationships are, what the results are, and what their significance is. Such interpretation usually takes place in the *conclusions* section of a technical report or the *discussion* section of a research paper.

For example, to arrive at conclusions about awarding a contract for a remote sensing survey in Portsmouth Harbor, the writer might compare the submitted proposals on such established criteria as experience with similar projects, time estimates for completion, expertise of project directors, and projected costs. Conclusions must often consider many requirements—and cost, though important, is not always the deciding factor. Figure 5-3 shows how such conclusions might look.

Notice that while these conclusions provide a summary of the key facts of the report, they also evaluate those facts and begin to guide the reader's thinking. Conclusions like these are especially appropriate with a thesis of evaluation or judgment.

Another conclusions section, for example in procedures or physical description, might be a simple summary of the key facts. See Figures 19-1 and 19-2 for examples.

Definition of Recommendations

In many documents, the writer will move beyond interpretation of evidence to make actual *recommendations,* telling readers what they should believe or what action they should take. Recommendations answer the question "What should I do about it?" Recommendations are also opinions, but because they are based on fact and arrived at through reasoning, they are informed judgments—much more valid than unsupported emotional claims.

Recommendations can be either in a separate section or combined with conclusions. Sometimes recommendations appear first to be supported by the conclusions; at other times the conclusions come first and lead into the recommendations. The method you choose will be determined by your purpose (what's the main thing you are trying to do?) and by your reader (what's most important to that reader?)

CONCLUSIONS

Four proposals were submitted for the remote sensing survey of Portsmouth Harbor, by

> Allied Marine Survey
> Excel Corporation
> PMS Co.
> Smith Oceanic Survey

Of these four, Allied Marine presents the best record of experience, especially with magnetometer and sidescan sonar. PMS and Excel Corp. have recently added new staff members, but their experience is in shallow water and therefore only partially applicable. All four proposals accept the three-month time limit, with October 1, 19__, as the completion date. Project costs by each company are

Allied Marine Survey	$67,000
Excel Corp.	72,000
PMS Co.	65,000
Smith Oceanic	67,500

Figure 5-3 Conclusions Section of a Report

Figure 5-4 shows how the recommendations might look for the Portsmouth Harbor study.

Requirements for Concluding and Recommending

If you are going to be successful in persuading readers to believe or act as you wish, you must meet these four requirements:

1. Provide a solid basis of fact in the body, including statistics, examples, evidence, and expert opinion.

2. Use logical reasoning to evaluate the evidence. Often you must establish criteria (standards) as a basis for judgment, and then test alternative products or courses of action against those criteria. To help set up criteria, you can ask questions like these:

- Will it work?
- Can it be done?
- Will it yield the results I want?
- How long will it take?
- Can I (or the company) afford it?
- Is this product or way of doing it better than others?

3. Be fair. To be fair, you must present *all* evidence, even that which promotes the other side. If you suppress alternatives, your readers won't trust your judgment. In

---

RECOMMENDATIONS

Allied Marine Survey should win the contract for the remote sensing survey of Portsmouth Harbor. It has the best record of experience in deep water, and its bid is second lowest of the four bids at $67,000.

Figure 5-4 Recommendations Section of a Report

the evidence-weighing process, therefore, you must consider *in writing* the implications of alternatives.

4. Use a positive, confident tone (see section 1.4 for more on tone). You cannot hedge or be half-hearted in your judgments because readers won't trust that either. Sticking your neck out with a firm decision may be difficult, but you should do it. Remember, though, that this is your *informed* judgment, so your tone should be serious and unemotional as well as positive and confident.

Figure 5-5 shows the conclusions and recommendations section of a report (Jarrett 1984) that analyzes stethoscopes for a group of home health nurses in order to find the one that best meets these requirements:

1. is priced at $50 or less
2. includes two chest pieces, a bell, and a diaphragm

CONCLUSIONS AND RECOMMENDATIONS

Recommendation Supporting conclusions

Based on this study, the Deluxe-Dual Head Stethoscope by Marshall is the best purchase. This stethoscope received high marks in all six categories.

Recommendation Supporting conclusions

The Lightweight Stethoscope by Littman is rated as the second choice. Although this stethoscope rated high in four categories, it does not offer the optional, flexible eartips. However, if the consumer is able to purchase these separately, this stethoscope would be a good choice.

Recommendation Supporting conclusions

The Double Head Color Coded Nurse Stethoscope by Ritter-Tycos is rated third. The cost of this stethoscope exceeds the $50 limit, and the optional, flexible eartips are not offered. However, if the consumer wishes to pay the cost of the stethoscope and to purchase the flexible eartips separately, this stethoscope is also a good choice.

Recommendation Supporting conclusions

The Sprague-Rappaport Stethoscope by JAC is rated as the fourth and last choice. The chest pieces on this stethoscope are heavy, resulting in an overall heavy instrument. The bell does not have a non-chill ring, which is important for patient comfort. This stethoscope is not recommended for purchase.

Figure 5-5 Conclusions and Recommendations Section of a Report

3. has screw-out eartips and optional, flexible eartips
4. is lightweight
5. has metal side pieces (binaurals) that can be adjusted to the angle of the ear canal
6. has a non-chill bell

After reading these statements, would you be willing to accept the author's conclusions and recommendations? Do they seem to be based on fact? Does she explain why she made the choices she did? In this case, the conclusions and recommendations are combined, but sometimes they will appear in separate sections.

For long and complex reports, conclusions and recommendations sections can be much broader in scope. In 1991, the Office of Technology Assessment (OTA) of the U.S. Congress issued a 212-page report titled *Complex Cleanup: The Environmental Legacy of Nuclear Weapons Production*. Figure 5-6 shows a summary of the conclusions, called Key Findings, in this report.

Following the conclusions, the report recommends to the Congress the Policy Initiatives contained in Figure 5-7. Notice that these recommendations for policy changes are written in language suggesting what Congress "could" do.

KEY FINDINGS

- The waste and contamination problems at the DOE [Department of Energy] Weapons complex are serious and complicated, and many public concerns about potential health and environmental impacts have not yet been addressed.
- DOE, other Federal agencies, and the States are trying to carry out their legally mandated cleanup responsibilities, but they presently lack the necessary personnel and infrastructure, and have yet to develop an effective process for public involvement in setting priorities and making important decisions. Despite laudable efforts at changing the DOE culture, substantial credibility and public acceptance problems continue to hinder progress.
- The environmental program now underway at the Weapons Complex is in the very early stages, and little actual cleanup has been done. It may be impossible with current technology to remove contaminants from many groundwater plumes and deeply buried soils within reasonable bounds of time and cost. Many sites may never be returned to a condition suitable for unrestricted public access.
- Despite DOE statements about lack of imminent off-site health threats due to contamination, possible public health effects have not been investigated adequately. The current regulatory process is not sufficient to effectively identify urgent health-based remediation needs or to comprehensively evaluate possible public health impacts. Among the missing elements are a coherent strategy for evaluating potential

---

Figure 5-6 Conclusions of a Long Report

off-site human exposure to radioactive and hazardous contaminants, a coordinated and scientifically sophisticated approach for evaluating potential health impacts from contamination, and an open process for public involvement in identifying risks and setting priorities for reducing risks.

- Because of the limitations of existing cleanup technologies it is prudent to invest in promising new developments; however, such efforts should not delay addressing situations in which containment and monitoring are warranted now. OTA finds that a technology development program will be most beneficial if it is focused on the most serious contamination problems identified by possible health risks.

- DOE's stated goal—to clean up all weapons sites within 30 years—is unfounded because it is not based on meaningful estimates of work to be done, the level of cleanup to be accomplished, or the availability of technologies to achieve certain cleanup levels. Neither DOE nor any other agency has been able to prepare reliable cost estimates for the total cleanup.

- DOE currently has large quantities of radioactive and hazardous waste in storage at all sites, often under marginal conditions. There will be an increasing need to store waste safely on-site for fairly long periods until disposal alternatives are available. Adequate and workable standards and criteria for improved storage and treatment on-site are urgently needed.

Figure 5-6 Continued

POLICY INITIATIVES

I. Increase congressional oversight of environmental restoration and waste management activities that require improved performance by the responsible agencies.

Congress could increase its oversight of DOE, EPA [Environmental Protection Agency], and other Federal agencies to develop and implement improved programs to deal promptly with the following matters:

- strengthen agency personnel,
- plan for safe waste storage,
- improve technology development,
- increase public access to information,
- coordinate and accelerate standard setting, and
- strengthen site monitoring programs.

Figure 5-7 Recommendations of a Long Report

> **II. Enhance the structure and process for accessing the public health impacts of Weapons Complex waste and contamination.**
> Congress could establish the following institutional mechanisms to evaluate potential off-site health impacts:
>
> - a new and separate health assessment office,
> - health "Tiger Teams" to conduct exposure assessments at each site, and
> - a national independent environmental health commission.
>
> **III. Develop a structure and process to provide public participation in key cleanup policy and technical decisions.**
> Congress could establish the following institutional mechanisms:
>
> - advisory boards with technical staff at each site, and
> - a national coordinating and advisory board.
>
> **IV. Establish a national mechanism to provide outside regulation of DOE radioactive waste management programs.**
> Congress could authorize an institution to regulate those aspects of radioactive waste management activities now subject exclusively to DOE's authority. These functions could be given to:
>
> - a new national body or
> - an existing body such as the Nuclear Regulatory Commission or EPA.

Figure 5-7 Continued

Placement of Conclusions and Recommendations

Where do you place conclusions and recommendations? The answer depends on your purpose and intended readers, as well as where the document will appear. In a journal article, for example, conclusions and recommendations usually appear at the end, but in a report for industry they are often better placed before the introduction, where busy managers, technicians, or specialists can easily find them. Exact placement in a document depends on the primary reader. Remember that not all documents require recommendations, and a few—like instructions—do not even require conclusions. For placement of conclusions and recommendations in specific forms of technical writing, see Chapters 15–23.

Checklist for Writing Conclusions and Recommendations

1. Are my conclusions supported in the body of the document?
2. Are the conclusions fair?
3. Is the tone appropriate?
4. Are the recommendations logically based on the conclusions?

EXERCISE

In a group of several students, study the following purpose statements and conclusions and recommendations sections from student reports. Decide whether the conclusions fulfill the stated purpose, and if not, where the problems are. Be prepared to share your analysis with the rest of the class.

1. Self-installed Burglar Alarms

Purpose

Burglary is big business today. Police and insurance company reports show that burglary costs the American public well over a billion dollars a year. This does not include the untold loss in human injury that cannot be expressed in dollars.

The purpose of this report is to inform the reader of the various ways of protecting one's residence by installing a burglar alarm. Four alarms will be researched and compared, and one will be recommended.

Conclusions

Each system researched had different advantages and disadvantages. The Seeker 1 control module possessed the best overall qualities, mainly due to its dual power supply and capability of supporting options.

The DTI's hardware circuit proved to be the most reliable. This was due to the fact that being a perimeter system, intruders were detected before entering any opening.

The Home Guard system would be the easiest to install. Most components were installed by simply peeling off an adhesive strip and placing them near entrances and openings.

On the point scale used to rate each product on predetermined criteria, the Seeker 1 accumulated the most points.

Recommendations

The Seeker 1 is recommended as the best alarm system of all that were researched, because of the control module's potential. It is capable of supporting the combined options that each of the other systems claims. These options include extra sirens, telephone dialers for phoning the police, and smoke alarms. A hardwire system can also be connected to the Seeker 1 control module. Adding a hardwire system would be very inexpensive and advantageous. Ease of installation would decrease considerably, but circuit reliability would be increased by a large factor (Bejar 1982).

2. Table Saws

Purpose

This is a comparative study of noncommercial 10-inch table saws. The different features of four saws are discussed in terms of quality and performance for use in a small woodworking shop.

Conclusions

The Shopsmith Mark Five is the best of the four saws compared in this report. The Shopsmith Mark Five is five tools in one and uses only one motor, one stand, and one work table. It does everything that five individual tools do, and has been engineered with accuracy and precision in mind. The Shopsmith Mark Five is not only a 10-inch table saw, but it also includes:

1. a vertical drill press
2. a horizontal boring machine
3. a 34-inch lathe
4. a 12-inch disc sander

These are all standard items when purchasing the single unit.

The next best saw is the Powermatic Model 66. This saw is excellent for both noncommercial and light commercial use. The DeWalt Model 7756 and Rockwell Model 34-710 lack the necessary features to produce high-quality wood products.

Recommendations

1. The Shopsmith Mark Five is the best choice for our needs and preferences.
2. The Powermatic Model 66 is perfect for light commercial use because of its construction.
3. The DeWalt Model 7756 and Rockwell 34-710 are not recommended because they are not as well made as the other brands.

The unit price for these saws can be found in Table 1.6 (Zertouche 1982).

5.4 Preparing an Introduction or Overview

You might want to put off writing the introduction of a long document until you have finished writing the body and the conclusions and recommendations. When the paper itself is written, you will have a good idea what to include in your introduction.

In a way, an introduction is an overview because it tells readers where they are going, the word *introduction* coming from two Latin words meaning "lead inside." Short papers generally do not have a separate introduction, but the opening paragraph will include introductory material such as the purpose and topic. In a very short paper, purpose and topic can be combined in a single thesis sentence. Introductions to longer papers should include these four items, which are often combined in a thesis sentence.

- a full statement of the topic
- the purpose (reasons for conducting the study)
- the scope (how much is covered)
- the plan (method of development)

Other introductory items include:

- definitions of key terms (with equivalent acronyms or abbreviations)
- a brief history or explanation of theory
- a description of the intended reader
- documentation of source material

In a very long or complex document, you might also include short introductions to individual sections. In this book, for example, each chapter has an introduction, as do many major sections within individual chapters.

The introduction in Figure 5-8 is from a report that compares surgical lasers for possible purchase by a hospital (Leonard 1985). Study the way the writer has organized the introduction and note the information that is included.

INTRODUCTION

This report is an evaluation of state of the art Nd:YAG (neodymium: yttrium aluminum garnet) surgical lasers suitable for Allied Hospital and available in 19__. Allied Hospital is a private, metropolitan hospital, which presently owns a mobile carbon dioxide gas laser and a fixed argon laser.

In this report, four surgical lasers are compared on the basis of five criteria:

versatility
cost
safety
mobility
ease of operation

I do not intend for this report to be the final word on available models. Rather, I hope to provide useful information and an objective recommendation that will be a valuable resource in final purchasing decisions by Allied Hospital staff.

I encourage hospital staff to inquire professional opinions of the specific models from the physicians and facilities listed in the Appendix. I consider this information vital to a purchasing decision, but understandably, it is not openly distributed to the public.

The report is directed to the reader who may have medical or other science-related experience but who is not an expert in the field of

---

Figure 5-8 Introduction to a Comparative Report

laser surgery. It is intended to aid Allied Hospital specifically, but much of the information contained could be useful to other facilities and to anyone with an interest in lasers and their use in surgery. For the convenience of the reader, terms defined in the Glossary, page 56, appear in capital type the first time they are used in a section of the report.

The report includes four major sections:

- an explanation of what surgical lasers are and how they work
- a description of the criteria used to choose a suitable laser and how those criteria are applied
- a technical description of the four models evaluated
- a recommendation for Allied Hospital.

Figure 5-8 Continued

In summary, an introduction should give a broad overview of the subject without going into detail. As a technical writer, you need to review your planning sheet before you write the introduction so that you are sensitive to the needs and knowledge of your primary and secondary readers.

Checklist for Writing Introductions

Does my introduction include each of the following?
___ a full statement of the topic
___ the purpose
___ the scope
___ the method of development
___ a description of the intended reader
___ definitions of key terms (with equivalent acronyms or abbreviations)
___ a brief history or explanation of theory
___ documentation of source material

EXERCISE

The following introductions are from two student reports. The first analyzes and compares cooking steamers for restaurant food preparation. The second examines the feasibility of establishing a child care facility for students who are also parents. With a group of fellow students, evaluate each introduction against the checklist to see if it is complete, and write a brief commentary based on your analysis. Be prepared to share your evaluations with the rest of the class.

1. Cooking Steamers (Li 1985)

1.1 *Purpose of Investigation*

During the past two months, the cooking steamer in the kitchen of the Spartan Food Service Company broke down at least once a week. This caused a lot of inconvenience to the staff of the kitchen as well as to the students who eat in the dining hall. It is necessary to replace the old steamer as soon as possible. The purpose of this report is to investigate the best model of steamer to replace the old one.

1.2 *Scope of Coverage*

There are many different types of cooking steamers on the market. This report only covers the investigation of one-compartment pressureless convection steamers. The models selected for investigation are listed below:

1. Cleveland   model CET-5
2. Ember Glo   model FS-2
3. Groen   model GG3-E
4. Market Forge   model 3005-GTE

1.3 *Sources of Information*

1. Cleveland Specification 1283
   Cleveland Range Company, 1333 E. 179th St.
   Cleveland, OH 44110
2. Continental Avard
   San Francisco, CA (415) 489-6850
3. Ember Glo model FS-2 operation manual EFS3-380
   Midco International, 2717 North Greenview
   Chicago, IL 60614
4. Groen catalogue and operation manual GG3-E
   Dover Corp., 1900 Pratt Boulevard
   Elk Grove Village, IL 60007
5. Ledyard Food Equipment Supplies
   1005 17th Ave., Santa Cruz, CA 95062
6. Market Forge specification sheet S-2258
   Market Forge, 35 Garvey St.
   Everett, MA 02149
7. Spartan Food Service
   Spartan University

2. Child Care at Central State University (Devere 1989)

Introduction

The existing child care facility at Central State University is not able to handle the number of children of students enrolled at the school. Presently there is a waiting list of over 100 student/parents that need child care.

Privately owned child care centers are expensive, and most students are unable to pay the costs.

The purpose of this project is to examine the feasibility of providing a child care center big enough to adequately handle the children of students presently enrolled. Four possible solutions are suggested in detail. Of these four solutions, one is strongly recommended based on the evaluation criteria used. Comparing cost, staff, capacity, building and land site, we will prove why the solution we picked is best suited for the described problem.

There is also a background section showing statistics statewide of all the campuses providing preschool child care. This information will tell how much these centers cost, what they provide, and how much money is necessary to maintain them. There is also information on centers in the Central City area, giving the cost that students would have to pay to enroll their children, along with the services these centers will provide for the cost.

 WRITING ASSIGNMENT

Study the planning sheet for your major project and revise it as necessary. Then study the main points from your outline to determine where you will begin writing. List those points on the planning sheet in order from easiest to hardest.

Choose one of the following:

1. If you conducted a research, feasibility, or comparative product study, combine your purpose statement and one major point into a thesis. Write the thesis in one sentence and submit it for review if requested. Use it to guide your writing of the report.

2. If you conducted an interview for a separate paper, combine your purpose statement and your one major point into a thesis. Write the thesis in one sentence and submit it for review if requested. Use it to guide your writing of the interview report.

3. If your project is collaborative, meet together to write your thesis based on your purpose statement and your one major point. Use it to guide all group members in writing their individual portions of the report.

When you have finished the body portion of your report, write conclusions and recommendations that are supported by the facts and analyses in the body. Make sure that the conclusions and recommendations are understandable even to a person who has not yet read the report itself, and that they are justified. If your instructor so directs, submit the conclusions and recommendations for review.

When you have finished the draft of the body and the conclusions, write an introduction that will cover as many of the items on the checklist as appropriate. Submit it for review if requested.

Combine the introduction, body, and conclusions or recommendations into a draft of your project report. If portions of the draft have not been reviewed earlier, this is a good time to seek review.

REFERENCES

Bejar, John. 1982. *Self-installed burglar alarms.* Unpublished report. San Jose State University.

Clayton, La Vada. 1985. *Technical evaluation of COM recorders.* Unpublished report. San Jose State University.

Devere, John, Rita Hertel, Gary Ursy. 1989. *Child care: A study of the feasibility of creating a program to meet the needs of students with children.* Unpublished report. San Jose State University.

Jarrett, Nancy. 1984. *A comparative study of four brands of stethescopes for home health nurses.* Unpublished report. San Jose State University.

Leonard, David. 1985. *An evaluation of Nd:YAG surgical lasers for Allied Hospital.* Unpublished report. San Jose State University.

Li, Wing-Kei. 1985. *The investigation of cooking steamers.* Unpublished report. San Jose State University.

U.S. Office of Technology Assessment. 1991. *Changing by degrees: Steps to reduce greenhouse gases.* Washington: GPO. OTA-O-482, February.

U.S. Office of Technology Assessment. 1991. *Complex cleanup: The environmental legacy of nuclear weapons production.* Washington: GPO, OTA-0-484, February.

Zertuche, Ralph. 1982. *The selection and purchase of a new non-commercial 10-inch table saw.* Unpublished report. San Jose State University.

Chapter 6

Making Information Accessible

When you finish writing a draft—whether it is a manual, a résumé, or a report—you want to ensure that the information is accessible to your readers. Most technical documents are not read from beginning to end; instead they are usually scanned and then read in discrete sections. You can help readers find information in two ways: with *content cues* and with *format* (or design) *cues*.

Content cues come from the internal context of the document: they are those guides you provide *within* the text through language and sentence construction to help the reader locate and understand information. *Format cues* are provided by the physical design of the document: they are those guides you provide by the way you make each page—and the whole document—look visually. At this step in the writing process, you can easily examine your draft for its content cues and add any that are needed. Therefore, this chapter covers matters such as titles, page and chapter markers, overviews, parallelism, transitions, and patterns of information.

Chapter 12, Designing the Document, presents detailed information on providing format cues: matters such as page design, type sizes and styles, white space, and lists. If you are designing the document as well as writing it, you might want to read that chapter now.

As you examine your draft for content cues to its information, remember that readers don't have the same experience with the material you do; you are now an expert, but they may find the information totally new. Think of your draft as a dense forest without a path. Your job is to lead your readers through that forest in such a way that they find the important features—the giant specimen tree, the swamp, the rare species of fern—that you have already noticed. Your readers have made their way through document forests before, so they have certain expectations, and you'll communicate best when you meet those expectations.

6.1 Composing Meaningful Titles, Subtitles, and Reference Lines

The first cue you give a reader about the purpose and content of any piece of writing is the title or reference line.

What a Meaningful Title Is

For readers of technical literature, the title or reference line needs to tell both what information is covered and how that information is handled. Before your readers plunge into the thicket of your writing, they deserve to know what's in store for them, and you cannot assume that they are familiar with your subject. For example, here are the titles of two technical articles and two books. What subject do you think each one covers?

> "How to Find the Right Software"
>   (Did you guess computer programs for stock market investors?)
> *The Two Faces of Management: An American Approach to Leadership in Business and Politics*

(Did you guess points of similarity between management in industry and management in government?)

"RAM MODS: Horns of Plenty"

(Did you guess modified Cessna airplanes?)

What Color Is Your Parachute?

(Did you guess a book on job hunting?)

Whether you are searching in a computer database, a periodical index, or a library catalog, you depend on cues to help you find what you want. In titles or the reference lines in memos, letters, and reports, the primary cues are *key content words* and *organizational indicators*. Key content words tell you about the subject of the document, while organizational indicators tell you the method of organization and the purpose. Look again at the titles listed above.

In the first example, the words *how to* provide a valuable cue indicating that you will get instructions. The key word *software* identifies the subject. But software (computer programs) can be used for many things. In this title, the key content words that would identify the specific use are missing. A title "How Stock Market Investors Can Find the Right Software" would supply that missing information.

In the second example, *The Two Faces of Management: An American Approach to Leadership in Business and Politics,* notice all the key words in the title: *management, American, leadership, business, politics*. The combination of all these key words gives a good general idea of the subject matter. In addition, the words *two faces* imply a comparison, as does the parallel structure of *business and politics.*

The third example, "RAM Mods: Horns of Plenty," could be very confusing to a general reader because it assumes familiarity with a name and an abbreviation. If you know computer terminology, you probably thought RAM meant "random access memory." In this case, however, RAM is the name of an airplane company that *modi*fies airplanes, and the article describes such a plane. The title is catchy and attention-getting, but it does not clearly indicate either the subject or the approach. To be fair, though, this article title appeared in an aviation magazine, whose readers would be more likely than general readers to recognize the name and the abbreviation.

The fourth example, *What Color Is Your Parachute?* is a fanciful, attention-getting title, but it gives no cue to the subject. Recognizing this, the author supplies a secondary title, *A Practical Manual for Job-Hunters and Career-Changers.* Notice that the secondary title contains both an organizational indicator, *practical manual*, and key content words, *job-hunters, career-changers*. The secondary title is the *real* title. The problem with secondary titles, however, is that they are not always indexed, and readers searching for job-hunting information might not be led to this book. On the other hand, shelved in a bookstore with other job-hunting books, the catchy title does attract book buyers (generalist readers). Usually, attention-getting titles are better for generalist readers than for others in the technical spectrum, who want information and are impatient with fancy words.

Subheadings within a document are as important as headings or titles; if they contain both key content words and organizational indicators, they can help a reader find information quickly. In a table of contents, good subheadings send a reader to the right section of the document; in the text itself, good subheadings direct the reader's eye to the right part of the page.

If you are writing a memo or a letter, you obviously do not have a title, but you do have a subject or reference line to guide your reader. If your memo deals with a proposal for a comparative study of music synthesizers, don't write a reference line like this:

Subject: Term Project

Instead, write one that tells the reader the task, the form, and the key content. Write one like this:

Subject: Proposal for a Comparative Study of Four Music Synthesizers

How to Compose a Meaningful Title or Subtitle

Notice that you can take two specific steps to create meaningful titles or subheadings. You can do the following:

1. Use *key content words,* which are words that tell the reader about the topic or content of the document. These words should be as specific and concrete as possible. For example, if you are writing about automobiles, *Ford* is more specific than *car,* and *Escort* is more specific than *Ford.* (See Chapter 8 for more information about concrete words.)

2. Use *organizational indicators,* which are words that tell the reader what the method of organization and purpose are. Such words include the following:

How to
An Explanation of
An Analysis of
A Description of
A Definition of
The Process of
    (or How _____ Works)
A Recommendation for
A Comparison of
An Evaluation of
A Justification of
A Report on
The Plan for
The Results of

What to Put on a Title Page

Long documents like reports and manuals have title pages that provide key information about the contents. Besides the title itself, the title page can include the authors' names, the number of the edition or report, and—if commercially published—the publisher's name. Company documents may include only the title on a title page because authors are not named and the document is self-published. Information about title pages in specific forms of writing (proposals or feasibility studies, for example) appears in Chapters 15–23.

EXERCISES

1. Read the following abstracts of technical articles, and write a meaningful title for each. Remember to choose words that show both the organization and the key content.

1. This study, using a set of six evaluation criteria, compares four-track cassette recorders in order to recommend one or more for use in a home recording studio. The products compared are Aria Studiotrack IIII, Clarion XD-5, Cutec MR402, Fostex 250, and Tascam 244. The evaluation criteria include the following: tape speed, pitch control, noise reduction, equalization, rack mount, and price. The study includes (1) a technical definition of a four-track cassette recorder, (2) a description of the principal components of a multitrack recorder, (3) a process description of the multitrack sound-recording process, (4) an explanation of the evaluation criteria and the point evaluation system, (5) individual product evaluations, (6) product comparisons, and (7) conclusions and recommendations. The study recommends the Tascam 244 for use in a home recording studio because it best satisfies the criteria set up for the study and combines all the features necessary for recording, processing, and mixing a high-quality four-track cassette tape (Neher 1984).

2. This report evaluates four Nd:YAG surgical lasers for use at Allied Hospital: Lasers for Medicine FiberLase 100; Cooper LaserSonics Model 8000; MBB-Angewandte Technology mediLas 2; Laserscope OMNIplus. Allied Hospital is a private hospital in a large urban community. Knowledge of surgical laser action, application, and market competition is needed to choose the right laser. The evaluation is based on five criteria: versatility, cost, safety, mobility, and ease of operation. Each model is analyzed by these criteria; the comparison emphasizes key differences among the models. The FiberLase 100 is recommended for Allied Hospital. It costs about $20,000 less than its nearest competitor, the LaserSonics Model 8000. Its audible fault indicators and self-calibration system combine to give it the best overall safety features of any model. Additionally, its broad delivery options and push button input panel give it creditable versatility and easy operation (Leonard 1985).

3. Assembly, operation, and disassembly of the Battelle Large Volume Water Sampler (BLVWS) are described in detail. Step by step instructions of assembly, general operation and disassembly are provided to allow an operator completely unfamiliar with the sampler to successfully apply the BLVWS to his research sampling needs. The sampler permits concentration of both particulate and dissolved radionuclides from large volumes of ocean and fresh water. The water sample passes through a filtration section for particle removal then through sorption or ion exchange beds where species of interest are removed. The sampler components which contact the water being sampled are constructed of

polyvinylchloride (PVC). The sampler has been successfully applied to many sampling needs over the past fifteen years. 9 references. 8 figures (Thomas 1985).

4. The technical aspects of using observations of animal behavior reported by the general public to predict earthquakes are discussed. The possibilities and problems inherent in two alternative approaches are explored. One approach would be to obtain daily reports of egg and milk production on farms along faults. The other approach would be to encourage a selected group of volunteer observers to report all observations of apparently abnormal animal behavior (Verosub 1985).

2. Bring to class three examples of misleading, ambiguous, or unclear titles from magazines, newspapers, or books that you have read this term. Be prepared to discuss what's wrong with them and how they might be improved.

6.2 Including Clear Headers and Footers

In long documents, readers rely on information at the top and bottom of each consecutive page to (1) remind them where they are in the documents, and (2) help them to move quickly to another section, either one that is referenced in the text or one they seek for added information.

Headers (often called running heads) are single lines of type across the top of most technical and scientific books and long documents. They are separated from the text by white space and are often in a different typeface. They typically include the chapter or section number, the chapter title, and the page number. They can also include a title that summarizes the contents of that particular page. However, many documents omit page content headers because they cannot be written until the document pages are in final form.

Footers (often called running feet) are single lines of type across the bottom of pages that can give similar information. Commercially published books often use only headers, but company-produced manuals and reports may use headers, footers, or both. The headers and footers help define the look of any document, so companies usually have guidelines that determine both the format and the information to be included.

6.3 Providing Overviews

Another good way to help your readers find information and understand your organization is to

- tell them at the beginning where they are going
- remind them at midpoints where they have been and what's ahead
- review at the end the path they have followed

Technical writers use specific locational guides called *overviews* (sometimes called advance organizers) to carry out these tasks. Among the overviews are the table of contents, list of illustrations, index, introduction, summaries (at the end of a document and at the end of sections), abstract, and executive summary. In a long document, even the body can contain overviews where necessary. Obviously, not all writing tasks need these specialized guides. If you are writing a one-page letter or a three-page memo, you can provide all the necessary organizational signposts through headings and subheadings, parallelism, transitions, and visual markers. But as your writing task becomes longer and more complex in documents like reports and proposals, readers will need overview statements. Here's a brief look at each one; details on how to write them appear in Chapters 15–23.

Table of Contents

Throughout Chapter 4, you organized your material more and more carefully in working toward a formal outline. One big advantage of having a carefully constructed outline is that you can easily turn it into a table of contents for a long document like a manual, formal report, or proposal. The table of contents shows your reader the main points and the method of organization. It also allows easy access to individual sections of the report. Since technical reports often do not have an index, a clear table of contents is especially important. Study the table of contents in the student report in Chapter 22. Notice how the table of contents reproduces the outline and shows where to find the information.

In a very long proposal, report, or book-length manual, you can also put short lists of contents at the beginning of each chapter or major section. As the table of contents provides an overview of the whole, so these lists give an overview of the individual sections. This book has such lists at the beginning of each chapter.

List of Illustrations

Like a table of contents, a list of illustrations helps the reader see at a glance what is included in the document, especially if you have taken care to illustrate all your key points with graphics. *List of illustrations* is a general heading that often includes separate lists—one for tables and another for figures. Robert Brown, an expert in rocket fuels (and a specialist reader), told me that he won't even read proposals or reports that don't have a list of illustrations at the front, because the data he wants is best presented in graphic form, and he wants to be able to find it quickly (1986). Chapter 11 explains how to use graphics to enhance your writing, and Chapter 22 explains how to compile the list of illustrations.

Index

Like the table of contents, the index provides an overview of the whole document. The difference is that while the table of contents shows the reader the document by the *way it is organized,* the index displays the document by listing its items of information, which are *arranged alphabetically* for easy access.

Because many companies realize that their readers want fast access to information, they require indexes in any documents that are more than 12 to 16 pages long. Most word-processing software now has an indexing feature that will compile and print alphabetized entries with the correct page number. However, the decisions about what words to include and whether words should be primary, second-level, or third-level entries are still the responsibility of the writer.

If you follow these four general principles, you can write good indexes.

1. *Select key nouns or noun phrases for entries.* Readers look for "what something is," and nouns provide more specific information than verbs or adjectives. In an index, verbs should be turned into gerunds, their noun form (*scrolling* not *scroll*).

2. *Provide entries for common terms as well as technical terms.* Some readers will know the technical terms, but others will need to be guided to the section where those technical terms are explained.

3. *Limit undefined page entries to no more than three.* Undefined page entries give only a key term with no further breakdown. If your reader looks up the term *formal reports* and it looks like this, your reader will be annoyed.

   formal reports 5, 78, 97, 105, 121

   To avoid undefined page entries, set up a second level of entries that detail the information and give the exact page numbers.

   formal reports
       abstract 97
       executive summary 105
       graphics for 78
       introduction 121
       writing of 5

4. *Provide frequent cross-references.* Cross-references show relationships and give readers an alternate access to topics. *See* references send readers to preferred terminology, while *See also* references direct attention to additional, related information.

Introduction

Within the text, another kind of overview (or perhaps *preview*) is provided by the introduction, which should give information about the topic, the purpose, the scope, and the way in which the information will be developed. Introductions are discussed in section 5.4.

Summaries

Papers of analysis or persuasion usually end with conclusions and recommendations. If the subject is complex, many writers include a *summary* of the report before such conclusions and recommendations to remind readers where they've been—to review the facts and add up the main points. Papers that simply present information without conclusions or recommendations can also end with a brief factual summary. In the same way, major sections of a long document or chapters of a book can end with intermediate summaries: paragraphs that review for the reader what was covered in that section or chapter. Some summaries are written as outlines or checklists rather than as paragraphs. This book uses such checklists at the end of each major section or chapter to sum up and apply the preceding contents.

Again, a summary should be written after the body of the paper is written. It should be concise and should contain only information that appeared in the document. Summaries can be included as a part of conclusions and recommendations. See section 5.3 for examples.

Abstract

An *abstract* is really a condensation or summary of a longer document. It has a different name because of its specialized purpose and its special location in a technical document. Like any summary, an abstract's purpose is to give an overview of the key information in a document. However, that overview is used by readers to determine if they want to read the report itself. Abstracts appear in two places:

1. Separated from the report itself in a database or journal of abstracts. Readers scan abstracts of technical reports, choose the ones in which they are interested, and print out or order copies of those reports.
2. Separated from the report itself, but attached to it on a cover sheet, on the title page, or on the first page after the title page. Readers determine from the abstract whether the report is of interest to them and in general what the report covers.

Abstracts are written for the *primary* reader of the technical report (usually a specialist or technician), and they must be understandable without reference to the report itself. They are not considered part of the main report, so information in an abstract is also repeated in the text. Abstracts are called either *descriptive* or *informative* depending on their purpose and content. See Chapter 20 for explanations and examples of these two types.

Executive Summary

An *executive summary* is another kind of overview. Like an informational abstract, it condenses the key information in a document, including major facts, conclusions, and recommendations. However, the intended reader is usually a nontechnical person like a manager; therefore, the executive summary often contains background information, explanations of technical terms, and discussion that will help the reader make decisions or see the entire picture. The executive summary is considered part of the main text, and many documents include both an abstract and an executive summary. See Chapter 20 for examples of executive summaries and information on how to write them.

6.4 Making Related Points Grammatically Parallel

One of the easiest and best ways to help readers find information and understand your organization is to write grammatically parallel points. Readers look for patterns; thus, whenever you want to show that points are equal or related, you should state those points in the same way. In your written document, those parallel points might become section headings, the first sentences in a series of paragraphs, sentences within paragraphs, or even the parts of an individual sentence.

A Definition of Grammatical Parallelism

Grammatical parallelism is the use of the same structure in sentences or parts of sentences for points that are alike, similar, or direct opposites. Parallelism gives writing balance and proportion, and it creates a pattern for the reader by saying, "Look at these things in the same way." As you polish your first draft, you should use parallelism (1) whenever you want to give the same importance to two or more items, as in a list, an outline, or a sentence, and (2) whenever you want the reader to see the way two or more items are related.

How Parallelism Helps the Reader Understand

Parallelism helps the reader understand because it

- communicates a pattern
- clearly indicates when one point has been completed and another begins
- emphasizes the relationship between ideas

For example, notice the repetition of structure in the following overview statement from the 1982 IBM Information Development Guideline called *Designing Task-Oriented Libraries for Programming Products*.

Operation consists of the following subtasks:

- starting a program
- checking and controlling a program
- recording the status of a program and its data
- reacting to abnormal events
- stopping a program

The following sections describe each subtask in greater detail.

The five subtasks are clarified by placement in a list, by a bullet, and especially by parallelism—each subtask begins with an *ing* word.

The following example is from an operator's manual (Varian 1986). Notice how the parallel structure of the sentences shows the difference between the two temperature ranges and directs the operator's attention to the colored areas on the dial face.

Operating ranges are shown on the dial face. When the cryopump is at room termperature, the needle should be in the silver-colored area. When the cryopump is at proper operating temperature, the needle is in the blue-colored area (i.e., the temperature of the second stage is then below 20K).

In contrast, here is a confusing job description for a software engineer:

Duties: Performing concept definition, software design and development, and to produce system software.

Can you determine what the job requirements are? Try listing each item, one under the other, to determine how many parts the job has.

- performing concept definition
- software design
- software development
- to produce system software

These are all job functions, but they seem to be different because they are different in form. The writer has given you one verbal, two nouns, and one infinitive. Try to clarify the sentence by using the same grammatical structure for each one. You may have to try several possibilities. Since the structure describes job functions, try verbs:

- define concepts
- design software
- develop software
- produce system software

This may not work because it's unclear what the writer means by "to produce system software." You might have to talk to the writer to find out. If you have guessed right, you could restate the job description like this:

Job requirements for a software engineer:

- define concepts
- design and develop software
- produce system software

Another option might be to use nouns:

- concept definition
- software design and development
- system software production

Still another choice might be verbals:

- defining concepts
- designing and developing software
- producing system software

Whatever option you choose, notice that in each case you can make the points understandable and equal by using grammatical parallelism. Remember too that the order you choose for items will influence the reader in a subtle way. You may choose to arrange items from most important to least important (descending order) or from least important to most important (ascending order). If no other order is important to you, follow the principle of emphasis by position: the most important position is last, and the next most important position is first.

In summary, here's how to make subpoints grammatically parallel:

1. List them, one under the other (even if later you will put them in a sentence).
2. Examine the list. Sometimes the order of items should be changed for clarity, especially if some are singular and others are plural.
3. Use the same grammatical structure to state each one. The structure can consist of single words, phrases, or clauses. Here are examples of each:

Single Words	The four operations are etching, masking, screening, and scanning.
Phrases	Three typical printing presses are listed below: blanket-to-blanket press in-line open press common impression cylinder press
Clauses in a List	During the Jan. 1–June 1 period, 5,230 rolls were processed 742 tons of waste were baled 12,470 pieces of freight were handled.

Usually technical writers will not break sentences with only three parts into a list unless they want to call attention to the individual parts, as do these writers.

Clauses in a Sentence	Facepieces fall into three types: the full-face mask, which covers the entire face including the eyes; the half mask, which covers only the nose and mouth and fits under the chin; and the quarter mask, which covers only the nose and mouth and fits on the chin.

Some of these examples use colons to introduce the lists. For more information on punctuation with colons, see the Handbook. To see how parallelism operates on a larger scale, look at the table of contents of this book. There you can see how chapters, sections, and subsections have parallelism in their titles.

EXERCISES

1. Complete the following sentences by adding a series of parallel words, phrases, or clauses.

1. (List by name) are four units of measure in the metric system.
2. A driver can increase a car's gas mileage by (give three techniques).
3. I can get a job as a technical writer if I (suggest three courses of action.)
4. You may find it necessary to (suggest two possible activities) in order to find a parking space near campus.

5. If you want to write a good résumé, you should (list four steps).
6. People who (list several undesirable forms of behavior) make poor employees.
7. Changes this campus needs urgently are (list three improvements).
8. The procedure for starting a car is as follows: (list at least four steps).
9. Three alternatives to oil as a home heating fuel are (list them).
10. (List four courses of action) will reduce air pollution.

2. Revise each group of sentences below into a single sentence by condensing and by putting groups of similar ideas into parallel patterns.

1. The vehicle is a three-door auto.
   It is a cooperative effort.
   General Electric participated in the effort.
   So did Chrysler Corporation.
   Also included was Globe-Union, Inc.
2. The ETV-1 is intended for short trips.
   It can be used for commuting.
   Grocery hauling is one aim.
3. At a constant 45 mph the ETV-1 provides an expected 97-mile range. Sixty-nine miles is the figure dropped to with the SAE J227a driving cycle, which more realistically includes stops and starts.
4. The auto weights 3300 pounds.
   The vehicle has a half-ton battery pack.
   A 600-pound payload can be carried by it.
5. Sixty miles at 60 mph is a good typical driving range for a practical electric vehicle.
   Today's electric cars can't claim an average velocity much better than 8 mph, however.
6. Vacuum-fluorescent displays have found more applications recently.
   They have low cost.
   Power requirements are modest.
   High readability is a good feature.

3. Rewrite the following sentences to make the equivalent items parallel in structure. Then meet with three to four fellow students to review your rewritten sentences. Discuss each sentence, and be prepared to defend your rewrites to the whole class.

1. The purpose of this section is to provide information about

   - our technical staff
   - performing other similar projects, and
   - qualifications

2. The equipment she worked with included camera use, developing negatives, metal plates, light composer, and rub-on lettering.

3. They found the weave of the silk open enough to allow paint to pass through easily, and it was also found that the strands were so fine that they left no permanent impression on the paint.

4. This knowledge must include, but not be limited to

> reading music quickly,
> being familiar with most musical instruments,
> an understanding of the range of each instrument,
> a knowledge of music literature and history, and
> he must be familiar with all kinds of music notation and proofreading marks.

5. Some of the many things they check for are listed below.

- The amount of wire allowed to be stripped of insulation should equal the wire width or be less.
- The component should be flush or level, with no more than $1/16$ angle.
- When the wire insulation melts into the solder, it contaminates the solder joint and must be reapplied.
- Cracked components are never allowed to be used.
- Metal should never touch an open trace; this will cause a short in the unit.

6. The chemicals are kept in two separate tanks and with connecting rods are shot into a container through a gun-like dispenser around the product.

7. Painted on the front side of the vase is Hermes carrying the infant Dionysus, and on the reverse the three Muses are painted.

8. III. Technical Definitions of Microwave Ovens

    A. What is a microwave?
    B. How microwaves operate within the oven chamber.
    C. Operation of microwave ovens.

6.5 Writing Good Transitions

Another way to make information accessible is to write good transitions. Because technical documents are often composed of a number of discrete parts, transitions can help your reader see the relationships among those parts. Transitions are like the string holding beads in a necklace. Individual beads may be very beautiful, but until they are strung, they are neither a necklace nor are they very useful. The same principle holds true for sentences and paragraphs. Individual sentences or paragraphs may be clear and well written, but until they are fastened to one another in a meaningful way, they will not make sense to the reader. You join sentences, paragraphs, and sections with transitions.

Definition of a Transition

A *transition* is a word, phrase, sentence, or paragraph that moves the reader from one idea to another. Transitions work by simultaneously looking backward to the previous idea and forward to the new idea. You can link both sentences and paragraphs by transitional words or by repetition of key words. You can also link paragraphs by sentences or short paragraphs that both summarize the previous ideas and introduce the new ideas.

Transitions between Paragraphs

The paragraphs in Figure 6-1 explain the process of color separation in printing (International Paper Company 1981). Notice how the fifth paragraph sums up the

Content paragraphs

> The process of color separation is analogous to the process of seeing by the eye, but the printing process introduces new concepts. The original is photographed using three filters, each corresponding in color and light transmission to one of the additive primaries.
>
> Placing a red filter over the lens produces a negative recording of all the red light reflected or transmitted from the subject. This is known as the red separation negative. When a positive or print is made from this negative, the silver in the film will correspond to areas which did <u>not</u> contain red but contained the other two colors of light, which are blue and green. In effect, the negative has subtracted the red light from the scene and the positive is a recording of the blue and green in the scene which is called <u>cyan</u>. The positive is the <u>cyan printer</u>.
>
> Photography through the green filter produces a negative recording of the green in the original. The positive is a recording of the other additive primaries, red and blue, which is called <u>magenta</u>. The positive is the <u>magenta printer</u>.
>
> The blue filter produces a negative recording of all the blue in the subject. The positive records the red and green which when combined as additive colors produce <u>yellow</u>. This positive is the <u>yellow printer</u>.

Transition paragraph

> These three colors, cyan, magenta, and yellow are called <u>subtracting primaries</u> because each represents two additive primaries left after one primary has been subtracted from white light. These are the colors of the process inks of process color reproduction.

New topic

> When the three positives are combined and printed, the result should be a reasonable reproduction of the original. Unfortunately, it is not. The colors, outside of yellow and red, are dirty and muddied. There is too much yellow in the reds and greens and too much red in the blues and purples. This is not a flaw in the theory but is due to deficiencies in the colors of the pigments used in the inks. . . .

Figure 6-1 Use of a Transitional Paragraph

Permission to reproduce this material from the 12th edition of POCKET PALS® was granted by International Paper Company.

previous information and then prepares the reader for a discussion of color reproduction. For more information on writing good paragraphs, see section 5.2.

Transitions between Sentences

Transitional words like *for example* and *therefore* are simpler links; to be most effective, you should put such words near the beginning of the sentence. In the following example, notice all the transitional words (italics added).

> In the event divers are carried into a trawl from which they cannot readily extricate themselves, they must cut an exit through the web. *Usually* trawls have heavier web in the aft portion (cod end). *Therefore* an escape should be cut forward in the top of the trawl body and a 3-foot long diagonal slit should be made in the trawl. *Another* similar slit should be made at 90 degrees to and beginning at the upstream end of the first slit. The water current should *then* fold a triangular flap of webbing back out of the way, leaving a triangular escape hole. The diver's buddy should assist the trapped diver through the opening to free any gear which snags on meshes (Miller 1979).

Transitional words and phrases show relationships. Table 6-1 shows how transitions function and gives examples of common transitional words and phrases.

Key Words as Transitions

Another kind of transition is repetition of key words, a technique that keeps readers on track by refreshing their short-term memory. To tie sentences or paragraphs together, you repeat the key term at the beginning of the new clause, sentence, or paragraph. Such repetition reminds the reader what has gone before and says that more on this topic is coming. In technical writing, it's important to use the same word to refer to an object or an idea from sentence to sentence and paragraph to paragraph; clarity is more important than variety. Thus, if you call something a *shaft* in one sentence, don't call it a *rod* in the next.

In the following paragraph (Neher 1984), notice how the italicized words link the ideas. The key words *engineer, performers, record,* and *tapes* are picked up and repeated (in varied forms) throughout the paragraph to provide the links (italics added).

Table 6-1 Transitional Words and Phrases

Function	Examples
add	also, in addition, second (etc.), moreover
enumerate	first, second
compare/contrast	likewise, similarly, however, on the other hand
illustrate	for example, that is, for instance
summarize	in conclusion, in summary
tell place	below, above, on the right
tell time	later, then, by noon
show cause	therefore, consequently, thus

Once the recording equipment is set up and properly adjusted, the master recording is made. The *engineer* turns on the recorder and signals the *performers* to begin. During the *performance*, the *engineer* monitors the *record* levels and tone adjustments to make sure they stay within acceptable bounds. When the *performance* is over, the *engineer* stops the *tape* and rewinds it. If, after playing back the *tape*, the *engineer* and the *performers* are satisfied with its quality, the *tape* is ready to be mixed down. Otherwise, the *recording* process may need to be repeated, or individual *performers* may do their parts over (overdubbing).

To avoid the stuttering of constant repetition, sometimes you must replace a key noun with an appropriate pronoun. If you use a pronoun, however, be sure the original noun is not too far away. Often it's better to add an adjective to the key noun instead of using a pronoun. The following paragraph (Lissaman 1980) uses both pronouns and adjectives with key nouns (italics added).

Among the greatest reservoirs of solar energy are the ocean gyres, or circulating current systems, generated by the prevailing global winds. The energy density in *these currents* is very high, because *they* are created by a twofold intensification of the direct sunlight: first, by the transformation of the sun's heat into wind, then by transfer of the surface wind stress into the wind-driven currents. *This* intricate energy *transfer* is part of the mighty global heat engine. In the Northern Hemisphere, the gyres rotate clockwise, and because of the Earth's spin, are strongest on the western shores of the oceans.

6.6 Supplying Patterns of Old-to-New Information

As you polish your first draft, you need to keep reminding yourself that the information and advice you are presenting to your reader is often *new* and *unknown* to that reader. Therefore, you want to do whatever you can to help readers fit that new information with what they already know; to make connections; to comprehend; to "take it" for themselves.

Research has shown that one of the ways to do this is to structure the flow of information so that new information is linked to old or known information, with the old coming first. In fact, at the sentence level, old information can be the subject, and new information the predicate. According to one expert, "After the new information appears in the predicate, it is old and can then appear as the subject. So the chain of information can be expressed abstractly as A + B, B + C, C + D, or old plus new, which becomes old plus new and so on" (Matalene 1989).

Related to this old-plus-new structure is the principle that the general should come before the specific, the permanent before the temporary, and information that affects many before information that affects few (Felker 1981).

Checklist for Making Information Accessible

1. Do my titles, subheadings, or reference lines contain both key words and organizational indicators?

2. Have I provided headers and footers in long documents?
3. What overviews are appropriate for this document?
   ___ table of contents
   ___ list of illustrations
   ___ index
   ___ introduction
   ___ summaries
   ___ abstract
   ___ executive summary
4. Are any points that are of equal importance also grammatically parallel?
5. Have I included transitional words, phrases, sentences, and paragraphs between sections?
6. Have I used a pattern of old-to-new information?

WRITING ASSIGNMENT

Write a meaningful title for your term project paper. Test the accuracy of your title by asking two fellow students to write three to four sentences explaining (1) the exact topic, and (2) the purpose and the method of organization of your report. If their responses are not accurate, rewrite your title. Then prepare a title page for your document.

When you have a good title for your term project paper, design the table of contents by adding page numbers to your final outline. After determining the placement of the figures and of the tables in your report, write a list of illustrations. Attach these overviews, along with your title page, to your report. See Chapter 22 for examples.

REFERENCES

Brown, Robert. 1985. United Technologies. Conversation with author. 15 Feb.

Felker, Daniel et al. 1981. *Guidelines for document designers.* Washington: American Institutes for Research.

IBM Information Development Guideline. 1982. *Designing task-oriented libraries for programming products,* No. ZC28-2525-2, 77.

International Paper Co. 1981. *Pocket pal: A graphics arts production handbook.* 12th ed. New York: International Paper Co., 79–81.

Leonard, David. 1985. *An evaluation of Nd:YAG surgical lasers for Allied Hospital.* Unpublished report. San Jose State University.

Lissaman, P. 1980. Tapping the ocean's vast energy with undersea turbines. *Popular Science,* September, 72.

Matalene, Carolyn B. 1989. *Worlds of writing: Teaching and learning in discourse communities of work.* New York: Random, 273.

Miller, James, ed. 1979. *NOAA diving manual.* 2nd ed. Washington: National Oceanic and Atmospheric Admin. Section 8-45.

Neher, Jonathan. 1984. *A comparative study of four-track cassette recorders.* Unpublished report. San Jose State University.

Thomas, V. W. and R. M. Campbell. 1985. Abstract of Assembly, operation, and disassembly manual for the Battelle large volume water sampler (BLVWS). *Energy Research Abstracts* 10, (15 April), 1605.

Varian Vacuum Products Div. 1986. *Cryopump operator's manual for CS8FA*, No. 87-400437. Santa Clara, CA: Varian Associates.

Verosub, K. L., D. F. Lott, B. L. Hart. 1985. Abstract of Use of volunteer observers to detect abnormal animal behavior prior to earthquakes. *Energy Research Abstracts* 10, (15 April), 1617.

Chapter 7

Reviewing and Revising the Document

One test of your planning skill comes when you have finished writing your first draft. The test is whether you have enough time before your due date to (1) carefully review what you have written, (2) revise it as needed, and (3) complete and edit the final draft. Students are often surprised at how long these steps take—and sometimes must work through the night to finish a major paper.

Both professional writers and professionals who write (engineers, scientists, and managers, for example) realize the importance of careful review. They schedule review and revision time and push themselves to prepare a first draft early enough to allow time for multiple reviews and careful revision. Karen Zotz, for example, gives this advice: "Give yourself plenty of time for preparation, thought, writing, review, revisions, *review, revisions,* **REVIEW**." Zotz works in the agriculture extension program of a large midwestern university; her job is to develop programs in nutrition education and family community leadership. Even though she is not called a professional writer, she spends 85 percent of her time writing. Antonia Van Becker *is* a writer for a large semiconductor manufacturer. She says, "I have found that the best way to review is to go through the document several times, each time looking at different elements. For example, I review once for technical information, once for graphics and graphic captions, once for chapter titles and headings, and once for grammar and syntax."

As you can see, no matter how hard you have worked on the draft, your memo, letter, or report will profit from revising. That's why it's called a draft; it is a beginning or preliminary version, and as such, is subject to revision. Revision may be needed in any or all of these three areas:

1. content: that is, purpose, reader, and topic, plus technical or factual information
2. organization and accessibility of information
3. diction, syntax, and punctuation of sentences; use of numbers and symbols

7.1 The Need for Review

Suppose that for the past year you have been building a single-engine airplane from a kit. You have put in countless hours building the frame, assembling the fuselage, building the engine, and attending to the details of wiring, control surfaces, and instruments. After you tighten the last screw, will you hop in the seat, rev the engine to its maximum, and take off at top speed down the runway? Probably not. Instead, you will run a series of tests, checking out each major component to see if it is doing its job and working properly with all the other components. You might even have it checked by an experienced mechanic—because only after careful checking will you feel ready for the sky. Your life might hinge on the outcome of those tests.

The writing situation is very similar. Even a short memo or letter has taken concentrated effort on your part. Before you release it, you want to ensure that each part is doing its job and working with the other parts to fulfill your purpose. A long proposal or report, like an airplane, has many parts—all the more reason for careful review and, if necessary, revision. Before you started writing the draft, you reviewed your outline to be sure that the organization you had chosen was the best way to fulfill your purpose and meet your reader's needs. Now you want to look again at what you have written to see if your document is technically accurate, clear, correct, and easy to read—because your reputation as a writer might hinge on the outcome of this second look.

If possible, let the draft cool for a day or two before you begin reviewing. You need to make the difficult shift from being a writer to being a reviewer. When the material is still fresh in your mind, it's hard to put yourself in a reader's place, but once you've been away from it for some time, you can be more objective.

7.2   The Three Phases of Review and Revision

When you review, you can do a better job if you break the review and revision into three phases, as shown graphically in Figure 7-1. To carry out the three phases, you should read through the document at least three times, each time looking for different things. By concentrating on one thing at a time, you can be more thorough. Content review and organization review are covered in this chapter because you will probably need to revise after each one. The third review—editing—is covered in Chapter 8, along with exercises to sharpen your editing skills.

Content Review

Phase 1 is a review of *content*; you are checking for technical accuracy and logic and the appropriateness of the purpose, reader, topic, tone, and form. Check your notes or your original sources to verify the accuracy of numbers, facts, dates, names, quotations, and conclusions. Check all the graphics for accurate presentation of data. Review documentation references to ensure completeness and correctness. Ask these questions about the content:

- Are the facts, figures, and computations correct?
- Is the purpose clear and appropriate?
- Is the thesis clear and is it supported by evidence?
- Does the thesis support the purpose for writing?
- Is all necessary information at hand?
- Is the information appropriate for the intended reader or readers?
- How can the tone be described?
- Is that tone appropriate for this document?
- Does this document need more or different graphics to make the point?
- Can anything superfluous be eliminated?
- Do the conclusions follow from the data presented?
- Do recommendations make sense in light of the conclusions?
- Does the form of this document best convey the content to the reader?

When you complete the phase 1 review, make any revisions or corrections.

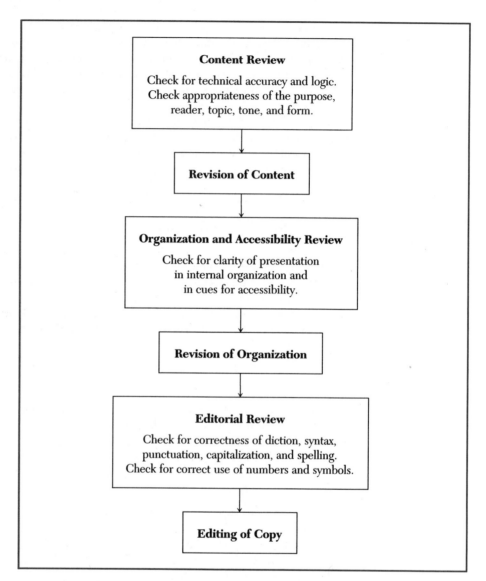

Figure 7-1 Flow Chart of the Three Phases of Review and Revision

Once you have words on paper, it may be difficult to acknowledge that you need to spend time writing a document (or at least parts of it) over again. But when you scheduled your project (following the advice in section 1.6), you actually allowed time for this step in the process. Industry planners know that review and revision often take up to 40 percent of the total writing time; professionals know that the review and revision step is at least as important as any other step in the process.

In revising content, make the most significant revisions first. Correct anything that is wrong, vague, or misleading. Find the specific support for weak or vague sections. In reports, revise conclusions and recommendations at once because they will be read first by nearly all readers. In instructions or procedures, correct the sequence of operations if it is wrong or misleading. Complete the content revisions, and then review the document for organization and accessibility.

Organization and Accessibility Review

Phase 2 is a review of *organization and accessibility*; you are checking for clarity of presentation, knowing that it does no good to be accurate if what is written is disorganized or inaccessible. Start this review by reading through only the headings in order; then read the subheadings in order. Verify that any numbering system is consistent, that the levels of headings follow an accepted format, and that the headings are grammatically parallel. Make sure that the headings in the table of contents are the same as those in the document. Ask if the method of organization clearly communicates the thesis. Many word-processing programs allow you to print out only headings and subheadings so you can evaluate the hierarchy. Others let you also print out first sentences in paragraphs to check for topic sentences.

After you have reviewed headings and subheadings, ask these questions about the document's overall organization and clarity:

- Does this document make sense? If not, why?
- Is the major point clear? What causes confusion?

Now study paragraphs for organization and coherence. Ask:

- Do the subheadings and the topic sentences support the thesis?
- Do the sentences in each paragraph support that paragraph's topic sentence?
- Is each section or paragraph linked logically and smoothly to the one before and the one after it?
- Can the intended readers find the information they want? Is it easy to find information by skimming? Do most pages highlight key words and phrases so they are easily found?
- Does this document need more or different definitions, examples, descriptions, instructions, details explaining who, what, why, when, or how?
- Does the document include aids that help the reader understand the organization? Are there advance organizers? Summaries or checklists? Is the document easy to read?

When you complete the phase 2 review, make any necessary changes or corrections. You may have to move or rearrange sections or paragraphs to achieve greater clarity. You probably already worked on this when you reviewed and revised your outline, but now you may have to organize once more. If you are working from a typed or handwritten draft, chop sections apart with scissors if necessary and reassemble the draft with tape. If you are working on a computer, add, delete, copy, and move blocks of text as necessary. Add headers or footers to long documents. If at all possible, type or print out a clean draft before phase 3, the editorial review.

Editorial Review

Phase 3 is a review of various *editorial* matters, including word choice, sentence construction, punctuation, capitalization, and spelling. This kind of review is sometimes called line-by-line editing or copyediting, and corrections are made on individual sentences. Chapter 8 and the Handbook give detailed information for carrying out editorial review and revision. Even though this is the last of the three phases of review, it must always be done. Errors of punctuation and spelling are not acceptable in the workplace.

7.3 Types of Review

In both the classroom and the professional world, several types of review are possible, and for any written piece you may choose one or more of these reviews. The longer and more complex the writing assignment is, the more likely you'll be to use more than one type of review. The three basic types are self-review, peer review, and instructor or management review.

Self-Review

Whatever other type of review you might use, you're sure to have to review your own document at each draft and then again in its final form. To help objectify your review, look at your original planning sheet and remind yourself of your intended purpose, readers, topic, tone, form, and style. Then check your document to see if your goals have shifted or if you have done what you intended to do. Follow the three phases of review, revising as needed after each phase. Don't be afraid to read your document aloud. You yourself should provide most of the changes, and you should review and revise the document before you submit it to others.

Peer Review

Most companies provide for peer review because colleagues can closely approximate your intended readers. In addition, your peers often have a stake in what you're writing, so they will review carefully. Peer review can be done internally or externally. In either case, it's important to give the manuscript to the reviewers in advance, so they have time to read it carefully before you meet.

Internal peer review on the job involves members of your department; in the classroom it involves fellow students in your writing class. In both cases, the reviewers will know something about what you're doing, and they may even be experts who can help with technical accuracy.

External peer review on the job involves people from other departments who have an interest in what you have written. They are concerned with what you say because it affects what they do. When they read, they can check for clarity as well as technical accuracy. Medical writer Ann Petersen says, "All our writing is reviewed by at least one other writer and as many as a dozen individuals outside the group (people in the marketing, regulatory, clinical, development, public affairs, quality assurance, quality control, and legal groups). These people rarely agree on every point."

When you are a student, external reviewers may be family members or friends whose critical abilities you trust. Often they won't know the subject matter at all, but they can provide a valuable check for clarity. If you are writing to generalists and your external reviewers don't understand, you know you need to revise. Peer review can be carried out one-on-one or in small groups. Often small groups work best because of the group dynamic: that is, one comment triggers another, or one observation encourages another person's new idea. Peer reviewers work best when they are given questions like those listed in the section on the phases of review.

Instructor or Manager Review

As a professional, you might well use all three types of review, the last reviewer being your manager. As the person who probably assigned you this writing task, he or she is in a unique position to assess how effectively you have performed. Sometimes managers see only the final copy, but more often they help in its preparation by reviewing for technical accuracy and clarity. You should also think of your instructor as a manager and seek interim reviews when you suspect problems. Most instructors will be glad to share their expertise with you if you don't wait until the last minute to ask.

7.4 Reviewing a Collaborative Document

When several writers work collaboratively on a long and complex document, their various writing styles and approaches must be knit together before or at the reviewing stage so the document will have unity and coherence. Typically, a lead writer or editor will cut, paste, and revise to blend the various portions. Then, perhaps, all the writers will participate in the review. If this is the situation, you must force yourself to be very objective, viewing your own writing only for its contribution to the whole. Many writers are defensive about their writing and find criticism hard to accept. Yet think how much easier it is to be criticized at the draft stage when you can revise, than it is to be blamed for faults in the final document.

As a critic of someone else's writing, remember to be positive as well as negative. All of us are vulnerable when we expose our writing for criticism; sometimes you will accomplish the most by focusing on the best parts of a document and then revising the rest to bring it up to that standard.

Whether you are writing as an individual or part of a team, it's important to have your draft written well before the due date. That's why early planning and frequent reference to your schedule are so important. The only way to benefit from a review is to have time for a thorough evaluation—and time for the revision that will follow.

Checklist for Content and Organization Review

Phase 1: Content Review

1. Are each of the following accurate?
   ___ figures
   ___ facts
   ___ computations
   ___ dates
   ___ names
   ___ quotations
   ___ conclusions
   ___ graphics
   ___ references
2. Does the thesis support the writing purpose? Is that purpose clear and appropriate?
3. Is the thesis clear and supported?
4. Is all information included?
5. Is the information appropriate for the intended reader?
6. Is the tone appropriate?
7. Are conclusions supported by data?
8. Do recommendations follow from conclusions?
9. Have I chosen an appropriate form?

Phase 2: Organization and Accessibility Review

1. Do headings and subheadings meet the following criteria?
   ___ accuracy of content
   ___ parallelism of form
   ___ correct numbering or indentation
2. Do paragraphs concentrate on one main idea?
3. Are topic sentences supported with specifics?
4. Are paragraphs and sentences logically joined with transitions?
5. Are key words and phrases highlighted to improve reader comprehension?
6. Does the document include cues to help the reader understand the organization? Are there advance organizers? summaries or checklists? headers or footers?

 EXERCISES

1. For practice in the kind of analysis needed for review, treat this chapter as a draft. Read it first for content and applicability to student readers like yourself. Ask the following: What is the purpose? What is the tone, and is it appropriate? Do the graphics contribute to your understanding? Do the conclusions and recommendations follow from the text and make sense? Write at least a paragraph based on your observations.

Now read the chapter again, looking at organization and accessibility of information. Are the four major divisions the best way to present this information? Are those divisions in the best order? For example, in an earlier draft, the order was

7.1 The Need for Review
7.2 Types of Review
7.3 Reviewing a Collaborative Document
7.4 The Three Phases of Review

Is that a better organization? Why, or why not? Look at headings and subheadings, headers, and topic sentences. Do they work? Could you improve them? If so, how? Write at least a paragraph based on your analysis and be prepared to discuss in class. If you have terrific ideas for improvements, ask your instructor to forward your suggestions to me; you can be my reviewer.

2. After letting the first draft of a major paper cool for some time, review it yourself for content and organization. Use the checklists and be meticulous in marking questions, unclear passages, and possible errors. Revise as needed.

3. Pair up with another student and exchange papers. Review the document for content and organization in two separate phases, making notes and suggestions on the manuscript. Apply the specific questions listed in section 7.2 and summarized in the checklists. Note positive as well as negative things. When you get your own paper back, analyze the comments and revise as needed.

4. Join a group of three to four fellow students. Submit a copy of your document to each group member. Review their documents for content and organization in two phases, making notes on the document and answering the questions from section 7.2 in writing. Make positive as well as negative comments. When your own document is reviewed, take notes on the comments and use them to revise.

5. Select a particularly complex or difficult part of your paper and give it to your instructor for review. Also give your instructor a set of comments, questions, or concerns to bear in mind while reading this section of your paper.

Chapter 8

Editing

The last review you give a document is an editorial review; you are checking for correctness of diction, syntax, punctuation, capitalization, spelling, and use of numbers. Marketing writer Jamie Oliff stresses the importance of this final review when she says, "Don't be sloppy. It could cost you and your company a lot of time and money. For example, words like *greater* and *better* can mean two different things; it takes time to check out those subtle differences." Oliff works for an international pharmaceutical company, but engineers, agricultural specialists, and packaging experts face the same demands for precision in word choice and sentence construction.

If you followed the suggestions in Chapter 7 and sought review of your writing, or if your instructor assigned peer review in or out of the classroom, you probably have rewritten and revised for content, organization, and accessibility of information. Now you will examine individual sentences—even words—to see if you can sharpen their focus. This kind of review and revision is called *copyediting* and *line-by-line editing*.

When you edit, you may find that you need to improve

- word choice. Your vocabulary must be simple, clear, and appropriate. If not, you'll need to revise.
- sentence construction. Your sentences must say what you mean efficiently and correctly.
- punctuation. Each sentence should be punctuated to *help* the reader understand.
- mechanics. Capitalization, spelling, and number use must be correct and appropriate.

In order to edit, you may need to improve your editing skills. Sharpening those skills will also help you comment intelligently on other people's writing. You can use this chapter as a guide to editing, a resource for information, and a place to practice those writing skills in which you are weak. When you find problems during editorial review, you may want to read the appropriate section and do some practice exercises. The Handbook in the Appendix contains a guide to punctuation, capitalization, spelling, and the treatment of numbers, and includes exercises to enhance those particular skills. This chapter and the Handbook assume a basic understanding of grammar. If you need help in basics, consult a college composition handbook or tutors at your school. Ask your instructor for advice.

8.1 Editing by Eye and by Computer

You can edit a document either by eye or by computer, and sometimes you will want to do both.

Editing by Eye

Students often find editing difficult because they know their own writing too well.

Here are some ways you can edit your own sentences more easily:

- Read the document aloud to yourself or someone else. Hearing the words often helps you spot awkward constructions, overly long sentences, misused words, or gaps in logic.
- Force yourself to read slowly by covering the bottom part of the page with a card and moving the card only when you complete each sentence.
- Read the document backward one sentence at a time, starting with the last sentence. This technique forces you to study individual sentences rather than skim whole paragraphs or sections.
- Have a trustworthy friend or colleague read the draft aloud to you while you follow on another copy. If the reader stumbles while reading, see if the problem is an unclear or awkward sentence.

Before you can edit for clear sentences, you need to know what kinds of errors you are looking for and how you can eliminate each one. Muddy writing is a result of poor word choice and poor sentence construction. A convenient term to describe word choice is *diction*: the choice of words and the force and accuracy with which they are used. The term for sentence construction is *syntax*: the pattern, structure, or arrangement of words in a sentence. An easy way to remember the difference is

Diction = Word Choice
Syntax = Sentence Construction

In the editorial review, you should look for problems in the following areas:

Diction (word choice)

- Are the words specific and concrete?
- Is each word accurate? Check especially for words commonly confused.
- Do the words avoid sexist bias?
- Is any jargon used? If so, is it appropriate for the intended reader? Can simpler terms be substituted? Do terms need to be defined?
- Can pompous words be eliminated to improve the tone?

Syntax (sentence construction)

- Are all the sentences complete, with a subject and verb and a single thought?
- Do subjects agree with verbs? Do pronouns agree with their noun antecedents? *Are* there antecedents for pronouns?
- Are tenses accurate and consistent?
- Are sentences pruned of unnecessary words and inappropriate expletives like *there are* and *it is*?
- Are most sentences in the active rather than the passive voice? Are passive sentences that way for a reason?
- Is the action of a sentence lodged in the verb?
- Is each sentence arranged to put the most important information either at the beginning or the end?

Punctuation

- Do the punctuation marks help the reader understand by showing the relationship of one part of the sentence to another?
- Does the punctuation follow generally accepted professional standards?

Capitalization and Spelling

- Are words capitalized following accepted standards?
- Is each word correctly spelled? When in doubt, check it out.

Number Usage

- Is the use of numbers and word equivalents consistent?

During the editorial review, use your dictionary and the relevant sections in this chapter and the Handbook to help you correct and revise. This chapter has information and exercises on diction and syntax; the Handbook has information and exercises on punctuation, capitalization, spelling, and number use. After editing and revision, you are ready to submit your document for another review or to produce your final copy. See Part 3 for help in preparing the final copy of specific documents.

Editing by Computer

Another way to edit a document is with editing software (a computer program). The most common editing function is checking spelling, but some programs contain other helpful features:

- a thesaurus that will suggest substitutes for words that aren't quite right in a sentence
- a set of rules about sentence length, constructions (like active versus passive voice), and diction problems (like unspecific words, sexist vocabulary, or clichés)
- a control to correct automatically for "widows and orphans" (single letters, words, or titles left by themselves at the bottom or top of a page)
- a punctuation check that will flag missing parts of pairs (like quotation marks and parentheses) and that will check for the position of periods, semicolons, and colons in relation to double quotation marks

Often the programs will highlight a suspect word and offer alternatives. However, even with computer editing programs, don't assume that your editing responsibilities are over. Spelling checkers will not flag correctly spelled words that are incorrectly used in a sentence (such as *principle* for *principal* or *two, too,* and *to*). Computer punctuation editors are bound by strict rules and can't make stylistic exceptions—like deliberately not using a comma in a short compound sentence in order to stress the continuity of two equivalent ideas. Therefore, in order to use a word processor's editing functions most effectively, you yourself need to know the standards for good writing.

Some companies use computer programs that assess readability by counting the number of words in each sentence and the number of short versus long words. These programs imply that shorter sentences and words make a document easier to read, but research in adult reading does not support that implication (Redish and Selzer 1985). In fact, short sentences may be harder to read because they do not indicate the relationships among ideas, and short words are no better than long words if your reader does not understand them. The problem with readability formulas is that they focus attention on words and sentences at the expense of larger concerns like organization and coherence.

The checklist for editorial review is located at the end of this chapter.

8.2

Using Clear Words: Diction

Writing is intended to communicate—to move facts and ideas from the mind of the writer to the mind of the reader. Too often in the technical and business world, though, that stream of communication is polluted: muddy or clogged with garbage and junk. Instead of communicating, the writing confuses; instead of educating readers, it enrages them. The lack of clarity in government writing is so serious that laws have been passed against it. In 1982 the California legislature, for example, passed a bill stating:

> Each department, commission, office or other administrative agency of state government shall write each document which it produces in plain, straightforward language, avoiding technical terms as much as possible, and using a coherent and easily readable style.

The idea of passing a law to eliminate muddy writing may seem absurd. However, to the reader looking for information and to the writer trying to make a point, muddy writing is not a joke but an obstacle. You should remove the mud from your sentences during editing.

Use Concrete Words

As a child, did you ever play a game of locating yourself in space? You were sitting, perhaps, on the bed in your room, doodling on a sheet of paper, and you began there with your name and the place, as shown in Figure 8-1. With each level, you moved outward from a place easily understood and easily visualized (your room) to larger places that were harder and harder to see and understand. You were creating something very similar to a "ladder of abstraction."

On a ladder of abstraction, the bottom rung holds the most specific term. This might be a model name or a part number. Thus, the bottom rung has the most *concrete* word: a specific place, object, action, or person. Concrete words help readers understand because they create pictures in the mind's eye. For example, if you are writing about fruit, a concrete term would be Golden Delicious, a specific kind of apple. With each level that you move up the ladder, the term becomes less concrete and more general, until you reach a rung where the term becomes too inclusive to be

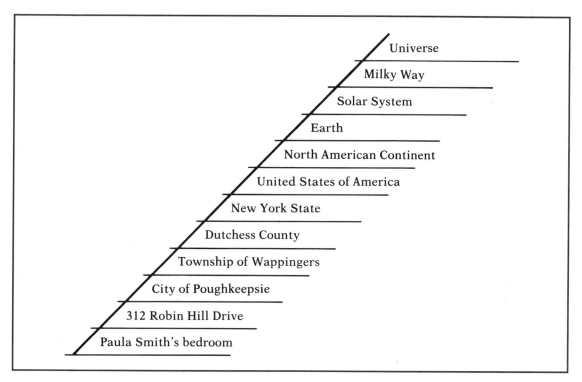

Figure 8-1 Locating Oneself in Space

visualized at all. Now you have an *abstract* word: a word naming a general condition, quality, concept, or act. Figure 8-2 shows the Golden Delicious apple on a simple ladder of abstraction. As you move up the ladder, the words become more general and less specific. When you read the word *sustenance*, do you *see* anything in your mind's eye? At what level on this ladder do you stop being able to visualize an object?

Objects (named by nouns) are not the only words that can be placed on a ladder of abstraction. Verbs (action words) can also be concrete or abstract. Consider the verb *perform*. What visual image do you have of this word? Do you see an actor on a stage? A pianist at a grand piano? A quarterback passing the football? Or do you see a 5-liter V-8 EFI engine accelerating smoothly and rapidly? *Perform* could be the verb you'd choose in each instance above, but you'd help your reader by choosing a verb to describe that specific action.

It's tempting to say that in technical writing you should always use words at the bottom of the abstraction ladder or concrete words. But it's not that simple, because you will also need abstract words like *economy, industry, perseverance,* and *liabilities.* When you edit for diction, you can, however, achieve clarity if you:

1. Choose the most concrete words (those at the lowest level on the abstraction ladder) whenever possible.

   Mazda RX7, not *automobile*
   rheostat, not *resistor*

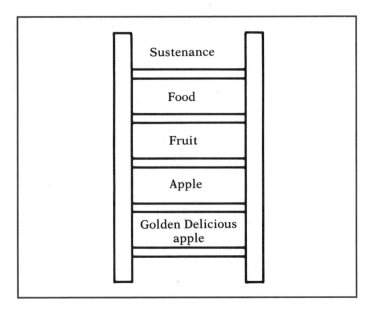

Figure 8-2 A Simple Abstraction Ladder

2. Clarify abstract words by giving concrete examples—several of them if you wish to indicate a range.

> Your *communications* [abstract word] must be clear and organized whether you are writing *request letters, status reports,* or *user manuals* [concrete words].

If you choose general or abstract words when concrete or specific terms are available, you will be guilty of vague or ambiguous usage. In other words, if your reader is not sure of your meaning, you have failed to communicate.

 EXERCISE

Each of the sentences below is vague because it uses too many general and abstract words. First, circle the general words and phrases. Then, replace them with words and phrases that are more concrete or specific. Make up your own details. Discuss and revise your new sentences with a small group of students and be prepared to present them to the class.

1. The substantial part of the body of the report rests on the information expected to be received from these companies.
2. The firm's excessive charges over a long period of time caused us to cancel their contract recently.
3. After flying at altitude for a long time, many pilots experience hypoxia.

4. Recent research indicates that substantial numbers of immigrants from the Far East experience feelings of disorientation and culture shock.
5. Adding a weak acid changed the color of the solution.

Avoid Clichés and Vogue Words

Clichés have been called "pre-fab" phrases. They drop into our minds without thought because we have heard them so many times. Once they were fresh and appropriate, but overuse has dulled their meaning. A cliché muddies a sentence by keeping the reader from focusing on a specific message.

Some writers of business letters seem to be still wedged in the nineteenth century. They use formal, archaic expressions that obscure their intended message. Study the list below, and think of an equivalent in plain English.

> Pursuant to your letter of May 3
> In re your letter of November 15
> I enclose herewith
> Thanking you in advance, I remain
> Please be advised that

Your business letters and memos will be shorter and cleaner if you replace pre-fab words with words of your own choosing.

If letter writers seem drawn to old-fashioned clichés, technical-report writers often favor vogue words and phrases. Today many vogue or "trendy" words come from the computer industry, and they have quickly become overused. In the computer world, terms like *interface, input,* and *user-friendly* have a specific meaning and a specific place. But when words like these are used to describe interpersonal relationships, or interactions in another technology, they can mislead and misinform.

Following are some vogue words that have become clichés in the last few years. Add others that you hear or read on the job or in the marketplace, and resolve not to use these terms unless they specifically apply.

glitch	state of the art	interface
parameters	utilizing	window
access	debug	bottom line
input	facilitate	networking

Many technical writers use clichés that are not vogue words at all, but simply old and tired expressions. Phrases like the following should be replaced whenever they appear in your first draft. To assure yourself that you do know other ways of stating the idea, think of an equivalent phrase for each cliché.

each and every	no sooner said than accomplished
in the final analysis	pave the way
needless to say	method in his madness
last but not least	better late than never
to make a long story short	to all intents and purposes
few and far between	get down to brass tacks
easier said than done	

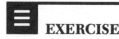

EXERCISE

Rewrite each sentence to eliminate clichés or vogue words. Make up details.

1. I enclose herewith sufficient postage to pay for shipping the new catalog.
2. Please be advised that we no longer manufacture photovoltaic collectors.
3. If you can access that information promptly and get back to me, I will do everything in my power to meet the schedule.
4. Needless to say, the relative humidity of the desert is low; average rainfall for a year is less than 125 mm.
5. Each and every person on earth is utterly dependent on water for survival.

Discriminate between Words That Are Easily Confused

Many of the words in your vocabulary have come to you orally; you've heard them on radio or television broadcasts, but you haven't seen them in print. Therefore, you don't have a visual image of the word, and you may have only a hazy notion of its spelling or its meaning. It's easy under those conditions to confuse words that sound alike. Likewise, words may have visual similarities that don't correspond to similarities of meaning. The only way to ensure accuracy is to make constant use of the dictionary.

For example, what is the difference between *replace* and *reinstall*? Both words mean to put something back, but what is put back differs. If you *replace* a faucet, you take one out and put a different one in its place. If you *reinstall* a faucet, you put the same one in again (perhaps repaired, but still the same one.) Can you discriminate in this way between *respectfully* and *respectively*? between *universally* and *generally*?

EXERCISES

1. To sharpen your understanding of the following paired words, look up each word in the dictionary. Write a sentence that explains the *difference* between the two words, and then use each word correctly in a sentence.

1. adjacent/contiguous
2. advise/inform
3. affect/effect
4. alternative/choice
5. among/between
6. amount/number
7. anxious/eager
8. assure/insure/ensure
9. bimonthly/semimonthly
10. capital/capitol
11. complement/compliment
12. credible/creditable
13. disinterested/uninterested
14. few/less
15. filtrate/filter
16. infer/imply
17. its/it's
18. liable/likely
19. oral/verbal
20. principal/principle

2. Edit the following sentences to change any inaccurate words or terms. Check your dictionary to be sure of the meanings. Be prepared to justify your changes.

1. Caution: Take care in handling the film since the resistance to damage has now been increased.
2. Management has carefully studied the sight for the new building but has not yet decided to build.
3. The conference speaker had a real flare for words.
4. The amount of people in the room far exceeded those allowed by the fire regulations.
5. He inferred to me that I was in line for promotion.
6. The critic complained that Tri-Level Company did not "go after its goal in a humbling way."
7. The list of special features is infinitesimal.
8. I want to be appraised continuously of your progress.
9. The assembly base requires a four-legged tripod.
10. I have been anxious all day long to tell you the good news.

Recognize Denotative and Connotative Meanings

The *denotative* meaning of a word is its direct, objective, and neutral meaning. The word *mother*, for example, denotes simply "female parent." The *connotative* meaning of a word includes associated meanings that are indirect, subjective, and often emotionally loaded. When you hear the word *mother*, what are your immediate associations? Probably, you first think of your own mother, and then your responses are triggered by your association with your mother—maybe warmth, caring, sharing, and helping or perhaps nagging, meanness, and smothering.

Theoretically, technical writing is objective and denotative, but in fact, much of it must be persuasive in forms like proposals, memos, letters, and reports. Thus, the *connotative* meanings of words become important because they affect the reader's attitudes. Often when the *tone* of a letter or memo is criticized, the offensive tone is caused by connotations of particular words. For example, a "white-collar worker" might be offended at being called a "worker," while a "blue-collar worker" would accept the title. "Rank and file" implies lower-level unionized workers and may have a negative connotation, while "front-office" implies management and thus may have a positive connotation. On the other hand, if one belongs to the "rank and file" and has a grudge against the "front-office," the connotations could be reversed. If a student tells me she "expects" my approval of her proposal, I am irritated. However, if she "requests" my approval, I am pleased.

Avoid Sexist Language

During the last 25 years, Americans have become sensitive to the subtle ways in which language can influence our attitudes toward persons of the opposite sex. Substantial changes have occurred in common usage in the following areas:

- job titles

flight attendant	*instead of*	stewardess
firefighter	*instead of*	fireman

mail carrier	*instead of*	mailman
cleaner	*instead of*	cleaning woman

• letter salutations and attention lines

Dear Supervisor:	*instead of*	Dear Sir:
Attn: Research Dept.	*instead of*	Gentlemen:

• references to people in general

the human race	*instead of*	man
workers	*instead of*	the working man

Still, writers have not yet found acceptable substitutes for the use of *he* and *his* to refer to both men and women. Most writers choose one of the following solutions to fix a sentence like this one:

Each technician *is* responsible for *his* own equipment.

1. Make the sentence plural.

Technicians *are* responsible for *their* own equipment.

2. Eliminate the pronoun or replace it with *the*.

Each technician is responsible for *the* issued equipment.

3. Change the third-person pronouns to first-person or second-person pronouns.

We are responsible for *our* own equipment.
You are responsible for *your* own equipment.

4. Use *he and she* or *his and her*.

Each technician is responsible for *his or her* equipment.

5. Alternate pronouns from chapter to chapter or paragraph to paragraph, using male pronouns in one and female in the next.

The five solutions in this list appear in order from the most graceful and inoffensive to the least graceful and inoffensive, so you should keep the ranking in mind when you choose alternatives. You can also see that each solution changes the meaning slightly. You need to think about the words you use and the effect you want to create. But above all, you need to keep your purpose and your readers clearly in mind.

EXERCISES

1. To sharpen your sensitivity to word choice, find five examples of connotative or sexist language (or, conversely, five examples of denotative and nonsexist language). Look in newspapers, magazines, books, or technical reports. Clip out or copy the examples and bring them to class for discussion.

2. For each paragraph, choose one of the nonsexist editing solutions and revise accordingly. Be prepared to discuss why you made the changes you did. If you decide not to change the paragraph, be prepared to explain why.

1. . . . one of the remarkable features of the human voice is that it conveys much more information than is necessary to distinguish the words spoken. We can ordinarily tell from a person's voice whether he is happy, sad, or angry, if he is asking a question or making a statement of fact. To a great extent, we can recognize a familiar voice and associate it with the person it belongs to (Denes and Pinson 1973).

2. Reports on the reactions of patients to localized electrical stimulation in parts of the brain make fascinating reading. One gets the feeling of being very close to the essential nature of the human intellect, although the way it works remains a deep mystery. A patient may respond by moving an arm or finger, without knowing why he did so. He may consciously want to say something, but be completely unable to set his vocal organs into action. Or, again, he may want to name an object shown to him, but not be able to recall the object's name while his brain is being stimulated electrically. Although he is completely unaware of the electrical excitation, he recalls the word he was seeking immediately after the stimulation is stopped (Denes and Pinson 1973).

3. Once you've decided to fire someone, and reviewed the legal and emotional problems that could result, do it as quickly as possible. Meet with the employee. In the meeting, which should take 15–20 minutes, tell the employee from the start about the firing. Let the employee talk. Don't debate. "A man in a key position with a company views his job as a serious covenant with that company," Finely says. It needs to be done in a gentlemanly way. Calling them on the telephone is not the way to do that (*San Jose Mercury News* 1989).

4. Competition fails to provide ideal motivation for learning, according to our four principles; worse, it has harmful effects. By focusing the spotlight on the very best performer, it damages the self-respect of the remainder. The boy whose craftsmanship is mediocre needs to handle tools around home and to have a belief in his adequacy in this respect, just as much as the boy who can win prizes for his skill. The student who learns that he is hopelessly outclassed in public-speaking competitions will avoid the speech activities, when the school should be developing his self-expression and his confidence that others are interested in his statements. Class discussions, panels, and non-competitive talks can help him develop. Competition to see who is best draws emphasis away from bringing each person to his full potentiality. Moreover, by emphasizing the false standard that one should take pride only in fields where he excels, it discourages the pupil from developing his lesser talents. When competition becomes a principal form of motivation in school, tension mounts. If a person is emotionally aroused by the threat of failure, his performance deteriorates; he simply cannot do his best because he is so tense (Cronbach 1954).

5. One of the chief causes of vocabulary difficulty is the almost universal habit of superficiality. Life has grown so complex, and the student has

so much to learn, that he does not have time to learn anything accurately. He learns to read too quickly, and reads too fast; he is content with approximate meanings, and hardly gets to know the exact meaning of any word in the language. He talks vaguely about socialism, democracy, art, poetry, jazz, beauty, education, communism, inflation, honor, capitalism, tolerance, futurism, progress, and so on; but if challenged he can hardly define one of those terms (Dolman 1954).

Clarify the Jargon of Your Specific Occupation

Every occupation develops its own terminology or jargon—abbreviations, symbols, words, and phrases that have specific meanings to the people in that field. Jargon can be useful shorthand for communicating with co-workers; it saves time and limits meaning. Airplane pilots communicate with other pilots using terms like *pitch*, *yaw*, and *sink rate*. Automobile mechanics talk about *universal joints*, *shackle bushings*, and the *front planetary gear set*. Crab fishers discuss *peelers*, *pots*, and *progging*.

Each of these workers is using jargon or shop talk, and using it effectively and properly. However, the word *jargon* has two very different meanings, and technical writers need to understand both.

Jargon 1   the highly specialized vocabulary of a particular group, trade, or profession. Call this *good* jargon.

Jargon 2   meaningless or pretentious terminology: doubletalk. Call this *bad* jargon, pompous language, or *gobbledygook*.

Ironically, the same words or phrases can be either good jargon or gobbledygook depending on the context and the reader.

Because good jargon is specialized language, you must know the terms of the field in which you work. Those terms can range all the way from mathematical symbols and acronyms to phrases and clauses. The key to using good jargon effectively in your writing is constant awareness of your intended reader or readers. You can use jargon *only* if you are writing to a technical audience (an audience of specialists and technicians) and *only* if you are positive that all of them know the terms you use. Otherwise, you should use common terms whenever possible, and carefully explain or define each jargon term the first time you use it.

EXERCISE

Write down as many of the jargon terms from your major field as you can think of. Then meet with other students in your class who are in the same major, and work together to define each term for a generalist reader. Add to the list when you encounter more jargon terms.

Eliminate Gobbledygook

The word *jargon* comes from the same root that gives us the word *gargle*, or "a noise made in the throat." *Bad* jargon is just that—gargling noise that doesn't mean anything. Other names for bad jargon are *pomposity* and *gobbledygook*. Texas Congressman Maury Maverick coined the word *gobbledygook* in the 1940s to describe bureaucratic writing that he said made as much sense as the gobbling of a Texas turkey. Gobbledygook is now a standard dictionary term; unfortunately, there's so much bad writing in government and industry that we need a term to describe it.

Gobbledygook has two primary ingredients: circumlocution and jargon. Circumlocution is a roundabout way of writing that uses many unnecessary words, and jargon is language that is specialized for a particular occupation. Add these two together, and you're writing gobbledygook. Many writers think that to sound "technical" and "professional" they must use long, fancy words. They add verb endings to nouns or adjectives to get words like *incentivize, prioritize,* and *finalize*. They take the word *solution*, add *-ing* to it, and say "solutioning a problem." Or even worse, they take an adjective like *ambiguous*, add a prefix to make it negative—*disambiguous*—and then add a verb ending to the whole mess to turn it into a verb. The result (I didn't make this up!) *disambiguate*.

Another way gobbledygook writers work is to inject fancy words in an ordinary sentence. Here are the directions for collecting student ratings at a university:

> **To avert any procedural irregularities which can adversely affect your results, it is essential that you appoint a student who can reliably conduct the collection process described on each packet.**

What does this sentence say? Something like this:

> **Ask a reliable student to administer and collect the rating forms. If they're not filled out correctly, your rating might go down.**

How can you avoid gobbledygook? Use the simplest and most accurate words you can, and don't try to impress your reader with the big words you know.

EXERCISES

1. For each of the inflated words below, write a simple equivalent. Keep this list near your desk and use the simpler words whenever you can.

1. circumvent	11. peruse
2. utilize	12. initial
3. attempt	13. implement
4. unequivocal	14. minimal
5. initiate	15. effectuate
6. unavailability	16. endeavor
7. facilitate	17. increment
8. subsequently	18. rescind
9. sufficient	19. prioritize
10. ascertain	20. terminate

2. Rewrite each of the following sentences in simple, clear language. Be prepared to discuss problems of interpretation in class.

1. From an ad for a technical writer/editor:

> The position will require a degree in English, Journalism or equivalent; 3-5 years experience writing in an engineering/technical environment, familiarity with document preparation and production processes, and experience in expediting commitments from those providing inputs to documents.

2. From a government memo to employees:

> . . . for those bonus goods or benefits that accrue to the Government, an agency may not set up internal procedures for the return of such goods for which it has been determined that the Government is unable to use such benefits or goods.

3. From an ad for ski boots:

> This revolutionary design incorporates a traditional fitting system, which closes the shell around the foot, with high-tech fit control systems. The C series is a modern, rear-entry boot and features a single TD constant pressure buckle system which has two fulcrums for ease of adjustment.

4. From a government request for proposal:

> If, in the view of the Contractor, due to specified Training System architectural, functional, and/or operational characteristics, it is technically reasonable and prudent to address the subject Software as comprising some number of essentially independent functional or operational units, each having clearly defined interactions with all other interfacing units, if any, each such unit shall be identified, assigned a specific nomenclature, and briefly described in terms of its functional characteristics and interfaces within the specified Training System.

5. From a course description:

> This course will provide practitioners with guidelines and specific materials to plan and implement activities for the older adult in community and institutional settings.

8.3 Constructing Effective Sentences: Syntax

Choosing clear words, as you learned in the diction section, is essential to good technical writing. Equally important is arranging those words in an order that will communicate your intended message to your reader. Look at the following list of words:

| boy | had | little | the |
| wagon | red | a | black |

By themselves, these words do not convey a message. You can arrange them into an order that will say something meaningful, but you cannot be sure what the order should be. How many sentences can you write with these words? Spend a few minutes writing down three or four possibilities.

Did you notice that you could combine the words in several ways that changed the meaning radically? You could also, of course, combine those words in ways that make no sense at all. In putting the random words together, you completed an exercise in *syntax*, which deals with word order, or what can be called sentence construction. To understand the importance of syntax to clear technical writing, look at the following sentence:

> This is the *only* tape rated "acceptable."

Suppose we move the word *only*.

> This is the tape rated *only* "acceptable."

Do you see how important the position of a word can be in an English sentence? By the simple act of moving one word, you have completely changed the meaning of the sentence. And remember, your job as technical writer is to communicate clearly what you mean and only what you mean to your reader.

Expectations of Readers

Technical writing often deals with complex ideas, and when the ideas are complicated, the writing must be crystal clear. A good technical writer will be like a meticulous groundskeeper who moves over the golf green to clip the grass and smooth lumps and bumps. When the green is free of impediments, the players can concentrate on their strokes. Just so, when the writing is free of distractions, the reader can concentrate on content.

One easy way to help readers is to meet their subconscious expectations. Because readers come to any piece of writing with much practice in reading, they subconsciously expect sentences to be built in certain ways, following specific sentence patterns. In English the most common patterns are the following:

> subject-verb-object (SVO)
> subject-verb (SV)
> subject-verb-complement (SVC)

For example:

> The unit [S] can dump [V] the module's local memory [0] in 20 seconds.
> The report [S] will be rewritten [V].
> The Model 620 [S] is [V] the fastest digitizer [C].

How do we know these patterns are common? In 1967, Professor Francis Christensen studied samples of 200 sentences from each of 20 well-known American writers—10 fiction writers and 10 nonfiction writers. He analyzed the kinds of sentences they wrote and counted the different sentence patterns. He found that 75.5

percent of the time these writers constructed subject-verb-object, subject-verb, or subject-verb-complement sentences.

Another 23 percent of the time the sentences opened with some kind of adverbial modifier—a single word, a phrase, or a clause—and the rest of the sentence was in S-V-O order. Most of the adverbial openers served as transitions from a previous idea to a new idea. For example:

> Otherwise, designers should stay with industry-standard parts.

Adding these two figures together, you can see that respected professional writers use nearly the same sentence pattern more than 98 percent of the time. The very few exceptions are sentences in inverted order (with a delayed subject) and sentences that open with verbals. For example:

> Included also is a projected cost analysis. (delayed subject)
> Elevating the beaker, he examined the precipitate. (verbal opening—present participle)

What can you learn from this as a writer or editor?

1. You communicate best when you meet your reader's expectations.
2. Those expectations are for sentences that follow the standard S-V-O, S-V, or S-V-C order most of the time.

That doesn't mean you have to write short, choppy sentences of the "Run, Spot, run" variety. It does mean that you don't have to write complicated sentences in order to communicate complex subject matter.

 EXERCISE

Choose a 200-word sample (about a page of typed, double-spaced material) from a major paper for your technical writing class. Count the total number of sentences. Analyze each sentence by marking subjects, verbs, direct objects, and complements to determine the type. Count the number of each type and record them on the following chart:

Total number of sentences in sample: _____
Number of sentences of S-V-O, S-V, or S-V-C type: _____
Number of sentences beginning with an adverbial
    modifier: _____
Number of sentences beginning with a verbal: _____
Number of sentences with delayed subject: _____

How does your writing compare with Christensen's results?

Problems with Syntax

When you look out your window in the morning, what prompts you to say, "It's really clear today"? Where I live, that means I can see all the way across my valley and the bay to hills that are clearly sculptured by sun and shadow. Beyond those hills, I can even make out the peaks of distant mountains. In contrast, what is it that blocks a clear view? It might be fog, or smog, or rain, or smoke—perhaps snow. Whatever it is, the something preventing clarity is made up of many small particles. By itself, each particle causes no problem, but when all the small particles are present together, they obscure the clear view.

You probably never look at a sentence and say, "What a clear sentence!" In fact, you don't usually notice such a sentence at all, because its clarity leads you to focus directly on the content. A sentence like that—as a sentence—is invisible; it's doing its job well by clearly communicating its message. Sometimes you *do* look at a sentence and frown or even groan. You pick your way through it again, or you back up to the previous sentences seeking help in understanding this one. Where is your focus now? on content? on meaning? No. Your focus as reader has shifted to the sentence itself because the pieces of the sentence have obscured the clear view.

Just as many snowflakes in the air at one time can obscure the view, so too many words in a sentence can obscure the meaning. You might think that the answer is to write short sentences. In fact, some people give just that advice. "KISSS," they say: "Keep it short and simple, stupid!" But short words and short sentences will not, by themselves, guarantee clear communication. To communicate clearly, you must know the material, understand the purpose, and meet your reader's needs. If you *never* use three-syllable words (even if they could accurately say what you want), and if you *never* write a complex sentence, your writing may be short and simple, but it may also be lifeless, and you will force the reader to figure out the relationships between ideas that you should make through syntactical constructions. For example, the very short sentences in the following description make the reader work too hard to relate one idea to another.

> The kilns to be compared in this report are electric kilns. They have a 7-cubic-foot capacity. The firing chamber is where the pieces to be heated are placed. It is a 23 3/8-inch decagon. The sides are 27 inches deep. The outside dimensions are 30 × 36 × 41 1/4 inches. The inside of the kiln is insulated. The insulation is refractory firebrick. This brick is made from a fine grade of clay. The clay will withstand temperatures of 2500 degrees and above. The firebricks are similar to the common red bricks used in construction. But they are much lighter. The firebrick has grooves cut into it. These grooves are to accept the resistance elements. These resistance elements encircle the interior of the kiln.

Syntax problems are not caused by length itself, but by these culprits:

- repetitious words
- long modifier strings
- packed sentences
- inappropriate expletives

Repetitious Words

One cause of unnecessarily long sentences is redundancy, or saying the same thing in more than one way. This sentence is from a technical description of a typewriter.

> The main process of the typewriter is its printing mechanisms procedure.

This writer uses extra words to "sound technical." The two related words—*process* and *procedure*—both describe the main function of a typewriter. But what is a *printing mechanisms procedure*? The key word seems to be *printing*. This sentence could mean

> A typewriter's main function is to print.

If so, the new sentence uses only half as many words to say the same thing.

Here are two more examples of wordiness caused by repetition. These sentences were finalists in a "Classic Clunker" contest IBM held among its employees in 1983 to call attention to bad writing within the company.

> You can include a page that also contains an Include instruction. The page including the Include instruction is included when you paginate the document but the included text referred to in its Include instruction is not included.

> For a priority system based upon fixed numerical ordering of requesters, following the servicing of any given requester, a request from a lower priority requester, if deferred due to a conflict with the just-serviced requester, shall be honored prior to honoring a second request from the just-serviced requester.

Long Modifier Strings

Another kind of wordiness is created when the writer tries to pack too many technical words in a phrase—particularly in a long modifier string. Such packing yields a sentence like this:

> *The elevator primary power servo electrohydraulic servo* valve accepts series inputs from the AFCS autopilot control function to compensate for engine thrust pitch moments, using the elevator, during manual approach and landing operations.

Some expert readers will accept a modifier string that long because they are familiar with the jargon. However, generalist readers find such a sentence very difficult to understand. Here's an example from a different subject area:

> In my view, because I am a *biological aspects of management financial analyst*, diversification by management produces both more growth and safety than diversification by industry.

In this sentence, the subject complement *analyst* is modified by six words.

How can you fix sentences like these? You need to eliminate or move some of the modifiers and break up the sentence. You could move some modifiers to a prepositional phrase in the first sentence in this way:

> The electrohydraulic servo valve is located in the elevator primary power servo. This valve accepts . . .

For the second sentence, you might move some modifiers to a clause:

> I am a financial analyst who believes in biological aspects of manage-
> ment. Thus, I believe that diversification . . .

In general, you can follow these principles for improving sentences with long modifier strings:

1. Put key nouns first in the sentence.
2. Use possessives and hyphens.
3. Turn nominals into verbals (*management* into *managing*).
4. Keep related ideas together.
5. Put prepositional phrases after important nouns.

As noted in item 2 above, technical writing uses hyphens in modifier strings, so you need to understand when and how to use hyphens correctly. The punctuation section in the Handbook gives both explanations and examples of common hyphen uses.

Packed Sentences

Some writers have so much information to present that they overload a sentence, stuffing in detail after detail until the sentence is bulging. Packed sentences overload the reader's short-term memory because there is simply too much information to process at one time. For example, here is a sentence explaining a rating system:

> Two points are given if all office software is available and if CAD software
> is available—because they are most important; one point is given if the
> software is UNIX based, if CAD software requires at least 512kb RAM, and
> the computer has the RAM, if CAD software needs 1Mb RAM or more and
> the system has the RAM, and if the computer has a co-processor; and a half
> point is given if the co-processor is available.

You can repair an overloaded sentence like this one in at least three different ways:

1. Make individual sentences out of independent clauses. Periods provide stronger resting places for the reader than do semicolons.

> . . . important. One point is given if the . . .

2. Rewrite the sentence as a list, assigning numbers or bullets to subpoints of equal value. (See Chapter 6 for more on parallelism.)
   - Two points are given if . . .
   - One point is given if . . .
   - A half point is given if . . .

3. Reorganize the information into a table or figure that will graphically present the data. (See Chapter 11 for more about tables and figures.)

≡ **EXERCISE**

Restructure each of the following sentences to present the data more clearly, choosing one of the three methods described in the section on packed sentences. Use each method at least once.

1. Until recently, the typical VCR offered an audio frequency response at its fastest speed of about 70 Hz to 10,000 Hz, dynamic range of 40 to 50 dB, wow and flutter of 0.15 percent, signal-to-noise ratio of around 40 dB, and harmonic distortion of 3 percent.
2. The *normal program type* will be shooting at 1/60 seconds and at f/4; to minimize camera shake, *program tele* will shoot at 1/250 seconds at f/2; and the *program wide* will shoot 1/30 seconds at f/5.6: the slow shutter speed is to have more depth of field.
3. The T70's shutter is an electronically governed, vertically travelling focal-plane shutter with stepless speeds on automatic from 2 to 1/1000 seconds and manual speeds from 2 to 1/1000 seconds plus B (Bulb).
4. The increased understanding of the printing mechanisms, the editing features including memory, the portability features, etc., made me realize the futility of the typebar printer, the versatility of the electronic typewriter, the great progress made in the last few years with the dot matrix, and the possibilities for increased productivity in the future.
5. Keeping in mind that many reading this report would be doing so to make the best use of their finances, to escape from the stressful life of the progressive, technological atmosphere, to determine whether they could build a log cabin themselves, to take the lowest risk possible in buying their own home, and to investigate the possible return on their investment, the following criteria were chosen: cost efficiency, esthetics, convenience, company reliability, and durability.

Inappropriate Expletives

In English syntax, expletives are not swear words but the introductory words *it* and *there* appearing in the usual position of a subject. In a way, expletives are throat clearers; they prepare the way for what the writer wants to say. However, because they usually appear where the reader expects the subject of the sentence to be, the real subject is delayed while the writer clears his or her throat.

> *It* is apparent that the shipment will be delayed for 15 days.
> *There* was a manager assigned to every weekend work shift.

Expletives often add wordiness because: (1) the introductory expletives themselves provide no content in the sentence—they're simply extra words; (2) when the usual subject-verb-object order is reversed, the subject must often appear in a clause or phrase, a process that adds extra words; and (3) expletives often combine with passive voice, which also adds extra words.

Why not say:

The shipment will be delayed for 15 days. (still passive)

or

Apparently, the shipment will be delayed for 15 days. (still passive but with an adverbial modifier)

or

The wholesaler will delay the shipment for 15 days. (active)

or

Apparently, the wholesaler will delay the shipment for 15 days. (active with an adverbial modifier)

Why not say:

A manager was assigned to every weekend work shift. (still passive)

or

Every weekend work shift has an assigned manager. (active)

Often you can eliminate expletives, but occasionally you will need an expletive for emphasis, to avoid an awkward construction, to move a long subject to the end of the sentence, or to lead the reader to a new topic. You need to assess each sentence individually and in the context of the paragraph to see if the expletive functions effectively or is superfluous. In the following sentences, the expletives work well:

- to emphasize

    There are 14 subassemblies included in the working drawings.

- to avoid an awkward construction

    There was no attempt to interrupt the speaker with questions.

- to move a long subject to the end of the sentence

    It is clear that the project will be delayed six months.

- to lead the reader to a new topic

    There are four possible reasons for the equipment breakdown: poor installation, inadequate maintenance, power surges, and defective parts.

The pronoun *it* can also begin a sentence. Then the sentence follows standard word order because the pronoun *it* refers to a specific noun in the previous sentence.

    It calls for a steel-alloy jacket.

 **EXERCISE**

Analyze the following sentences to see if the expletive can be eliminated. Rewrite if appropriate, and underline subjects and verbs in the new sentence. Be prepared to discuss.

1. It should be pointed out that there are other things to consider in attempting

to purchase an office communication system, which an analysis of this sort does not cover.

2. There are many different ways to compare software.
3. It should be determined whether or not a tether will be required. Tethers must be bolted to the frame of the car and are very difficult to install. Lastly, it should be determined whether the safety seat will require an especially long car seat belt.
4. There are at least two basic tuning controls on the front panel of all oscillators: a switch to set the basic range and a vernier control knob to fine-tune the pitch.
5. It is mandatory that all employees who manually fill out their own timecards view this film and receive a copy of the appropriate timecard preparation brochure. It is asked that supervisors divide their groups into two sessions for each shift to minimize impact on production operations.

Enhancements to Syntax

Writing well involves more than avoiding errors and problems. Fortunately, you can also take positive actions in constructing your sentences. You can, for example:

- choose voice for effect
- verbalize the action
- emphasize by position

Choosing Voice for Effect

"Objectivity" and "detached abstraction" are often mentioned as the primary goals in technical and scientific writing. Writers used to think that to achieve objectivity they had to remove any mention of people. Forty years ago, for example, some engineering schools *required* their students to write weekly lab reports in the impersonal third person (*one, it, he, they*), the past tense, and the passive voice. As a result of such teaching, much technical writing removed itself from the reality of the workplace. The writing became detached and hard to read because the life had been sucked from it.

Today, the emphasis is on communication. In the introduction to this book, I said that technical writing succeeds only if readers understand it. That's quite a different goal from objectivity or detached abstraction. To communicate effectively, you need to adapt your style to your goals. One way to change your style is to choose either active or passive voice for effect.

First, some definitions:

- *voice*: the verb form that shows whether the subject is acting or acted upon
- *active voice*: the sentence construction in which the subject of the sentence *performs* the action of the verb

  T. Smith crated the machine.
  The manager moved the project to Dallas.

- *passive voice*: the sentence construction in which the subject of the sentence *receives* the action of the verb

The machine was crated by T. Smith.
The project was moved to Dallas by the manager.

Advantages of the Active Voice   Today most writing books and writing teachers advise using the active voice whenever possible, even in scientific and technical writing. Here are its advantages.

1. You can write shorter sentences. An active verb is generally only one word, and the doer of the action, as the subject, can also be one word. In the passive voice, you need a verb phrase; you also need a prepositional phrase if you intend to include the doer.

> Oil companies lease offshore oilfields from the federal government. (active voice, 9 words)
> Offshore oilfields are leased from the federal government by oil companies. (passive voice, 11 words)

Admittedly, this is not a large difference, but it's one that becomes large when passive voice sentences mount up in a paragraph or passage.

2. You can be more forceful. An active doer at the beginning of a sentence gets things going. The action then moves through the verb to the object, just as the batter connects with the ball and sends it flying. This makes the active voice especially useful for instructions.

> *Not*   The paint should be stirred thoroughly.
> *But*   Stir the paint thoroughly. [You stir.]

3. You can make the reader's task easier. Active sentences link doer, action, and recipient in a logical progression like a chain link fence. Passive sentences either omit the doer (a crucial link) or force the reader (once that doer is produced) to reread the sentence and insert that information.

> *Not*   A job-application letter should be addressed [by you?] to a specific, named person. (passive)
> *But*   You should address a job-application letter to a specific, named person. (active)

4. You can be more personal. Putting the doer "up front" in a key position lets you stress individuals instead of things. Most of us are more concerned with (and more interested in) other people than we are in objects. The active voice is especially useful in letters and memos, which are directed to specific persons. You can communicate better with those readers in the active voice.

> *Not*   Fourteen heat shields were ordered on September 20, and your order is being shipped today.
> *But*   You ordered 14 heat shields on September 20, and we are shipping your order today.

5. You can assign responsibility.

> *Not*   All costs will be paid if the unit fails.
> *But*   We will pay all costs if the unit fails.

Not   The overload circuit was designed by me.
But   I designed the overload circuit

Advantages of the Passive Voice   The passive voice also has its uses. Sometimes it's worth adding the extra words, being less forceful and personal—not assigning responsibility. You need always to ask where you want the *focus* to be. Here are the advantages of the passive voice.

1. You can emphasize the receiver, events, or results of an action.

Price-determination analyses were performed at a 0-percent DCFROR.

Who did the analyses is not important here, but the analyses themselves are, so they become the focus of the sentence. Emphasis on the receiver or results may make the passive voice useful in writing procedures or physical descriptions.

2. You can avoid assigning responsibility or soften the responsibility by removing the doer.

The T-14 air scrubber was badly designed.

This sentence avoids the finger pointing that you'd get in an active voice sentence like this:

You designed the T-14 air scrubber badly.

When you must convey bad news to a customer or a boss, you may find the passive voice useful.

Not   We are reducing your discount to 8 percent.
But   Although the discount has been reduced to 8 percent, you will notice
        that by paying within 30 days your actual cost is less.

3. You can avoid the first person. Using the first person in technical writing is no longer considered a bad thing, but when you want to remove yourself as the doer, you can shift to the passive.

Not   I took samplings of the precipitate at 60-second intervals.
But   Samplings of the precipitate were taken at 60-second intervals.

In this example, *who* took the samplings is not important, so the emphasis rightly shifts to the samplings.

4. You can avoid a long phrase as the subject, changing it to a prepositional phrase and moving it to the end.

Not   MINSIM, a computer-based comprehensive economic evaluation sim-
        ulator, performed the DCFROR analyses.
But   The DCFROR analyses were performed by MINSIM, a computer-based
        comprehensive economic evaluation simulator.

I have listed almost equal numbers of advantages for active and passive voice, so you might think that one is as good as the other. Remember, though, that you want to communicate as crisply and clearly as possible. When you want to be crisp and clear, you will use the active voice most of the time—for its shorter and more forceful sentences. Only in special circumstances will you use passive constructions. Wrongly used, passive constructions can confuse the reader, and overused, they can numb the brain. Look at this sample (Davidoff and Hurdelbrink 1983).

> Each hypothetical mineral operation was evaluated at three different predetermined profitability levels, referred to in this study as economic, marginal, and subeconomic. These three profitability levels are defined by the ore feed grades of the primary commodity that would yield, in constant terms, DCFROR's of approximately 18, 10, and 5 percent, respectively, under the Montana tax structure. The relative effects of taxation in each State were measured for each property type at each profitability level by comparing the derived tax payments and rates of return. Profitability levels are also discussed in chapter 2.

How could you make it easier to read?

EXERCISE

Analyze each sentence to determine if the passive is used to advantage or disadvantage. Rewrite if necessary. Explain why you made the changes you did.

1. Analysis and evaluation of various machines that will be cost effective in the long run will be undertaken.
2. Local models were expected to be visited by September 22. None of the models is close enough to have been personally examined within that time period. They require weekend trips and are expected to be visited by November 11.
3. A study of the problem was conducted early in May, but the results were not released until the end of June.
4. The reliability of the sewing machines being studied will certainly not be ignored by me.
5. Detailed analysis of macro-economic impacts is omitted because of the character of the results obtained in this part of the research.
6. Oil price controls were in part intended to prevent increased inflation that, in the absence of controls, it was thought would result from the increase in the world price of oil.
7. The area of specialty assigned me was to document an entire department and its interaction at every level inside and outside the company.
8. The flow of fluid in the air brush and the amount of fluid released are regulated by two separate actions.
9. It was indicated that the Civiche is a good quality product, even though the organization of the company is poor, which could account for the lack of replacement parts.
10. The SLR (single lens reflex) 35 mm camera was designed to combine viewing and picture-taking functions into one lens, so the image seen in the viewfinder is, in most respects, the image that the film will record.

Verbalizing the Action

Most companies and individual employees want to be "action oriented"; they see themselves as aggressive and innovative—always moving forward. It's also good business to project that image to customers and competitors. Companies that succeed in fostering an action image succeed partly because of the care they take with all aspects of their public and internal relations. Writers need to take the same care when editing, studying the "internal relations" of each sentence to see if the sentence moves the communication forward. To produce effective action sentences, you should "verbalize the action."

To *verbalize the action* means to put the action of a sentence in the verb itself. I usually avoid words ending in *-ize* because they can easily lead to pompous diction. In this case, though, *verbalize the action* says very directly what you want to accomplish. First, you want to find the action in a sentence and then lodge that action in the action word—the verb. If you don't put the action in the verb, you will often have a sentence problem called *nominalization*, the result of making nouns from verbs or adjectives.

Sometimes you must hunt for the action; it can be well hidden by writers who think they are sounding technical by using long nouns that end with *-ment, -tion, -ance*, and *-ence*. Here are the four most common places writers can hide the action:

1. in the *subject*

The supervisor's *intention* is to consolidate all rework requests.

Notice that placing the action in the subject weakens the verb. If you move the action into the verb, the sentence becomes lean and purposeful.

The supervisor *intends* to consolidate all rework requests.

2. in the *direct object*

C. Rasch made the *motion* that funds be approved for the pollution study.

Again, notice the colorless verb *made*. Put the action into the verb, and you can shed excess words.

C. Rasch *moved* that funds be approved for the pollution study.
C. Rasch *moved* that the pollution study be funded.

3. in a *verbal* or *prepositional phrase*

Downstream water users must take *into consideration* the sewage treatment facilities of the cities upstream.

If you move the action to the verb *consider*, you can cut two dull words from the sentence.

Downstream water users must *consider* the sewage treatment facilities of the cities upstream.

4. in a *noun that follows an expletive*

It is a *requirement* of the federal granting agency that résumés be submitted with the proposal.

Turn the noun into a verb by dropping the *-ment*, and you can erase the ineffective expletive as well.

> **The federal granting agency *requires* that résumés be submitted with the proposal.**

Notice in each case that by verbalizing the action, you have also tightened the sentence by eliminating extra words.

Figure 8-3 shows a three-step flow chart you can follow to put the action in the verb.

EXERCISE

Analyze each sentence to find the hidden action. Mark the word that contains the action. Then rewrite, putting the action into the verb and eliminating the excess words.

1. Since quality of work is a direct reflection of employee morale, this memo proposes that an investigation should be made of installing vending machines.
2. Your résumé has had a careful review by interested members of our staff; it is unfortunate that no openings exist at the present time.
3. An example of the sophistication level of the system is the ability of the instruction builder to discard corrupted memory information that appears in the cache without deviating from the execution stream in progress.
4. There is a possibility of the reception by students of monetary awards from other agencies such as Vocational Rehabilitation or Social Security.
5. The occasion arises to manipulate the cursor when there is a need to branch to any point in a text to perform an editing operation.

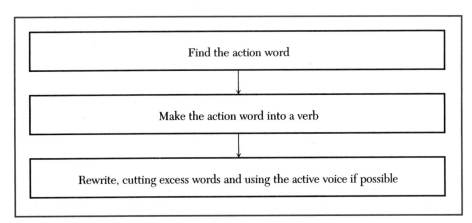

Figure 8-3 How to Verbalize the Action

Emphasizing by Position

Why is it that in an argument, most people try to have the last word? They do it because they expect that's what their listener will remember. Why do lawyers save their most persuasive points for the end of their speech? They do it because they know the jury will remember best what it heard last. These examples illustrate an important principle for technical writers: where a word is placed in a sentence often determines the amount of emphasis it receives. Technical writers should know that:

- The most important position in a sentence is the end.
- The next most important position is the beginning.
- Words in the middle of sentences tend to be skipped over.

These statements may seem arbitrary, but if you think about your readers (and even yourself as you read), you will understand why position is so important. Reading forces us to keep a number of ideas in our short-term memory while we process the words that supply new ideas. Those new ideas must then be related to the old ones for understanding to occur. Because short-term memory can efficiently hold only up to nine items, those items need to be rearranged and re-emphasized constantly as new information is introduced. A sensitive writer or editor will revise a sentence to ensure that in most sentences the beginning refers back to previous information, while the end introduces new information.

However, sentence revising is never quite that simple. As a writer, you must balance the principle of emphasis with other demands of the sentence. You need to think about

- the advantages of subject-verb-object order
- the choice of a passive or active construction
- parallelism
- the location of the action in the sentence
- emphasis by position
- clarity

And you must balance all these enhancements at the same time! In order to emphasize certain words, you may have to disregard other principles of good syntax. You might, for example, put a sentence in the passive voice in order to move the receiver to the front of the sentence. You might begin a sentence with an expletive so that the new information you want to stress will come at the end. Whenever possible, you should play with sentence arrangement at the editing stage, shifting words around for various effects until you find an order that pleases you and fulfills your purpose.

After weighing the other things you want to accomplish, you should try to use the end and the beginning of sentences for your key words. This may mean:

1. moving a dependent clause

   We are providing a steadily increasing number of jobs, although we have an unemployment problem.

Where does the emphasis seem to fall in that sentence? On the jobs? No, it seems to fall on the problem. But if you move the clause to the beginning, the emphasis shifts.

> **Although we have an unemployment problem, we are providing a steadily increasing number of jobs.**

2. moving modifier phrases or words

> **We have seen several years of increased per capita income after a decade of no growth.**

In this sentence, not only are the emphasized words negative, but the word *we*, now in an emphatic position, is probably not important to the message. If you shift the prepositional phrases, however, the sentence becomes more positive, and *we* is moved to a less important position.

> **After a decade of no growth, we have seen several years of increased per capita income.**

3. making the sentence passive, using an expletive, or both

> **In Japan the private sector has carried the burden of research and development, even in basic research areas.**

What are the key words in this sentence? Probably not *even in basic research areas*, although these words appear at the end. More likely, the key words are *research and development* and *private sector*, and those key words should not be buried in the middle of the sentence. If you rewrite the sentence in the passive voice, you might achieve the emphasis you are seeking.

> **The research and development burden in Japan, even in basic research areas, has been carried by the private sector.**

Note how the two emphatic positions are now filled by the key words *research and development* and *private sector*. In considering this last option, however, be sure not to distort the other things you are trying to do in the sentence. See earlier sections on passives and expletives.

 EXERCISES

1. Determine the key words in each of the following sentences. Then rewrite if necessary, moving those key words into positions of emphasis. Be prepared to justify any changes you have made.

1. Because the list price of the D200 is relatively high, two different distributors are cited as sources of the D200 instead of the manufacturer.
2. Some of the older, traditional wood rackets have virtues for beginner or intermediate players who are not strong-wristed yet.
3. It has been determined by Michelin test sites that synthetic rubber is the key to longer tread wear.

4. Representing the tire's ability to stop on wet pavement are the traction grades A, B, and C, which are measured under controlled conditions on surfaces of asphalt and concrete specified by the government.

5. We feel that the Cromalin process, which is also a screened process, is too costly and requires too much labor to get desired results.

2. Complete the following steps to practice all your editing skills.

- Bring to class a sample of your own writing, preferably something technical, of at least three pages.
- Exchange writing samples with a classmate.
- Choose at least five sentences from the sample and rewrite them for clarity and economy.
- Return the sample and rewritten sentences to the writer.
- When you receive your own edited sentences, read them carefully and consider their effectiveness as rewritten.
- Write a response to the editor, commenting on each rewritten sentence and then on the editor's work overall.
- Turn in all the papers to the instructor.

Checklist for Editorial Review

1. Check diction.
 ___ Are words specific, concrete, and accurate?
 ___ Are words neither sexist nor biased?
 ___ Is jargon appropriate to the reader or defined?
 ___ Are words the simplest that accurately project the meaning?
2. Check syntax.
 ___ Do all sentences have subjects, verbs, and a single thought?
 ___ Do subjects and verbs agree in number?
 ___ Are tenses accurate and consistent?
 ___ Do pronouns and antecedents agree in number?
 ___ Are sentences pruned of unnecessary words and inappropriate expletives?
 ___ Are most sentences in the active voice? Are passive sentences justifiable?
 ___ Is the action of the sentence in the verb?
 ___ Does the most important information appear either at the end or the beginning of the sentences?
3. Check punctuation (See Handbook).
 ___ Do my punctuation choices help the reader understand the relationship of ideas?
 ___ Does the punctuation follow accepted professional standards?
4. Check capitalization and spelling (See Handbook).
 ___ Are words capitalized according to accepted standards? Is capitalization consistent?
 ___ Is each word spelled correctly?
5. Check number usage (See Handbook).
 ___ Is numeral and word use of numbers consistent?

WRITING ASSIGNMENT

Edit your term project document by exchanging the last revised draft with a fellow student. Follow the checklist to edit for diction, syntax, punctuation, capitalization, spelling, and number usage. Use the information in this chapter and in the Handbook to verify your decisions. When you get your paper back, evaluate the suggestions and incorporate those that will improve the document.

REFERENCES

California Senate. 1982. Bill No. 2051, Section 1, Chapter 3.3.

Christensen, Francis. 1967. *Notes toward a new rhetoric*. New York: Harper, 41–51.

Classic clunkers. 1983. *Think* magazine. White Plains, NY: March/April.

Cronbach, Lee. 1954. *Educational psychology*. New York: Harcourt.

Davidoff, R. and R. Hurdelbrink. 1983. Taxation and the profitability of mineral operations in seven mountain states and Wisconsin. *Mineral Issues*. Bureau of Mines, May.

Denes, P. and E. Pinson. 1973. *The speech chain*. Garden City, NY: Anchor.

Dolman, Jr., John. 1954. *A handbook of public speaking*. New York: Harcourt.

Redish, Janice and Jack Selzer. 1985. The place of readability formulas in technical communication. *Technical Communication* 4th Q, 46–52.

San Jose *Mercury News*. 1989.

Part Two

Techniques in Technical Writing

Chapter 9

Defining and Comparing

A major task in technical writing is to explain what you did or what you know to your readers; in other words, to teach them something new. When you teach, you have to understand where those readers begin, meet them at that point, and lead them to greater understanding. That's why it's so important to identify your potential readers at the beginning of a writing project. (See section 1.2 for information on reader identification.) Once you've identified the readers, you need to ask "Does my reader understand what's going on here?"

One way to help readers understand is to define your terms. Medical writer Ann Petersen says, "A courteous writer defines terms, thereby allowing those in the know to feel very smart and preventing those still learning from feeling uneducated." In other words, by defining terms, you meet readers where they are. Petersen continues, "By defining terms in pieces I write, I save my readers not only the risk of misinterpretation, but also the chore of confirming something they may already know." Thus, definitions also clarify the content for the reader.

In addition, defining terms can help writers and subject-matter experts clarify their own thinking. Antonia Van Becker, a senior writer in the semiconductor field, reports, "In every project, we discuss definition of terms at some point. Often we do a page at the beginning of the project that defines the new or changed terms. If terms are not defined at the beginning, the issue comes up during a content review when the technical experts disagree about the meaning of a sentence. Then, of course, it's not easy to define the required term, but once the definition is agreed on, the whole process of reviewing speeds up."

Van Becker's comments underscore the importance of defining terms early in a project. It's also wise to incorporate analogies as you write the draft; analogies are especially helpful for generalist readers. This chapter discusses defining terms and using analogies—both techniques you can use to explain new material. It also helps you decide where to place these explanations in a document.

9.1 Defining Words and Phrases

You can often explain what you mean by defining words and phrases, but in technical writing, there are three reasons why you can't simply send your reader to a dictionary. First, technical readers are busy, and they often won't take the time to go to another source. Second, standard dictionaries seldom adequately define technical terms. Third, defining the terms yourself lets you stipulate the meaning—that is, specify what you want a word to mean in a particular context.

To decide which terms need definition, you must put yourself in the primary reader's place. Here are some questions you can ask:

1. Does the reader need this term, or can I avoid definition by using a simpler word? For example:

From the air, the field appeared to be striated (striped).

In this case, the simpler word would say what you mean, so you could eliminate the definition.

2. How many words do I need to define? If you are writing to your peers (fellow architects, scientists, or technicians, for example), you might not have to define many or any words. When you do, you can use jargon of the field in your definition. More often, though, you'll be writing for generalists, or you'll have more than one type of reader. For general and multiple readers, you need to define more often, and you must use simple language in the definitions. A definition like the following one is appropriate for generalists because, even though the defined word is formidable, the explanation is simple:

> **Epikeratophakia is a surgical procedure that involves sewing a machine-tooled donor lens onto the patient's cornea.**

But the next definition (Syva n.d.) would be acceptable only to medical specialists and technicians, who could understand the terms used to define the test.

> **A MicroTrak™ test is a monoclonal antibody-based immunofluorescent culture confirmation test that enables rapid detection of all known human seriological variants of *C. trachomatis*.**

3. If I'm new to this subject, am I defining too much? Generally, it's better to define too much than too little. Sometimes, though, when you first approach a subject, you are both overwhelmed and fascinated by its vocabulary. In an attempt to share your new knowledge, you may overload the reader with detail. Keep in mind your reader's needs as well as your own purpose, and ask yourself "What does the reader want out of this?" Do *not* write a definition for a triangle like this:

> **A triangle is a closed plane figure having three sides and three angles. A plane in mathematics is a surface generated by a line moving at a constant velocity with respect to a fixed point. A figure is a combination of geometric elements dispersed in a particular form or shape.**

Unless you are teaching geometry or these terms form the basis of your discussion, this much detail is smothering. Most adult readers will already have in mind the basic concept of a plane and a figure.

When you seem to need more than about five definitions in a single document, consider consolidating them in a glossary and highlighting the words in the text to send the reader to that glossary. You will find an example of a glossary in the student report in Chapter 22. Depending on your context and your reader's needs, you can define terms in three different ways:

1. informally, with a word or phrase
2. formally, with a sentence
3. by expansion

Informal Definitions

An informal, or parenthetical, definition is useful when a close synonym exists for the word you are defining, or when you want to stipulate (specify) the particular meaning you want to use. (Notice that I have just used an informal definition.) Informal definitions provide adequate explanations of the unfamiliar words in the following sentences:

- Nearly all species have clocks that regulate their metabolism in circadian (daily) rhythm.
- Psittacosis, so-called parrot fever, is transmissible to humans, causing high fever and infecting the lungs.
- Geologists today almost universally accept the theory of plate tectonics, which evolved from the phenomenon formerly called continental drift.
- Form in this book means the way a document looks on the page based on organization and layout. [a stipulated definition. This definition of form is specific to this book; in other contexts, form can have other meanings.]

To write an informal definition, you can simply enclose the new word in parentheses, commas, or dashes, or you can add a sentence or phrase of explanation.

Formal Definitions

A formal definition is so named because it follows a set form and is stated in a sentence. A formal definition follows the method of organization called classification and division; you first put the word into a larger class, and then you add the details that distinguish your word from the other words in that class. The order is *word to be defined + verb + class + distinguishing characteristics*. Table 9-1 shows the pattern to follow in writing formal definitions.

You have to think carefully about a word in order to write a formal definition because a dictionary definition by itself usually won't give you the precision you need.

Table 9-1 **Formal Definitions**

Word	Verb	Class	Distinguishing Characteristics
Gravity	is	a fundamental force	that operates between any two objects having mass, causing them to attract one another (Rensberger 1986, 176).
An enzyme	is	a molecule	that causes other molecules to undergo specific chemical reactions, breaking them into smaller pieces or making them combine with other substances (Rensberger 1986, 140).
A cave	is	a natural opening in the ground	that extends beyond the zone of light and is large enough to permit human entry (Davies and Morgan 1980).
Asphalt	is	a mixture of hydrocarbons	that is used, when mixed with gravel or crushed rock, for roadbuilding.

Putting the word in a larger class helps the reader see relationships; pointing out distinguishing characteristics narrows the focus. If you also keep the class reasonably narrow, you can limit the number of comparisons the reader must make. For the word *harrow*, for example, *implement* is too wide a class; *agricultural implement* narrows the field considerably.

For good formal definitions, you also need to remember the following guidelines:

- Do not repeat the term you are defining either in the class or among the distinguishing characteristics.

  Not A harrow is an agricultural implement used to harrow the ground.
  But A harrow is an agricultural implement that uses spiked teeth or upright disks to level and break up clods in plowed land.

- Maintain the form of *word* + *verb* + *class* + *distinguishing characteristics*. If you use either *when* or *where* in your definition, you are telling time, place, or circumstances, not what something is.

  Not Hatching is when you draw a series of parallel lines to shade or to differentiate one part from another.
  But Hatching is a series of parallel lines used in a drawing to shade or to differentiate one part from another.

- Use simpler language for the definition than the word itself. If you define with words the reader doesn't know, you have defeated your purpose, which is to explain or teach. Readers are seldom insulted by simple language, but they are easily angered by long, pretentious words.

Descriptions and introductions often begin with formal definitions; that way the reader knows at once what the key terms mean.

 EXERCISES

1. From the list below, choose two terms that would be suitable for an informal definition and define each one in a sentence. Then choose two terms that require a formal definition. Think about the terms first and then consult regular and specialized dictionaries to find a suitable class and to write a list of clear distinguishing characteristics. Be prepared to discuss your choices and your definitions.

hard copy	MIDI	asbestos
pronation	oscilloscope	escrow
fiscal year	amine	modem
head gasket	catalytic combustor	espalier
hurricane	grounding strap	kerning
ground water	radial tire	tear gas
	fibrin	

2. Select two words that need to be defined for a short report you are writing. Write clear formal definitions for each one and submit them for review.

Expanded Definitions

Useful as a formal definition is, sometimes you need to explain a word even further. You can do this by choosing one or several of these six methods of expanding a definition: describing, providing examples, comparing, classifying or dividing, showing the etymology, illustrating. In all these methods, you begin with a formal or informal definition and then add to it.

Describing

Since technical writing usually deals with either physical objects or processes, one of the best ways to expand a definition is to describe how the object looks or how the process works. Description can include weight, size, color, composition, and component parts. Such descriptions can be simply a sentence or two added to a formal definition. For example, you might add this description to the formal definition of asphalt given in Table 9-1:

> Asphalt cement is a brownish-black substance that can range from (1) a liquid during mixing, transporting, and placing, to (2) a plastic during compaction and service under traffic, to (3) a solid under low temperatures.

Description can also be the major purpose of a document; description of objects and places is covered in detail in Chapter 10, and description of procedures is covered in Chapter 19.

Providing Examples

Giving one or more examples helps the reader understand a word by making it more concrete and easily visualized.

> Two of the most famous caves in the United States are Mammoth Cave in Kentucky and Carlsbad Caverns in New Mexico.

> Examples of nitrogen fixers in the grasslands are lupines and clovers, both in symbiosis with bacteria of the genus *Rhizobium*.

Comparing

Like examples, simple comparisons help the reader visualize something unknown by relating it to something that may be known. Comparisons can point out likenesses or differences (contrasts).

> Asphalt cement is like concrete in being a mixture of bonding matter with sand or gravel. However, asphalt and concrete differ in their type of bonding material: in asphalt the bonding material is bituminous, and in concrete the bonding material is a calcined mixture of clay and limestone.

Synonyms, or words that mean almost the same thing, are a kind of comparison.

> Asphalt is also called blacktop or tarmacadam.

Comparison is also a method of organizing material; you can group items by the ways in which they are alike or different. See section 4.3 on organization.

Classifying or Dividing

You can expand a definition by grouping related items into a single class or dividing a larger class into related items.

> Caves can be classified into four main types: solution caves, lava caves, sea caves, and glacier caves.

Like comparison, classification and division are also methods of organization. You could, for example, structure a portion of a report on caves based on the four types listed above. See section 4.3 for more on classification and division.

Showing the Etymology

Etymology tells where a word comes from and what its history is. Etymology is particularly useful to explain a word combined from two or more source words. For example:

> *Edentate* means toothless and refers to mammals such as armadillos, sloths, and anteaters. The word comes from the Latin prefix *e-* (or *ex-*) meaning "out of," and the Latin root *denti* meaning "tooth."

Illustrating

You can illustrate meaning either with words or with graphics. Usually when you illustrate with words, you give examples. But drawings, diagrams, and photographs can effectively expand the meaning of a word. Figure 9-1 from a report on skis (Roulin 1985), combines a drawing with a definition:

> Torsion is the twisting that occurs sideways in a ski when a force acts on only one of the bottom edges of the ski. In Figure 1, the front of the ski encounters an irregularity in the terrain, but the ski is able to twist slightly without disturbing the skier's balance. The middle section under the skier's foot remains stable.

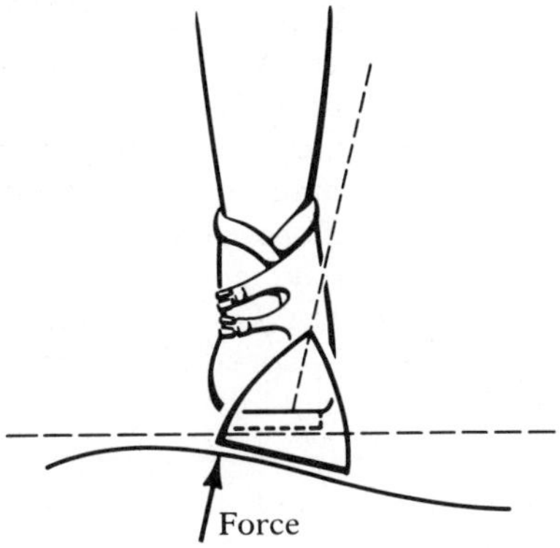

Force

Figure 1

Figure 9-1 Illustration Used as Part of a Definition

EXERCISES

1. Choose one of the terms you defined with a formal definition and expand the definition in two of the following ways:

description	classification or division
example	etymology
comparison	illustration

2. Study the following definitions and decide which expansion devices have been employed. Write a paragraph discussing what the devices are, how well they work, and whether others are needed. Be prepared to defend your conclusions in class.

1. A camera is a picture-taking device usually consisting of a light-tight box, a film holder, a shutter that allows a certain amount of light in, and a lens that focuses the image. Cameras have similar ways for viewing and recording images as do human eyes. The eye and the camera both have a hard outer covering, are light-tight, and have a sensitive surface capable of registering an image projected by light (eye = retina; camera = film). Each also has a lens mechanism for measuring the amount of light that enters. Muscles surrounding the lens of the eye change the curvature of the lens to make it focus on subjects at different distances. The camera focus is adjusted by moving the position of the lens. In both the eye and the camera, the image is registered upside down and laterally reversed on the light-sensitive surface. The brain corrects this for the eye, and developing and printing corrects this for the camera. The word *camera* comes from the Latin word *room* (Zicovich 1984).

2. A mountain bike is a bicycle designed to withstand the rigor of off-road riding. The mountain bicycle is a successful meshing of three bicycle designs:

   1. The Beach Cruiser. The mountain bike borrows the wide 1.75 by 26-inch tires of the beach cruiser, as well as the upright frame design. The tires and the frame design increase the bike's stability and handling in dirt and on bumpy trails.

   2. The Moto-cross Bicycle. The mountain bike adopts the pedals, brake levers, and handlebars from the rugged little moto-cross bicycle. The pedals, with their sawtoothed edges and wide platforms, offer a good grid and stability for the foot. The handlebars and brake levers are very similar to those of the off-road motorcycle. They offer greater control and stability.

   3. The Standard 10-to-15 Speed Bicycle. The mountain bike uses a 15-speed gear range to allow for easy pedaling up steep trails and hills (Cording 1985).

9.2   Comparing with Analogies

Comparison can be used in several ways to explain what words mean. You can expand a definition by comparing two objects or processes (showing their likenesses) or contrasting them (showing their differences). You can supply synonyms and then discuss how the two words are similar and/or different. But the kind of comparison called *analogy* is especially useful when you want to explain a complex phenomenon or object that can't itself be visualized. In an analogy, you compare what is unknown to something that is known, usually something belonging to a different class. For example, a scientist writing for generalists might explain that DNA (deoxyribose nucleic acid) serves as a pattern or model for the synthesis of proteins much as a dressmaker's pattern serves as a model for a garment maker. Just as the pattern notches of two garment pieces can be matched and sewn together, so the chemical markings on the DNA molecule match those on amino acid molecules, arranging them in a specific sequence. Analogies can help readers understand objects, processes, or abstract ideas.

To Explain Objects

The following analogies explain new or unfamiliar objects by comparing them to familiar objects.

> Carbon filter elements look like solid or broken charcoal briquets, but seen under a microscope, the carbon granules look like sponges, with little canals and pockets that trap suspended solids (Cochran 1985).

> A simpler definition would be to compare a varistor to the volume control knob on a radio. A person manually adjusts this knob (to raise and lower volume) regulating the amount of voltage passing through. A varistor automatically adjusts to the amount of voltage entering it. As a surge enters its circuitry, the "knob" is turned down to compensate for the increased voltage, and the output remains constant (Cook 1985).

> No expression describes the typical comet better than "dirty snowball." Basically, that's what comets are—chunks of ice and frozen gases mixed with rocky debris. The snowballs range in size from 100 yards in diameter to 50 or 60 miles (Rensberger 1986, 96).

To Explain Processes

Terms for unfamiliar processes can be clarified with simple analogies, as the following examples show.

> A disk duplication center is much like a book printing company. Just as an author might employ a printing company to print a required number of books, so would a programmer employ a disk duplication center to duplicate a number of programs. He or she would send the duplication company a copy of the program on a floppy disk, and the duplication company would send back the required number of exact duplicates (Chandler 1985).

> Lamination is the process of making something by bonding different materials together. Snow skis are laminated. A ski cut in half would look like a sandwich that has been cut in half. The fiberglass base of the ski is the bottom piece of bread, the layered wood core is the meat of the sandwich, and the fiberglass top "skin" is the top piece of bread (Roulin 1985).

To Explain Abstract Ideas

Eminent scientists have often used analogies to explain difficult or abstract concepts. Albert Einstein, for example, was a master at helping others visualize his abstract theories through simple comparisons. Biochemist Harold J. Morowitz, in a 1985 essay titled "Mayonnaise and the Origin of Life," used an analogy of homemade mayonnaise to explain the components of a living cell. Here are the first two paragraphs of that essay.

> The volume in my kitchen on the art of French cooking has a basic mayonnaise recipe that lists only three constituents: vegetable oil, egg yolk, and vinegar. A recipe for mixing up a living cell would be vastly more complicated, but the microstructures of both cell and salad dressing depend on a class of molecules of the greatest importance in all living processes. These ubiquitous chemicals are designated the amphiphiles. Since that noun occurs neither in the Oxford English Dictionary nor in the Webster's New Collegiate Dictionary, I feel safe in assuming it is not a household word. Actually, it was necessary to go to Chemical Abstracts to confirm the usage. What is surprising about the relative obscurity of the amphiphiles is the fact that as requisites for life they are just as important as proteins or DNA.
>
> The word itself is a compound of *amphi*, meaning "both kinds," and *philos*, "to love." The two kinds referred to here are oil and water, for these molecular structures have one end that is attracted to oil and another end that is attracted to water. The presence of amphiphiles in the egg yolk is what makes the smooth, homogeneous mayonnaise sitting in my refrigerator belie the old adage that oil and water don't mix. The egg material contains lecithin, a phospholipid molecule with a water-soluble phosphate moiety and a lipid portion that dissolves in oil. Lecithin therefore occupies the interface between the oil droplets and the surrounding vinegar and thereby produces a smooth emulsion—pleasant, tasty, and fattening.

Notice that within the analogy is an expanded definition: the etymology of the word *amphiphiles*.

You must use your imagination to develop useful analogies. Whenever you want to explain something for which no synonyms or easy comparisons exist, see if you can write an analogy. I often use brief analogies in this book to help you understand the writing process; watch for them as you read, and see if they clarify the explanations.

9.3 Placing Definitions and Analogies in a Longer Document

Depending on the type of document you are writing, you can place definitions and analogies in any of the following four places:

1. *In the introduction or body*. If a term is one that underlies the whole paper, the definition might go in the introduction. Analogies may also work well in the introduction, especially for generalist readers. Other informal definitions and analogies are most likely to go in the body of a document where they help explain key terms in the discussion.

As an example of how definition works in the context of a larger report, Figure 9-2 shows the definition of surgical lasers (Leonard 1985). This definition forms part of

SURGICAL LASERS DEFINED

Surgical lasers are medical instruments that use radiant energy to vaporize, coagulate, or photoradiate tissue. Consequently, the general term *laser surgery* involves any of these three laser operations.

Using a laser to vaporize tissue is specifically called *laser surgery*. In this application, a very fine, dense beam of radiant energy is converted to intense heat at the target tissue. The response is identical to water boiling: intracellular liquid vaporizes. Directing the beam along a path produces a fine line of vaporized cells: an incision. In this case, the surgical laser is like a scalpel except the blade is made of light instead of steel.

Using a laser to coagulate tissue is called *photocoagulation*. The process is similar to vaporization except that the thermal response is not as intense in photocoagulation. Before liquids can vaporize, protein rearrangements cause clotting. The same process occurs on a larger scale when an egg is cooked.

Using a laser in *photoradiation* is quite different from either vaporization or photocoagulation. Photoradiation therapy is a cancer treatment in which a nontoxic chemical (a hematoporphyrin derivative) acts selectively on malignant cells. Laser radiation reaching the chemical stimulates a toxic product (singlet oxygen). Thus, the laser treats a specific area with a local toxin.

Surgical lasers, then, are the tools that enable these valuable medical procedures.

Formal definition

Expansion of definition: vaporizing function

Analogy to a common process

Analogy to a common instrument

Expansion of definition: photocoagulation

Analogy to a familiar process

Expansion of definition: photoradiation function

Figure 9-2 Definitions in the Introduction of a Report

the introduction of the comparative report discussed in Chapter 22. Notice that the writer begins with a formal definition giving three functions of surgical lasers: to vaporize, to coagulate, or to photoradiate tissue. Then he expands the definition by the division method—devoting each of the next three paragraphs to one of these functions. Within the paragraphs he uses analogies: "identical to water boiling," "like a scalpel," and "when an egg is cooked."

Other terms needed to understand this report are defined in a glossary, but the term *laser surgery* is central to understanding the report itself, so it is defined in the introduction.

2. *In a footnote or endnote.* If you are writing to a mixed group of readers (specialists and managers, for example), too many definitions or analogies will slow the pace of your explanation. In this case, you should put a definition or analogy in a footnote or in an endnote, making it available to the group of readers that may want it but unobtrusive to the other.

3. *In a glossary.* If you need to define more than four or five terms in any one document, you should include a glossary. In addition, you should define the word within the text the first time it's used; the glossary makes the definition available the next time the word appears in case the reader has forgotten the meaning. Terms in glossaries are alphabetized for easy access. In a formal report, the glossary is usually in an appendix. (See Chapter 22 for the form of a report.) Occasionally, you will include a glossary as part of an introduction, especially if all the terms defined are essential to understanding the report.

4. *In an appendix.* Long analogies or extended definitions (etymologies, for example) may simply be of added interest—not central to the content of the paper. In this case, you should put the information in an appendix, even though you may not have a formal glossary.

Checklist for Writing Definitions and Comparisons

1. For this document, what kind of definitions do I need?
    ___ informal
    ___ formal
2. What method or methods shall I use to expand the definition?
    ___ description
    ___ examples
    ___ comparison
    ___ classification or division
    ___ etymology
    ___ illustration by graphics
3. Could I effectively use analogy for one of these tasks?
    ___ explaining an object
    ___ explaining a process
    ___ explaining an abstract idea
4. What is the most effective place for my definition or analogy?
    ___ introduction
    ___ body
    ___ footnote or endnote
    ___ glossary
    ___ appendix

EXERCISE

Choose a technical term or object from your major field that would probably be unfamiliar to a generalist reader. Define it by using an analogy, keeping the explanation simple.

WRITING ASSIGNMENT

Choose two or more terms that need to be defined for your term project paper. Consult specialized dictionaries and other resources to help you write formal definitions. If necessary, expand the definitions or include analogies. If you are working on a project as part of a team, be sure that one person is responsible for defining terms. Review those terms in the group so you agree on the definitions.

REFERENCES

Chandler, Richard. 1985. *Definitions of disk duplication terms*. Unpublished document. San Jose State University.

Cochran, Marian. 1985. *Evaluation and recommendation of drinking water sources for residents of Spartan City*. Unpublished document. San Jose State University.

Cook, Ronald. 1985. *Definitions of surge suppressor terms*. Unpublished document. San Jose State University.

Cording, Richard. 1985. *Off-road mountain bicycles*. Unpublished document. San Jose State University.

Davies, W. E. and J.M. Morgan. *The geology of caves*. U. S. Geological Survey, 311–348/34. Washington: Government Printing Office.

Leonard, David. 1985. *An evaluation of Nd:YAG surgical lasers for Allied Hospital*. Unpublished document. San Jose State University.

Morowitz, Harold J. 1985. *Mayonnaise and the origin of life*. Woodbridge, CT: Ox Bow Press, 27–28.

Rensberger, Boyce. 1986. *How the world works*. New York: William Morrow, 96, 140, 176.

Roulin, Kim. 1985. *Physical description of a ski*. Unpublished document. San Jose State University.

Syva. n.d. Microtrak™ test data sheet STD011 10M. San Jose, CA: Syva Company.

Zicovich, Robin. 1984. *The 35mm single-lens reflex camera*. Unpublished document. San Jose State University.

Chapter 10

Describing Objects and Places

A description is a picture in words: a way of calling on the reader's picture-making sense in order to present information. In the workplace, description is often used to introduce readers to new products or unfamiliar locations. "When you describe verbally," says technical writer Antonia Van Becker, "be very specific. Build the description from a general, overall view to specific sections and minute detail." Van Becker works in the semiconductor industry, where, she says, graphics are an essential part of description. "If you can use a drawing or other illustration as part of your description, do it. Because there are so many ways that people learn, we must appeal to their visual skills as well as their verbal skills." In technical writing, description can be of objects, of places, or of processes. Because procedure description has a distinct form, it will be discussed in Chapter 19. This chapter covers physical description of objects and places.

10.1 A Definition of Description

A *description* is a verbal picture of an object or place. As Van Becker points out, that word picture often is combined with an illustration like a drawing or photograph, giving the reader even more help.

Technical description can be either generic or specific; that is, you can describe a "typical," or generic, compact disk player or a specific one, such as Sony's model CDP-610. Both kinds of description are useful depending on the specific purpose, the type of reader, and the reader's needs. If your purpose is to evaluate four potential locations for a helicopter landing pad, you might need to begin by describing the "ideal" (generic) landing pad before going on to describe each location specifically. Then your evaluation can compare each site to the generic one in order to draw conclusions. On the other hand, if you are writing an operator's manual for a vacuum cryopump (which pumps gases at very low temperatures), the operator will want a specific description of that model and make of cryopump.

You must also consider types of readers and their needs. Generalists and managers may need generic descriptions to help them understand the thing or place itself, while specialists and technicians may want specific descriptions to help them visualize a particular variation. Each time you write, you need to analyze your specific readers and their needs.

In technical writing, as in most writing, a description is usually not a separate document but forms part of a longer whole, whether report, manual, or proposal. Description can also be a part of a letter, memo, or set of instructions or procedures. In addition, as noted in Chapter 9, description is sometimes used to expand a definition. As a student, you may write a separate description paper as a way to learn the skill, but most of the time on the job you will incorporate description into some larger document. Frequently you will combine description with graphic (visual) representation. Pictures and words together give a reader solid information about a thing or place.

10.2 Sources of Information

Before you can describe an object or a place, you must have detailed and accurate information about it. You can obtain that information yourself by examining, weighing, and measuring the object, or you can find the details in written material. Obviously, the best source is the object or place itself and related explanatory material like advertisements, catalogs, manuals, and specification sheets. If you use sales literature as source material, though, be careful to strip away biased or persuasive sales language from the factual data. Technical description ought to be objective, so your facts must be accurate and your language carefully chosen.

For example, Figure 10-1 contains the description of the 1992 Pontiac Grand Am®, taken from the text of an ad in a national magazine (GM 1991). How much of this description is factual? Strip away the emotion-laden words, and you are left with the following:

- anti-lock brakes (ABS VI) with a patented brake control algorithm, and an isolated modulation circuit
- wider track suspension
- optional variable-effort assist for standard rack-and-pinion steering
- three available engines
  2.3L Quad OHC
  16-valve 2.3L Quad 4®
  160hp 3.3L V6

Notice that in even the above list, the descriptive details are somewhat ambiguous. What precisely are a "patented brake control algorithm" and "an isolated modulation circuit"? If the car has "wider-track suspension," you need to ask "wider than what"?

As you can see, advertisements often do not supply the information needed for accurate description. In addition, ads like these are loaded with persuasive words: *most advanced, impressive as they are,* and *even greater road feel.*

Besides sales literature, sources of descriptive information can include technical dictionaries and encyclopedias, textbooks, handbooks, and catalogs. You can find most of these in your library; ask a reference librarian for help. In addition to written sources, you can interview specialists, technicians, and operators. You should, if possible, get hands-on experience with whatever you want to describe—seeing it and using it, measuring and weighing it if necessary.

10.3 Informal Descriptions

Informal descriptions are generally short and may appear almost any place in a long document. Frequently, though, you will find them in the introduction or in that portion of the body dealing with definitions, materials, or the explanation of a problem. An informal description may be only a paragraph, yet its purpose is the same as a more formally designed long description: to enable the reader to picture the object or place mentally and understand its function.

You saw it here first. We slammed the brakes on our usual high-performance prose to bring you a public service announcement.

Every 1992 Pontiac Grand Am® has **anti-lock brakes** as standard equipment.

Known to its creators as ABS VI, it's the simplest yet one of the most advanced anti-lock braking systems available in a production vehicle. By using a patented brake control algorithm and an isolated modulation circuit, ABS VI minimizes the pulsating pedal feedback of less advanced systems.

On the road, it's the same story. In emergency situations ABS VI modulates brake pressure to help you stop as straight and steer as precisely as the laws of physics and the unpredictable impulses of nature will allow. Impressive as they are anti-lock brakes are only one dimension of the all new Grand Am's character. There's also a new **wider-track suspension** for added stability. Optional variable effort assist for the standard power rack-and-pinion steering for even greater road feel. And the power and performance of three available engines, including a new standard **2.3L Quad OHC**, a 16-valve 2.3L Quad 4® or a newly available 160 hp 3.3L V6.

The last point? After all, this is a Pontiac, and we couldn't resist. Because now you can step on the pedal on the left with just as much confidence as the one on the right.

The all-new 1992 Pontiac Grand AM, with standard anti-lock brakes.

The excitement stops only when you say it does.

Figure 10-1 Advertising Description

Reprinted by permission of Pontiac Division.

For example, here is an informal place description written for pilots:

Thomasville Municipal Airport (on the Jacksonville Sectional) is located 7 mi. NE of the city of Thomasville. It has two 5,000 ft. asphalt runways, a rotating beacon, runway lights (available on request until 10 p.m.), hangars, tiedowns, fuel, and Unicom (AOPA 1965).

If this description were more complete, it would give the altitude of the airport and the compass orientation of the runways; with those added facts, a pilot would have a good mental picture of how the Thomasville Airport looked from the air and how it would function.

Just as pilots need descriptions of airports, so gardeners need descriptions of those garden pests called insects. The informal description in Figure 10-2 comes from the introduction to a chapter that teaches gardeners how to recognize various insects (Mother Earth 1989).

The paragraph in Figure 10-3 informally describes the feed system of an Elna SU68 sewing machine (Bernucci 1981). Coupled with an illustration, the description gives the reader a good picture of the feed mechanism and leads to an understanding of how it works.

All adult insects have two qualifying characteristics:

- They have three body divisions: the head, the thorax, and the abdomen.
- They have three pairs of legs, all attached at the thorax.

The head of an adult insect contains its antennae, mouth and (compound) eyes. The thorax bears the legs and wings (if the insect has wings). The abdomen contains the digestive and reproductive organs. An adult insect also has an exoskeleton, a hard outer shell, instead of an internal backbone. (It's an invertebrate, not a vertebrate.) And it breathes through little holes in its abdomen called spiracles. But the truly defining characteristics—the ones that say this, and only this, is an insect—are those three body parts and the six legs.

Figure 10-2 Physical Description of an Insect

One of the features that distinguishes the Elna from the Riccar, Pfaff, and Singer is its feed system. The feed system passes the fabric under the needle with feed dogs, a series of moving teeth that come up through tracks in the needle plate. The feed dogs must pull the material just enough so that it does not bunch up, but not so much that it catches and snags on the teeth of the feed dogs. Elna's feed dogs have diamond-pointed teeth instead of the traditional slanted saw-tooth as shown in Figure 1.

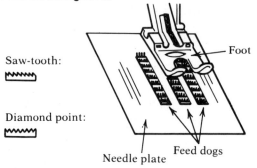

Saw-tooth:

Diamond point:

Foot

Needle plate   Feed dogs

Figure 1

The diamond points do not snag sheer fabric such as chiffon. Slanted saw-tooth feed dogs have a tendency to "eat" sheer fabric and the lining on bonded wool by pulling it down into the needle hole in the needle plate. Because of their gentle pull, diamond-pointed feed dogs also keep stitches from stretching out of shape on bathing suits and T-shirts.

Figure 10-3 Description That Includes Graphics

10.4 Formal Descriptions

Formal descriptions are generally organized in three parts: (1) an overview, (2) a breakdown into parts with part descriptions, and (3) a concluding discussion of how the object works or is used.

Overview

In a formal description, the overview should begin with a definition, as does this example (Davis 1985).

> A catalytic combustor is a device mounted in the firebox or flue of a wood-burning stove that causes fuel to be burned more completely than in a conventional stove.

Following the definition, you should tell the purpose of the object.

> Its purpose is to increase the energy efficiency of the woodburning stove by enabling smoke to burn at a lower temperature and to decrease the amount of creosote-forming gases and pollutants released into the air.

Next, as does Figure 10-4, you should describe its general appearance, including, if possible, a cutaway or exploded drawing. (These drawings are explained in Chapter 11.) In this section, you should also discuss the object's material composition, and number or list its major components. This list of components will serve as a transition to the breakdown of parts in the second section of the description. In this example, the catalytic combustor does not itself have component parts; instead, it works with other parts of the woodstove to produce efficient combustion. Those parts might be named next: the air intake, the baffle, and the bypass plate.

A catalytic combustor, as shown in Figure 1, is typically a ceramic cylinder 6 to 8 inches in diameter and 2 inches high. It has a porous interior that is coated with catalytic metals such as platinum, palladium, or rhodium. These metals cause a chemical reaction to take place with the incoming fuel molecules.

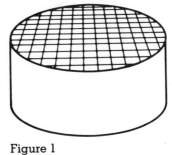

Figure 1

Figure 10-4 Line Drawing to Enhance Description

Breakdown and Description of Parts

The parts description of any object or place can be organized in several ways:

- *Spatially.* You can organize the description from left to right, top to bottom, or outside to inside. Spatial organization is especially good if there are no moving parts.
- *Chronologically by order of operation.* This organization is related to the function or process of the object or place. If you were writing a description of a golf course, for example, the parts (each hole) would be covered from Hole 1 to Hole 18—the way the course would be played.
- *Chronologically by order of assembly.* This organization is more appropriate for an operator's manual or as a description intended for a technician who must build or reassemble an object made up of a number of parts. If you were writing a description of a sink trap, for example, the parts could logically be discussed by the order in which they would be assembled: sink drain, nut, washer, trap, nut, washer, drain pipe. You could also include an exploded drawing showing that relationship, as does Figure 10-5.
- *From general to specific or complex to simple.* This order works best when some parts of a mechanism or place are clearly more important or more complex than others.

As you describe each of the parts, you should cover the same points as in the overview: definition of the part, its purpose, its general appearance (perhaps with an illustration), its composition, and if necessary, its parts (and those parts may need to be described as well). In what is called "top down processing," you continually move from more complex parts to simpler ones so the reader can see the larger picture first and then the details.

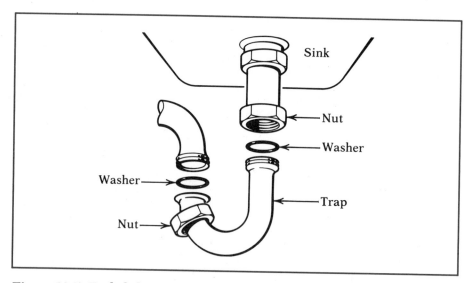

Figure 10-5 Exploded Drawing to Accompany a Description: Sink Trap Components Showing Order of Assembly

COMPUTER LOCKS

Formal definition

A *computer lock* is a device that prevents an unauthorized user from moving a computer or its components. There are several types of computer locks, including keyboard locks, power switch locks, and system locks, which lock a computer to its desk. Without computer locks, a computer and its components are very easy to steal.

General appearance

The Kensington Apple Security System (KASS) is a system lock consisting of a series of plates, cables, locks, and screws that together will secure an Apple Macintosh computer, monitor, peripherals, and keyboard to a desk. The KASS can be used on the following Apple systems: Macintosh, Macintosh 512k, Macintosh Plus, Macintosh SE, Macintosh II, and Macintosh IIGS.

COMPONENTS OF THE SYSTEM

List of major components

The KASS components include two large security plates, three small security plates, one keyboard loop adapter, one 8-ft wire cable, one padlock with two keys, one custom screwdriver, and eight tamper-resistant screws. See Figure 1.

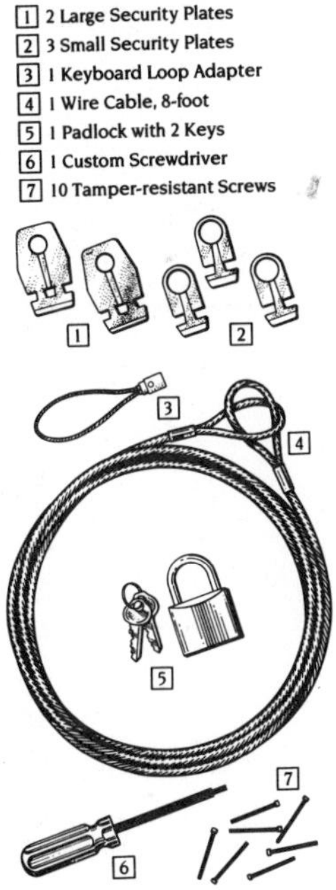

Figure 1. Components of Kensington Apple Security System
Reprinted by permission of Kensington Microware Ltd.

Figure 10-6 Example of a Formal Description

*First major
component, discussed
by order of assembly*

Security Plates

The computer systems on which the KASS can be used have security slots (marked with a chain symbol) on the back of each piece of equipment. The two large security plates (2.125″ × 1.375″ × 0.187″) are inserted into appropriate slots on the back of the monitor and on the back of the keyboard. The three small plates (1.75″ × 0.8″ × 0.175″) are inserted into slots on the monitor, keyboard, and external hard disk, if appropriate. Once in place, the plates can never be removed without damaging the computer housing.

*Second major
component*

Keyboard Loop Adapter

If the keyboard does not have a built-in security slot, the keyboard loop adapter, a 0.64″ twisted wire cable with a 2.5″ loop, plus an adapter (0.765″ × 0.382″ × 0.122″) can be used to secure the keyboard. The loop is pinched together and inserted into the slot under the left foot of the keyboard. The loop inserts from front to back so that the rectangular end of the adapter is captured beneath the keyboard, and the loop extends out behind it.

*Third major
component*

Cable

The cable is a 0.2″ diameter plastic-coated twisted wire and is 90″ long with loops on both ends. When installed, one end of the cable is wrapped around a secure immovable object (see Figure 2), and the other end is threaded through the holes provided in the security plates.

Figure 2. Placement of Cable

Figure 10-6 Continued

Fourth major component

> ### Padlock
> The padlock (1.25″ × 1.0″ with 2 keys) attaches to one end loop of the cable, thus securing the equipment to the desk, yet allowing parts to be removed easily for repair or replacement.

Other components

> ### Tamper-Resistant Screws and Custom Screwdriver
> The KASS expands to secure peripherals by using eight tamper-resistant screws (1.375″ × 0.108″). Peripherals used with the Macintosh system are connected to the computer with cables ending with connectors; these connectors are attached to the computer with screws. These screws are removable and replaceable with the KASS tamper-resistant screws, which have a special head. When screwed in with the custom screwdriver, the tamper resistant screws recess into the cable connector. They can only be removed with the custom screwdriver.

Figure 10-6 Continued

Conclusion

Some descriptions do not require a conclusion; once you have described the individual parts, you can stop. In other cases, you might add a short section—called a *process description or procedure*—that describes how the object works or how it is used. Notice how easily one kind of description merges into the next. Process descriptions are discussed in Chapter 19.

The description in Figure 10-6 is part of a collaborative study by five students that proposes a security system for new campus computer-equipped classrooms. This section of the proposal describes computer lock systems (Johnston 1990). The general method of organization is by order of assembly, and this description does not require a conclusion.

10.5 Hints for Writing Good Description

As in all types of technical writing, you must shape description to fit the particular needs of each assignment, remembering your purpose and reader. The computer-security system described in Figure 10-6 provides good information for generalists, but for technicians or specialists it might be inadequate. For them the description might need to be more like a specification sheet, with detailed measurements as well as descriptions of each part.

Your description assignments will vary depending on your field, your purpose, and your reader, but these general hints will help you write good description:

- Record measurements, dimensions, and weight accurately. A small discrepancy in a measurement might mean that the object won't fit where your reader needs it, and weight might be a factor in your reader's decisions.

- Be specific and concrete. Use terms at the bottom of the abstraction ladder (see Chapter 8); choose words that can be visualized easily. Give parts a name (even if your source doesn't name them) so they can be identified more readily. For example: the rubber plug, the C-shaped insert, the 6-inch cork gasket.
- Use simple language and analogies to help generalist readers visualize (see Chapter 9 for information on analogies). We visualize in the following five ways, and you can use many of them in a description.
  1. Size. In addition to giving measurements, use comparisons to common objects: a football field, a paper clip, a nickel.
  2. Shape. Use terms like *circular, rectangular, concave*, and *square*. Compare to common objects: U-shaped, saw-toothed, threadlike.
  3. Texture. Describe the texture as accurately as you can: sandy, honeycombed, spongy.
  4. Color. Describe both hue (red, blue, green) and tint (amount of white added).
  5. Position. Fix the position of your object or place relative to other objects or places. Use terms like *above, below, parallel to*; use directions like *northeast, west southwest*.

To see how a writer focuses on technical description within the context of a long report, study Figure 10-7, a description of the Nd:YAG Surgical Laser (Leonard 1985). This description is part of the introduction to the comparative report discussed in Chapter 22. Notice how the writer lists eight components in the overview and then organizes the rest of the description by detailing each of those components. The components are arranged chronologically by order of operation, thus also providing information about the process of producing and using the laser beam in surgery. The description concludes with a brief process description. Terms that appear in italics are defined in the glossary of the report.

SURGICAL LASERS DESCRIBED

Nd:YAG surgical lasers represent a specific class of surgical lasers which use a solid Nd:YAG crystal *lasing medium*. To a large extent, this medium determines the applications and the structure of the laser system.

The applications of the Nd:YAG laser and its relation to other surgical lasers are discussed in section 2.1.3.2. The structure of Nd:YAG systems is described here. The Nd:YAG system has eight basic components:

Overview lists major components by order of operation

  1. optical energy source
  2. optical resonator
  3. optical wave guide
  4. surgical handpiece

Figure 10-7 Example of a Formal Description

5. lens system
6. electronic shutter system
7. water cooling system
8. control panel

General description

These components are typically housed in a rectangular cabinet approximately 2 feet wide, 1.5 feet deep, and about waist height. The laser head may be separate and is often situated above the main unit. Different models vary somewhat in shape and size. The units range in weight from about 400 pounds to 800 pounds. See Figure 1.

Photograph enhances the general description

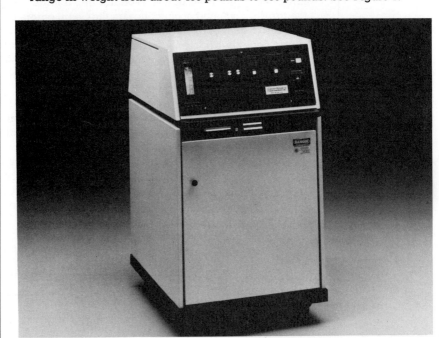

Figure 1. Photograph of the Laser Sonics Model 8000: A Representative Nd:YAG Surgical Laser
Reprinted by permission. © 1983 Cooper Lasersonics, Inc.

First component

The *optical energy source* consists of one or two high power xenon or krypton *flashlamps*. The lamps supply the optical energy required by the next component, the *optical resonator*.

Second component

The *optical resonator* is contained in the laser head. It is the critical component in laser production. The resonator consists of a cylindrical cavity with plane parallel mirrors at each end. The Nd:YAG crystal is contained within the cavity. When optical energy reaches the crystal, laser light starts resonating between the mirrors producing the beam. This process is described with more detail in section 2.1.2.

Figure 10-7 Continued

Third component

The laser beam leaves the resonator and enters the *optical wave guide*. The wave guide is usually a flexible *optical fiber*, but it can be rigid for special applications. The wave guide transports the beam from the resonator to the *surgical handpiece*.

Transition sentence

Fourth component

The *surgical handpiece* directs the beam onto the target tissue. A Nd:YAG system typically has interchangeable handpieces; each is designed for a different application.

Fifth component

The lens and shutter systems regulate the effect the beam has on the target tissue. The *lens system* concentrates the energy of the large beam produced in the resonator into a minute spot directed at the target. *Focusing* occurs once as the beam enters the wave guide and again as it exits the handpiece. The *electronic shutter system* controls the duration and frequency of laser exposure by repetitively opening and closing an aperture to the wave guide. Most systems allow *continuous wave* or *pulsed mode* with *pulse length* adjustable from 0.1 to 10.0 seconds.

Sixth component

Seventh component

The *water cooling system* prevents the resonator from overheating. Ordinary tap water is circulated around the resonator removing excess heat. The flow rate ranges from less than 2 gallons per minute to more than 6 gallons per minute among different systems.

Eighth component

The *control panel* is generally positioned near the top of the cabinet and angled for easy hand operation. It provides the interface between the operator and the other components in the system. Controls allow adjustments in power and pulse characteristics, while meters indicate the status of system components.

General description of the procedure

When the Nd:YAG laser is operated, optical energy flows into the resonator producing a laser beam concentrated and pulsed according to the operator's specification. The system is generally self-contained; the operator can devote his or her attention to the surgical procedure.

Figure 10-7 Continued

Checklist for Writing Description

1. What is my purpose?
2. Who is my primary reader and what are his or her needs?
3. Do I need a generalized description as an overview or a specific description?
4. Where can I find the information needed for this description? If I use sales literature, have I deleted any biased language?
5. Should my description be informal or formal?
6. If I choose to write a formal description, what should my overview contain?
   ___ a definition
   ___ a purpose statement
   ___ general appearance
   ___ illustration
   ___ a list of components

7. How is the breakdown and description of parts organized? Is this a logical organization? Have I thoroughly described each part and broken it into components if necessary?
8. Do I need a conclusion? Should I add a process description?

 EXERCISES

1. Choose an ad from a trade magazine in your field and make a copy of it. In one or two paragraphs, write an analysis of the ad discussing what parts of it are biased and what parts are objective. Give examples from the ad to back up your points. Be prepared to discuss in class.

2. Choose one of the following objects or places and write a generic description of two to three pages. Follow the three-part structure explained in section 10.4. Specify the purpose and the reader.

garage door opener	blow dryer	satellite dish
child auto safety seat	electric razor	umbrella
air popcorn popper	pressure cooker	smoke detector
roller skate	tennis racket	Bunsen burner
baseball glove	digital thermometer	the human ear
the human eye	local sewage treatment plant	the football stadium
local golf course		the nearest park

3. Choose one object or place from the list in exercise 2 and write a paragraph-long informal description.

4. Study the following description (Bejar 1985), analyzing it for language (is it specific and unbiased), for organization (does it have an overview, analysis by parts, process description), and for completeness of content. Determine whether it would be improved with added illustrations, and suggest what the illustrations should include. Be prepared to discuss in class.

> The installation of a burglar alarm in a home is now easy enough for one to do without professional help. A self-installed home alarm system, usually referred to as a local alarm, simply sounds a siren on the premises whenever an intruder is detected by the system.
> A typical alarm system works similar to an automotive dome lamp. When all openings are closed, the system is off and armed: ready for detection. When any door or window is opened, the circuit is broken. Then the siren sounds, just as the dome light in a car lights up. Another type of alarm system uses an energy field to detect intruders. It works similar to the electric eye used in many stores. When the beam of light is crossed, the alarm sounds. In home applications, the field takes on a bubble shape, protecting a large portion of the interior. The first system is usually referred to as perimeter protection, while the second is called specific-area protection. Both have advantages and limitations which will be discussed in the main text.

A typical alarm system consists of four main parts: sender, circuit path, control module (receiver), and siren. The sender is the component that initially detects an intruder. It notifies the control module via the circuit path.

The circuit path can be either in wire or wireless form. The wire form uses small gauge wire, similar to speaker wire, for connecting the components. Wireless systems use a noise-emitting box for a sender. It is similar to a cigarette box in size and shape. Instead of wires, the circuit is made complete by sound or radio waves inaudible to the human ear. The circuit works similar to a TV remote control.

The control module resembles a small, wood-grain clock radio with indicator lights on the front and terminals at the rear for connecting the system's components. The control module receives the signal from the sender when an intruder is present and activates the siren, usually through a wired circuit.

The siren is usually in loudspeaker form and resembles the hand-held speakers used by police. When an intruder is detected, the siren is set off, emitting an extremely loud noise throughout the exterior perimeter of the premises.

 WRITING ASSIGNMENT

Study the outline for your term project to see where you have included (or should include) a description of an object or a place. Write a draft of a formal or informal description (or both). Be sure to consider your purpose and reader. After editing, incorporate the description smoothly into the appropriate section of your report. If you are working on a collaborative project, be sure that one person is responsible for writing any needed descriptions.

REFERENCES

Aircraft Owners and Pilots Assn. 1965. *Places to fly*. Washington: AOPA, 73.

Bejar, John. 1985. *Description of a burglar alarm system*. Unpublished report. San Jose State University.

Bernucci, Julie. 1981. *Comparative study of top-of-the-line sewing machines*. Unpublished report. San Jose State University.

Davis, Mary. 1985. *An evaluation and comparison of four wood-burning stoves*. Unpublished report. San Jose State University.

GM Corporation. 1991. For once, a Pontiac ad that talks about slowing down. *Parade*, 25, October.

Johnston, C., R. Galindo, J. Shi, B. Jennings, B. Godfrey. 1990. *Proposal for a security system for CAI Classrooms*. Unpublished report. San Jose State University.

Leonard, David. 1985. *An evaluation of Nd:YAG surgical lasers for Allied Hospital*. Unpublished report. San Jose State University.

Mother Earth News Partners. 1989. *The healthy garden handbook*. New York: Simon and Schuster, 72.

Chapter 11

Illustrating with Graphics

"The role of technical writers has changed greatly over the last 10 years," says Jeffrey Vargas, a learning-products specialist. "Writers are expected to do more than just write. They must become editors, graphic artists, and human factors engineers. It's especially important to know how to present conceptual information graphically." Vargas's company manufactures computer systems and testing and measurement equipment, but his advice about graphics holds true for all writers dealing with scientific or technical information.

When you are a writer seeking to present information clearly and efficiently, you should ask yourself two important questions: "What kind of graphics will best achieve my purpose in this piece of writing?" and "How can I ensure that the graphics I choose do their job?" Just as various forms of technical writing are suitable for certain tasks, so different forms of graphics are suitable for specific purposes. This chapter will explain graphics from a task-oriented perspective, presenting each specific graphic in light of what it can do for you as a writer. In addition, you'll learn how to design each type of graphics most effectively.

11.1 Definitions of Graphics Terms

In this book the word *graphics* means any visual form of presenting information: that is, pictures or arranged numbers as opposed to sentences. Graphics is thus a general term, as are the related words *visuals* and *illustrations*. Graphics increase understanding by appealing to the eye. Graphics are not considered simply "aids" to the printed text but central elements along with sentences for conveying information.

Graphics are divided into two large categories: tables and figures. *Tables* present data (often numbers) in columns for easy understanding or comparison; the presentation is visual, but the data is set in type. Tables are of two types: *informal* and *formal* (see Figure 11-1).

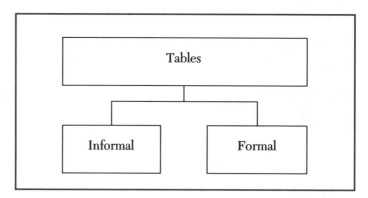

Figure 11-1 Two Types of Tables

All graphics that are not tables are called *figures*. The main divisions of figures are graphs, diagrams, photographs, drawings, examples, and icons, as shown graphically in Figure 11-2. In a formal report or proposal, tables and figures are numbered separately in the text—there can be both a Table 3 and Figure 3. Following the table of contents, you place a list of tables and a list of figures.

11.2   Advantages of Using Graphics

Graphics are widely used in technical literature to enhance and supplement the written text. While graphics can seldom totally replace sentences, they offer several advantages over printed words:

1. Graphics increase reader interest and understanding. They appeal to the reader's visual sense and pattern-making capabilities, and they usually are easier to understand than sentences saying the same thing. Tables and figures also help break up a page of solid words, providing rest and focus for the reader's eyes. (See Chapter 12 for more on document design.)

2. Graphics display and emphasize important information. In his 1990 book *How to Write and Publish Engineering Papers and Reports*, Herbert B. Michaelson suggests that when organizing, writers should (1) decide on the key points, (2) sketch a figure for each key point, and (3) then plan the manuscript around those figures. Certainly it makes sense to focus the reader's attention on key information through graphics.

3. Graphics summarize and condense information, pulling together many isolated bits and displaying them in one place. This capability makes them especially good in conclusions and recommendations sections of reports.

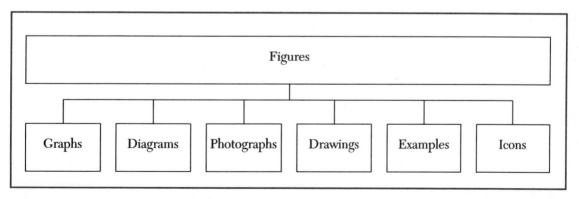

Figure 11-2 Six Major Types of Figures

4. Because of their visual presentation, graphics convey quantitative relationships—likenesses, differences, percentile rankings, trends—better than words.

5. Graphics are effective with international and multicultural audiences. Line drawings or icons, for example, can convey information to readers of all languages without translation.

6. Increasingly, graphics can be designed using a computer. In many industries 80 to 90 percent of graphics are designed by computer. As more low-cost computer-aided design (CAD) systems become available for microcomputers, students can produce tables, diagrams, and drawings of professional quality.

7. Graphics can be borrowed from other sources and incorporated into your own reports, as long as you *identify* and *give credit* to your sources. (You will, of course, have to secure written permission from copyright holders if you are going to publish or sell this borrowed material.) You can copy graphics, reduce or enlarge them on a copy machine if necessary, paste them on your typed page at appropriate locations, and then copy the whole page. This process will give you a professional-looking page. But be careful, because color photographs do not copy well, and reduction may destroy details.

8. Many graphics can be traced or hand-drawn with a ruler and pen; you need not be artistic. You can type captions, or you can use rub-on letters.

11.3   Choosing Graphics by Purpose

Your central question about graphics at the draft-writing stage is "What kind of graphics will best accomplish my purpose in each portion of what I am writing?" To answer that question, this section presents seven general purposes and suggests appropriate graphics choices for each purpose. Remember that you must also evaluate the type and background of your reader: specialists often want different graphics than do generalists. Although sometimes only one type of graphics will work, at other times you will have a choice. Remember, too, that you are not limited to these suggestions. Just be sure that the graphics you choose will accomplish your purposes. Table 11-1 summarizes these seven purposes and appropriate graphics choices, but for details about how to use each one you should also read the explanations that follow.

Showing Exterior and Interior Views

Photographs

When you want to show the *exterior* surface of an object or place, you can use photographs, which provide absolute reproduction; they can show damage, record phases of a phenomenon such as growth, identify or locate parts, or show how equipment functions.

For these reasons, photographs—whether color or black and white—are widely used in technical writing, especially in documents like brochures, annual reports, and

Table 11-1 Choices of Graphics for Specific Purposes

Purpose	Tables	Graphs	Diagrams	Photographs	Drawings	Examples	Icons
Showing exterior and interior views				X	X	X	
Displaying part/whole relationships		X	X	X	X		X
Explaining processes or operations			X		X		X
Summarizing	X		X		X	X	
Comparing	X	X					
Providing examples	X	X	X	X	X	X	
Representing ideas or functions							X

newsletters intended for general readers. Photographs are also used on the covers of manuals or reports or with introductory material; photographs like these usually show the outside of a machine, building or object, and they are called *external* photographs.

Photographs can also show *interior* views with closeups, disassembled parts, or through a "blowout": a telescoped view of a small part or assembly "blown out" of the interior of a machine, enlarged, and displayed. Whether showing exterior or interior views, photographs must be clear enough to reproduce well, and they should be cropped to show only the desired portion. Parts should be labeled carefully, and the title and figure number should be placed below the photograph. To show scale, you can include in the photograph a ruler or familiar object such as a coin. Photographs are especially good for introductory sections because they provide an overview. The disadvantages are that surrounding areas may clutter the photograph, or you may be unable to highlight the part you wish. See Figure 11-3 for an example of a photograph of gears used in a report for the mining industry. Notice how the center of interest is on the gears.

Drawings

Another way to show an object or place is with a drawing, which can be simplified to show only the important features. Exterior drawings, like photographs, are good for introductory and overview sections of manuals and reports.

Sometimes called "overall" drawings, *line drawings* show the outside of something. They are like photographs in presenting a total picture, but because they can be simplified to show only important aspects, they are sometimes more effective than

Figure 11-3 Photograph of Gears for a Mining Report

Courtesy of Engineered Systems & Development Corporation. Source: Final
Report, *Design, Fabrication, and Testing of a Hopper-Feeder-Bolter*, submitted
to the U.S. Bureau of Mines, July 1987. Reprinted with permission.

photographs. Line drawings are often shaded perspective drawings that give a three-dimensional impression. Sometimes line drawings are simply titled without explanations, the figure number and title placed below the drawing. At other times, parts are labeled on lines or arrows extending from the parts, as in the line drawing in Figure 11-4.

When you want to show the interior of an object or a machine, you can use photographs and cutaway or exploded drawings. Cutaway drawings show what is under the surface, and exploded drawings show the order of assembled parts.

In a *cutaway drawing*, the view is either of a cross-section or a part from which the outside covering has been removed. "Phantom" drawings are similar except that the outside covering is drawn in but appears to be transparent. Cutaway drawings are excellent for showing how interior parts are arranged and how they relate to the exterior. Figure 11-5 shows a cutaway view of a condenser.

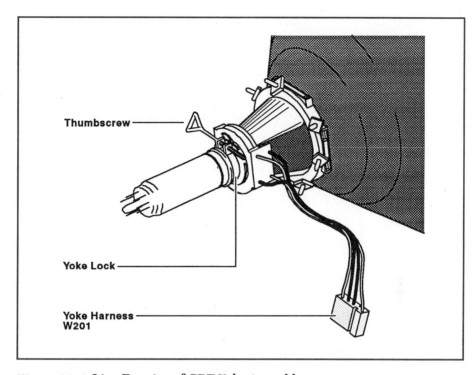

Figure 11-4 Line Drawing of CRT Yoke Assembly

Courtesy of Wyse Technology Inc. From *WY-160 Maintenance Manual*, copyright 1990.

An *exploded drawing* shows the correct order of assembly of a complex mechanism, and does it better, more easily, and in less space than a written explanation could. In an exploded drawing, the parts are spread out along an imaginary line or lines that indicate the axis. Each part is drawn in perspective where it would fit in the assembly. Parts are either labeled for identification or numbered, with the numbers identified in a key. The figure number and title usually appear at the bottom, source references in the lower left corner. Figure 11-6 shows an exploded drawing of a drive coil assembly.

Displaying Part/Whole Relationships

Readers often need to know what the parts of something are and how those parts are related to one another and to the whole. When you want to show an overview of part/whole relationships in material things, you can use pie charts, segmented bar graphs, and cutaway or exploded drawings. When you want to show part/whole relationships in concepts, organizations, or geographical entities, you can use organizational and hierarchical diagrams or maps.

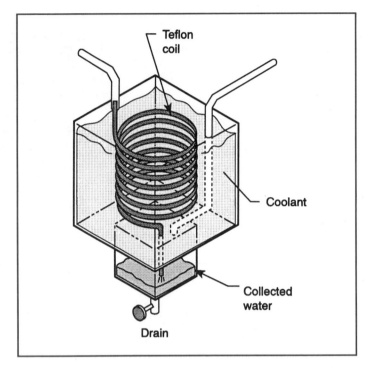

Figure 11-5 Cutaway Drawing of a Condenser

Courtesy of Jahnke, J.A. 1992. *Continuous Emission Monitoring Systems*. Van Nostrand Reinhold. New York. S. Peeler, artist.

Pie Charts and Segmented Bar Graphs

For showing a whole budget or allocation, pie charts or segmented bar graphs are ideal graphics. Both appeal especially to generalists and managers and are good additions to conclusions and recommendations sections or the analysis sections of a report.

A *pie* or *circle chart* is a good example of a diagram of parts; it easily shows both the parts of a whole and the relative size of each part. It uses a circle to represent 100 percent, so that each percentage point equals 3.6 degrees. The first radius should be drawn at 12:00, and the segments should be drawn clockwise around the circle, usually starting with the largest segment. Pie charts are most effective if they contain no more than seven segments, and if each segment is large enough to be seen easily. When you draw a pie chart, check that your percentages total 100 percent. Make labels horizontal, and indicate the percentages or parts of the whole with numbers.

A *segmented bar graph* shows the same relationships as a pie chart. Pie charts and segmented bars work well to represent large sums like a whole budget or one year's tax revenues. Figure 11-7 is a pie chart showing the relationship of parts and wholes in a city's residential waste collection by percentage of weight. Figure 11-8 shows the same relationship in a segmented bar graph.

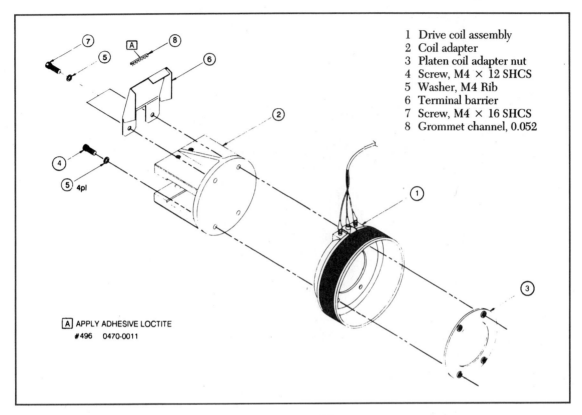

1 Drive coil assembly
2 Coil adapter
3 Platen coil adapter nut
4 Screw, M4 × 12 SHCS
5 Washer, M4 Rib
6 Terminal barrier
7 Screw, M4 × 16 SHCS
8 Grommet channel, 0.052

A APPLY ADHESIVE LOCTITE
#496   0470-0011

Figure 11-6 Exploded Drawing of a Drive Coil Assembly

Reprinted courtesy of Tandem Computers, Incorporated. Copyright © 1990.

Drawings

Cutaway or cross-sectional drawings are especially effective in showing part/ whole relationships in descriptions of something like a geological formation, where the various layers can be shown both in size and in relationship to one another. An exploded drawing, which shows each part in the order in which it should be used, is excellent for instructions of assembly or repair because it shows how each part fits with each other part to make the whole.

Organizational or Hierarchical Diagrams and Maps

Diagrams and maps provide another way to show how the parts of something relate to each other and to the whole. When you want to show how a company is organized and who reports to whom, an organizational diagram is ideal. Likewise, a hierarchical diagram shows how ideas and concepts are related. In the same way, maps—whether of the sky, sea, or land—effectively show the relative location of moons around a planet, reefs around an island, or sales offices within a region.

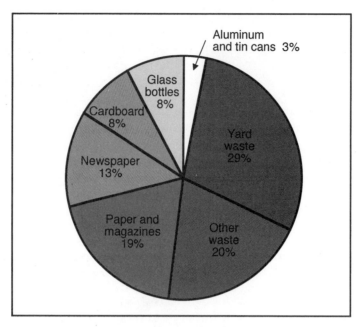

Figure 11-7 Pie Chart Showing Weekly Residential Waste by Percentage of Total Weight

Source: City of San Jose, 1991

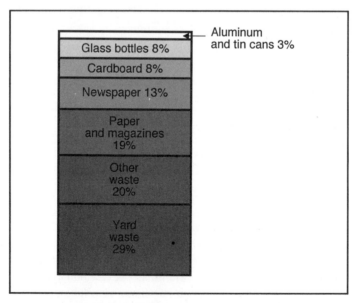

Figure 11-8 Segmented Bar Graph Showing Weekly Residential Waste by Percentage of Total Weight

Organizational and hierarchical diagrams and maps meet the needs of all types of readers because they provide an overview of relationships. They can be placed in the introduction of a report as an overview, in the conclusion as a summary, in the body as supporting data, or in an appendix as added information.

An *organizational diagram* is designed to show areas of control and responsibility; a *hierarchical diagram* shows the relationships among the parts of a complex subject. As in a pie chart, a whole is divided, but this time what is divided is a department, a company, or a complex idea with many subpoints. These diagrams consist of a series of connected boxes and are read from the top down or from left to right. The title and figure number appear below the diagram. Diagrams are unequalled for showing areas of responsibility and the chain of command when they refer to people and positions. When they refer to ideas, diagrams effectively outline hierarchical organization. By placing items of equal importance along the same horizontal, they show the equality of main points and the relationships among subpoints. Figure 11-9 shows a typical organizational diagram for a company division; Figures 11-1 and 11-2 at the beginning of this chapter are hierarchical diagrams showing the relationships among the types of graphics.

To identify and locate places within a larger geographical area use *maps*. Like diagrams, maps show the relationships among parts and how the parts make up a whole, but the whole represented by a map is located on land, sea, or in the sky. All types of readers use maps at one time or another, but operators—whether pilots, hikers, or traveling sales representatives—are the most frequent users. Aeronautical and nautical maps are usually called *charts*. A map will be most effective if it presents only information relating directly to the reader's purpose and needs. Thus, maps concentrating on rivers may not include cities, and road maps do not usually show changes in elevation. Label those items you want emphasized, and below the map give a title that clearly explains what the map shows. Figure 11-10 shows a simple map.

Explaining Processes or Operations

Some readers of technical documents must operate equipment; build, install, test, or repair complex machines; or understand how equipment or a natural process like digestion works. When you write for these readers, you want to choose graphics that effectively reveal processes or operations.

Diagrams of operations include flow charts, block diagrams, schematics, and wiring diagrams. A *flow chart*, sometimes called a *flow diagram*, can greatly simplify the description of a procedure because the reader simultaneously sees an overview (the whole process) and the individual steps making up that process. *Block, schematic*, and *wiring diagrams* all show how each element in a larger system functions and how it relates to other elements.

Flow Charts

In writing procedures or instructions, you will often want to show how one step leads to the next, if a step branches, and how many steps make up the process. For such purposes, a *flow chart* is a good choice: on one or two pages it can show movement and

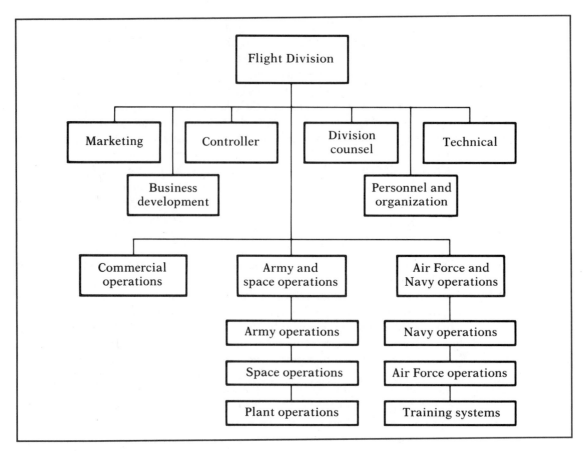

Figure 11-9 Organizational Diagram of a Company Division

relationship in ways that would take many pages of prose. Flow charts are often used in overview sections of reports or procedures; they may be broken into sections and detailed in subsequent parts of the report.

Flow charts visually represent the steps of procedures. They can be constructed with a series of labeled blocks, drawings, standardized symbols, or photographs. The steps are connected by arrows showing the direction of flow or movement from one step to another. Usually flow charts are read from top to bottom or left to right; if you want to show a repetitive process, you might design a clockwise flow. For clarity, give the flow chart a descriptive title, placing it and the figure number below the chart. Label the parts clearly, and spend some time planning the most effective presentation. Figure 11-11 is a simple flow chart; see Figure 19-5 for an example of one that is more elaborate.

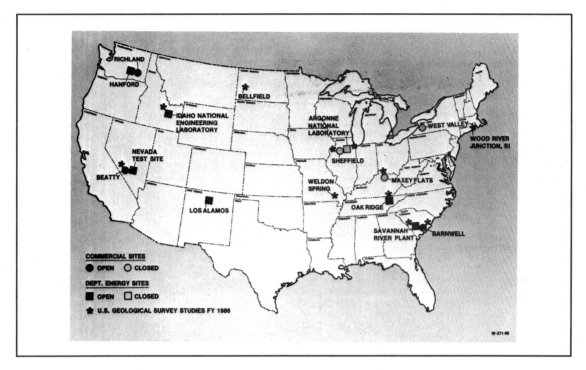

Figure 11-10   Map of Low-level Radioactive Waste Sites and U.S. Geological Survey Studies
Source: U.S. Geological Survey Circular 1005.

Block Diagrams

You can help a reader understand a complex subject if you provide an overview of the "big picture" before you plunge into detail. Block diagrams do this well; they simplify the description by "chunking" information into a few large categories. Block diagrams are also good for managers, who may want only the big picture.

Like flow charts, *block diagrams* join blocks or symbols, but those symbols show relationship rather than movement. When you must present very complex circuitry to inexperienced readers, it often pays to use a block diagram first to give them a simplified overview. Block diagrams are well placed in overview sections of manuals and also at the beginning of (for example) a maintenance section. They can also be used for managers who need a basic understanding without detail. See Figure 11-12.

Schematic and Wiring Diagrams

Once you have introduced the concept with a block diagram, you can use schematic and wiring diagrams to show specialists, technicians, and operators the

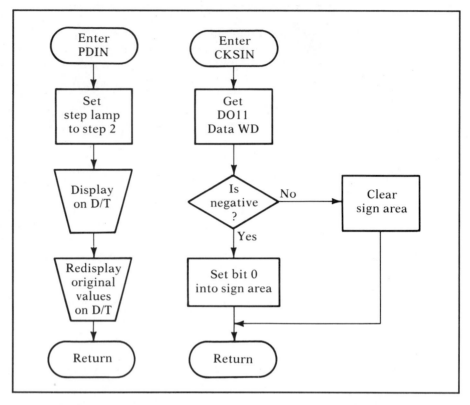

Figure 11-11 Flow Chart of Subroutines in a Computer Program

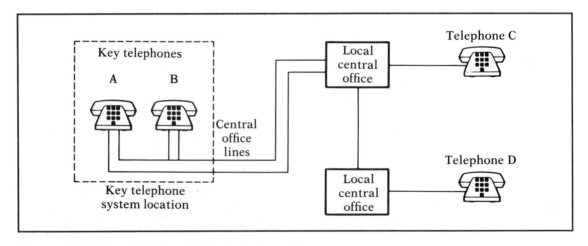

Figure 11-12 Block Diagram of Key Telephone System

details. Schematic and wiring diagrams are often needed in manuals that explain instructions and procedures.

Schematic diagrams are detailed maps of circuits showing the *logical* connections and current and signal flow. *Wiring diagrams* show the *actual* point-to-point connections. The American National Standards Institute (ANSI) issues guidelines and lists of the standardized symbols for schematics and wiring diagrams. Be aware, though, that the symbols and layout may vary from industry to industry. Figure 11-13 is a simple schematic. Figure 11-14 shows a wiring diagram.

Action Drawings

In description of mechanisms or in manuals of assembly, installation, or repair, you may want to use an action drawing, which captures an action in time and thus shows the reader what is happening. Figure 11-15, for example, effectively shows the sequence of actions in a four-stroke gasoline engine. The drawing is both attention-getting and explanatory. Action drawings are useful for all types of readers, but especially for operators and technicians, who must follow procedures or carry out instructions.

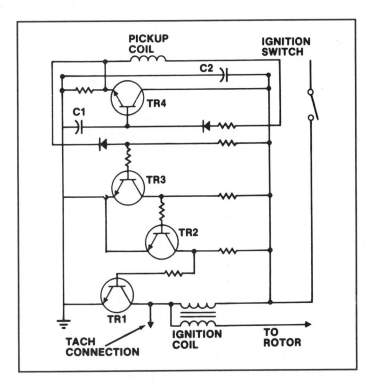

Figure 11-13 Schematic Diagram of a Module in an Electronic Automobile Ignition System

Reprinted from *Automotive Electrical Systems* with permission of Chek-Chart Publications, a division of H.M. Gousha.

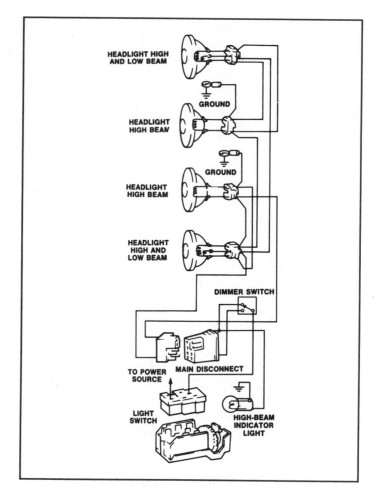

Figure 11-14 **Wiring Diagram of Automobile Headlights**

Reprinted from *Automotive Electrical Systems* with permission of Chek-Chart Publications, a division of H.M. Gousha.

In the way they help explain what happens, action drawings are related to flow charts. However, flow charts indicate sequence, while action drawings may simply show the part in motion. Action drawings can be three dimensional or stylized.

Summarizing

Technical writers often need to *summarize* or *condense* large amounts of information and make that information easily accessible to the reader. Several kinds of graphics summarize exact numbers, measurements, percentages, and locations; those that do it most effectively are tables, diagrams, maps, examples like printouts, and drawings.

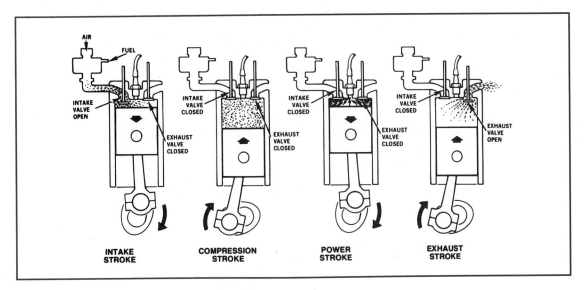

Figure 11-15 Action Drawing of a Cylinder Firing

Reprinted from *Automotive Electrical Systems* with permission of Chek-Chart Publications, a division of H.M. Gousha.

Tables

Tables are usually the best choice for condensing and referencing data or for giving exact numbers, measurements, and specifications. The advantage of a table is its ability to present a large amount of specific information in a small space, thus making comparisons and summaries easy. The disadvantage is that a table does not point out or stress relationships; either the reader must make comparisons from the data presented, or you must discuss the relationships in your comments on the table.

Tables of summary are best placed in the conclusions and recommendations section or in the appendix of a report. Tables of detail appear in the body or appendix. Specialists appreciate tables; they like the specific numbers and don't mind making the comparisons themselves.

Tables are arrangements of numerical or factual data by rows and columns; they can be either informal or formal.

Informal tables are lists that are part of the text itself, but have been pulled out of the paragraph and put into column format. For example, I can break out a list of reader types and appropriate graphics for that type into column format, like the following, as long as I explain within the text what I mean by the list.

Reader Type	*Especially Effective Graphics*
specialists	tables, line graphs, schematic diagrams
technicians	line graphs, flow charts, block diagrams, schematic diagrams
operators	bar graphs, flow charts, block diagrams, schematic diagrams, maps

managers	bar graphs, block diagrams, organizational diagrams
generalists	bar graphs, block diagrams, organizational diagrams
all types	pie charts, maps, photographs, line drawings, examples, organizational and hierarchical diagrams, exploded drawings

This list gives general guidelines for choosing graphics by type of reader; however, the categories overlap considerably, and many graphics are suitable for all readers. Informal tables like this are not titled or numbered because they require additional text to be understandable.

A *formal table*, by contrast, must be both understandable *without* additional text and referenced or explained *within* the text. A formal table has a table number and a title, which appear on a line above the table, either flush to the left margin or centered. Table 11-2 is an example of a formal table from a student report.

To keep the table free of clutter, use a minimum of lines; use white space for separation whenever possible. Place one horizontal line above and one below the column heads to form what's called a "boxhead," and one line at the bottom to separate the table from the following text. The first column on the left is called the "stub"; it contains the list of items, and it may or may not have a head. If you need footnotes for explanation, use lowercase superscript (above the line) letters; put footnotes at the bottom of the table.

Applying what you know about short-term memory, keep the information load in a table manageable. Sometimes it's better to design several tables than to jam a single table with every possible item of information. For example, generally it's best to keep

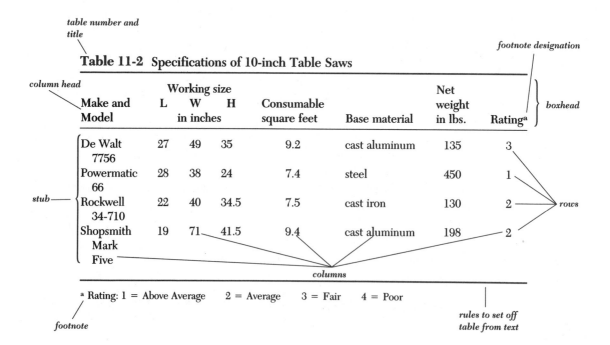

table number and title

footnote designation

Table 11-2 Specifications of 10-inch Table Saws

column head

boxhead

Make and Model	Working size L	W	H in inches	Consumable square feet	Base material	Net weight in lbs.	Rating[a]
De Walt 7756	27	49	35	9.2	cast aluminum	135	3
Powermatic 66	28	38	24	7.4	steel	450	1
Rockwell 34-710	22	40	34.5	7.5	cast iron	130	2
Shopsmith Mark Five	19	71	41.5	9.4	cast aluminum	198	2

stub

rows

columns

[a] Rating: 1 = Above Average   2 = Average   3 = Fair   4 = Poor

footnote

rules to set off table from text

both the number of items in the stub and the number of columns following the stub to fewer than 10.

Because the information in any table can usually be arranged in two or three different ways, you may need to test several options to find the most effective one. Use these suggestions to design an effective table:

- Confine the table to one page if possible and design it to fit vertically on the page.

- Place items that use names in the stub and place items that use numbers under appropriate column heads in the rest of the table.

- Write column heads that clearly identify the items and the units of measure used.

- Align decimal points and limit decimals to two places if possible.

- Add footnotes to any items that need further explanation. Use lowercase letters to label the footnotes.

Diagrams and Maps

Diagrams that display many parts of an organization all in one place and *maps* that display many locations simultaneously can also summarize large amounts of information. Such diagrams have the advantage over tables that they can also indicate relationships (equal job importance for example, or relative geographical location). However, because of space limitations, diagrams often do not include numbers, so they may be less accurate than tables.

Examples

Printouts of computer screens, EKG monitors, oscilloscope screens, and the like can both summarize information and attract the reader's eye to what is important.

Comparing

Consider *graphs* whenever you want to compare amounts; to show relationships and relative quantities; to show trends, distributions, cycles, and changes over time. Graphs work well in comparing both because they appeal to the reader's eye and because they *show* changes instead of simply listing the numbers. Thus, in a way tables do not, they help the reader see how the items compare.

A graph is a pictorial presentation of numerical data. In other words, graphs can give the same information as tables, but they picture comparisons and trends, making the reader's job easier. The major types of graphs are line graphs and bar graphs.

Line Graphs

Trends, distributions, and cycles are best shown with *line graphs*. Standard usage calls for plotting variables of distance, time, load, stress, or voltage on the horizontal axis (called the abscissa or *x* axis) and variables such as money, temperature, current, or strain on the vertical axis (called the ordinate or *y* axis). Figure 11-16 shows an example. The figure number and title usually appear below the graph, and the source line appears at the lower left, although placement may vary depending on each company's style. In a line graph, start your scales at zero, if necessary showing a break in the scale by a jagged line interrupting the tick marks on the axis. Take care not to distort increments by making the scales too wide or too narrow. Percentages should always go from 0 to 100. Include a key if needed, and make the lettering horizontal, not vertical.

Bar Graphs

To compare amounts and show proportional relationships, use *bar graphs*. The bars should be of the same width but can be either vertical or horizontal. Unless they are tied to time, bars often are arranged in increasing or decreasing length. Figure 11-17 is a typical bar graph showing how computing time in a laboratory is reduced by

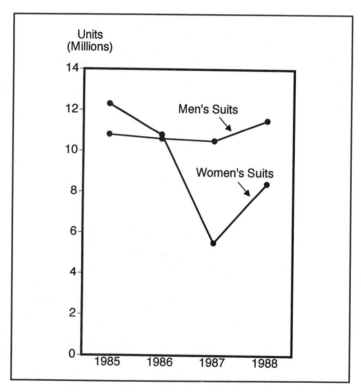

Figure 11-16 Line Graph Showing U.S. Production of Suits (millions of units)

Source: U.S. Bureau of the Census

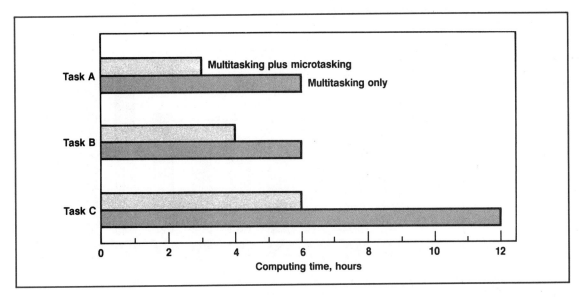

Figure 11-17   Bar Graph Showing How Computing Time Is Reduced by Multiprocessing

multiprocessing. The bar graph has labeled vertical and horizontal axes, and as with other figures, the title and figure number are usually located below the graph. For more precision, bar graphs can include exact numbers—a good location for numbers in a vertical graph is at the top of each bar. All graphs should be plotted on a computer or on graph paper with a horizontal and a vertical axis. Once the graph is plotted, omit the grid lines, which clutter the page. Show the scale with tick marks on the axis. Bar graphs can be truncated to show a break in the scale, as in Figure 11-18, showing the enrollment trend at a university over 10 years.

Segmented Bar Graphs and Pictographs

Two special kinds of bar graphs are *segmented bar graphs* and *pictographs*. In a segmented bar graph, a single bar represents 100 percent. It is divided into parts to show proportional relationships, as shown in Figure 11-8.

A pictograph uses a single simple drawing to represent one unit and then repeats those drawings to assemble the bars, as shown in Figure 11-19. Pictographs are not as accurate as standard bar graphs, but they add interest and are therefore good for generalists. Be careful when using pictographs not to increase the size of the drawing to show a greater quantity; a larger drawing will increase by more than a single unit.

Different kinds of graphs meet specific needs in presenting comparisons. Line graphs show trends, distributions, and cycles; they are especially effective in showing changes over time. Bar graphs are better for comparing amounts and showing proportional relationships. Visually, bar graphs also make a stronger statement; however, they may not be as accurate as either line graphs or tables. Pictographs are especially good for generalist readers because they add interest to the page.

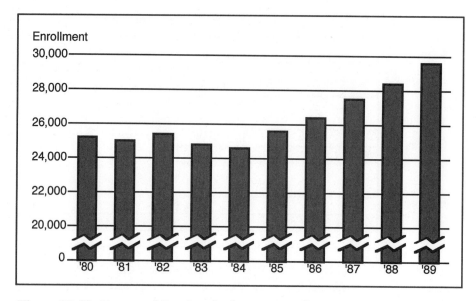

Figure 11-18 Truncated Bar Graph Showing Enrollment at Central State University from 1980–1989

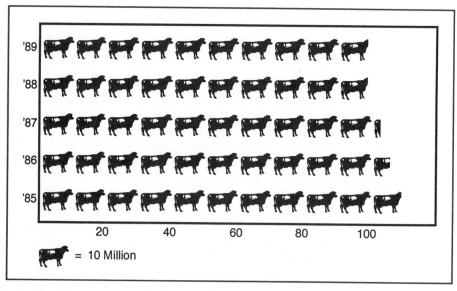

Figure 11-19 Pictograph of U.S. Cattle Production 1985–1989

Source: U.S. Department of Agriculture

The advantages of graphs lie in their visual appeal; they are more interesting than tables, and they show comparisons better. Generalists and managers appreciate bar graphs for the quick and easy view they give of relationships. Specialists and technicians like line graphs for their detail. However, unless they also give specific numbers, graphs may be less accurate than tables, and they can easily distort information if the scales are not properly proportioned.

You may want to experiment with tables and graphs to determine which type will best meet your readers' needs. Some computer graphics programs will allow you to display the same data in several different forms so you can decide which is most effective. For example, following is a list of raw data about the cost of various home appliances (PG&E 1989). Read through the data to see what information is given. Then study Table 11-3 and Figures 11-20 and 11-21 to see how a graphics program can display this data in a table, a line graph, or a bar graph.

Gas furnace	$0.51 per hour
Gas water heater	$0.02 per hour
Gas range burner	$0.04 per hour
Gas oven	$0.06 per hour
Electric furnace	$0.70 per hour
Electric water heater	$0.07 per hour
Electric range burner	$0.13 per hour
Electric oven	$0.14 per hour

Providing Examples

All types of readers and listeners want examples to help them understand technical material; thus, as you write, you should plan to include many kinds of examples in your documents. In a computer manual, an example might be a reproduction of the terminal screen, as in Figure 11-22; in an engineering report, examples might be photographs of the damage caused by excessive temperature; in a scientific report, examples might be drawings or photographs of microscope slides. In this chapter, note that I have provided you with an example of each major type of graphics.

In speeches and oral presentations, you will communicate most effectively if you appeal to both sight and sound—emphasizing the main points listeners *hear* with graphics (called visuals in this case) or with projected key words that they can also *see*.

Table 11-3 Cost of Home Appliances per Hour (dollars)

Appliance	Gas	Electric
Furnace	0.51	0.70
Water Heater	0.02	0.07
Range Burner	0.04	0.13
Oven	0.06	0.14

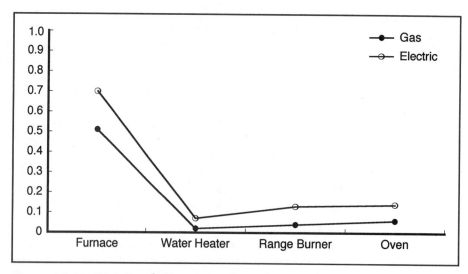

Figure 11-20 Line Graph Comparing Cost of Home Appliances (per hour in dollars)

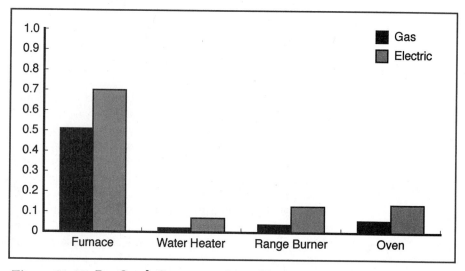

Figure 11-21 Bar Graph Comparing Cost of Home Appliances (per hour in dollars)

All the graphics mentioned in this chapter can be used in speeches, but most of them need to be simplified for projection. Many professionals use transparencies (also called *viewfoils* or *foils*) to project simple tables, line or bar graphs, diagrams, and the like. Others use flip charts or slides on which they write or draw appropriate visuals. All audiences respond well to visuals, but you seldom can include as much detail as in a printed document. Chapter 24 gives more detail on using visuals in speeches.

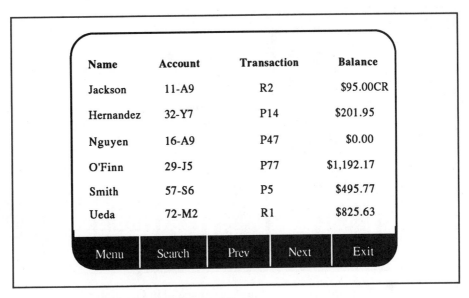

Name	Account	Transaction	Balance
Jackson	11-A9	R2	$95.00CR
Hernandez	32-Y7	P14	$201.95
Nguyen	16-A9	P47	$0.00
O'Finn	29-J5	P77	$1,192.17
Smith	57-S6	P5	$495.77
Ueda	72-M2	R1	$825.63

| Menu | Search | Prev | Next | Exit |

Figure 11-22 Example of a Terminal Screen

Representing Ideas or Functions

As you design a document, you always want to provide signposts that will help readers find their way through the sentences and paragraphs. You can provide such signposts with numbers, with type size and style, and with hierarchical organization. But increasingly, writers are also turning to *icons* as signposts. The visual impact of an icon is stronger than that of a number, thus attracting attention. Icons also add interest to a page and take up little space. If you decide to use icons, choose them carefully: each icon must be visually distinct, easy to understand, and repeated throughout the document.

Many icons in contemporary usage are a product of computer technology and are often used in computer-related documents. Others are used to communicate without words to international or multicultural readers. The term *icon* simply means picture or image; in the last 25 years, icons have appeared as small stylized drawings that are used to represent functions or ideas. For example, icons can be used:

1. in place of commands in word form ("Simply click on the trash can icon.")
2. as identifiers for sections ("The section on video monitors is always indicated by a screen icon.")
3. as attention-getting devices for inserts like warnings ⚠
4. as universal symbols easily understood by international and multicultural readers

Jonathan Price, in *How to Write a Computer Manual*, calls icons one of the common elements used in designing a manual (1984). In this view, icons must be chosen early in the writing process so that the manual can be designed around them.

Figure 11-23 Types of Icons Used in Packaging and in Computer Software

Icons can either be used alone or with their word equivalents; they help readers move through the text by supplying interesting and vivid pictorial signposts. See Figure 11-23 for examples of icons used both in packaging and in computer software.

11.4 General Rules for Using Graphics

Follow the general rules listed below to present all your graphics clearly and effectively:

1. Remember your readers and their needs. As you choose or design graphics, ask yourself "What does the reader want to know?" For different kinds of readers, you may have to use different graphics to present the same information.

2. Keep your options open. Most information can be displayed by more than one kind of graphic, so don't assume that your first choice is necessarily the best. As you review and revise, be willing to revise graphics as well as paragraphs.

3. Keep graphics simple and uncluttered; keep lines and wording to a minimum. Design a table or figure large enough to be read easily. If you reduce it for better fit on the page, be sure that the reduced numbers and letters are readable.

4. Insert graphics immediately after their first mention in the text. Referring to them in the text specifically by number (Table 6, Figure 1), tells the reader when to look at them and why, especially if conclusions are involved. A major fault in technical writing is using graphics without referencing them. Also, keep your terminology consistent from text to table or figure and from one graphic to another.

5. Make each graphic self-sufficient: that is, intelligible without the explanation in the text. The keys to self-sufficiency are (1) a title that has both key content words and function words like "How the Four-Stroke Cycle Works," (2) labels or column heads that clearly identify all parts, and (3) simplicity, so the reader is not bogged down by extraneous information.

6. Unless your company style guide has a different system, place the table number and title *above* the table. Tables are usually read from top to bottom, so this arrangement is sensible. For figures, place the figure number and title *below* the figure. Label the parts clearly with lines or arrows to the part (these are called "callouts"). Provide a key either on or below the figure.

7. Identify the source of the table or figure or the source of information in small print at the bottom of the graphic—for example: "Source: U.S. Dept. of Labor." Sources are often put at the lower left corner, but they may be placed elsewhere as long as you are consistent within a document. *Always* identify your sources; otherwise, you could be accused of plagiarism.

8. Separate the graphics from the text either with thin lines or with a frame of white space. Keep graphics on one page if possible, and try to avoid printing them sideways and making the reader turn the document. Make all wording read horizontally (left to right, not top to bottom).

9. Use graphics only if they are functional. If your table or figure won't directly help the reader understand your point, don't use it.

10. Include a list of illustrations in the front of a long report or proposal—especially if you use more than three graphics within the document. If you have both tables and figures, make two separate lists, since tables and figures are usually numbered separately. Some readers are more interested in the graphics than in the text itself, so make it easy for them to find the graphics.

 Checklist for Choosing and Using Graphics

1. As I consider my document, do I have places where graphics can do the following?
   ___ increase reader interest and understanding
   ___ display and emphasize important information
   ___ summarize and condense information
   ___ convey relationships
   ___ communicate with international and multicultural audiences
   ___ be designed by computer
   ___ be borrowed from other sources or hand-drawn
2. What purpose do I want to accomplish with any specific graphic choice?
   ___ to show exterior or interior views
   ___ to display part/whole relationships
   ___ to explain processes or operations

___ to summarize information
___ to compare, show relationships, indicate trends
___ to provide examples or reproduce what the reader might see
___ to represent ideas or functions
3. Who is my primary reader?
___ generalist
___ manager
___ operator
___ technician
___ specialist
4. What specific type of graphics should I use to meet that reader's needs?
___ tables (formal or informal)
___ graphs
___ maps
___ diagrams
___ photographs
___ drawings
___ examples
___ icons

EXERCISES

1. Search in textbooks, magazines, and newspapers for examples of at least four different kinds of graphics. Notice the purpose and type of readers for which each is used and analyze how effectively each one is presented. Clip out or photocopy the graphics and bring them to class. Be prepared to discuss their good and bad points.

2. For each of the following situations, choose the most appropriate type of graphics. If more than one would work, note that. List the reasons for your choices, and the type of reader. Be prepared to discuss your choices in class.

1. Your state has (or is considering) a lottery to raise money for education. The graphic aid should show the breakdown of revenues, showing the percentage to go to each educational level.
2. A company manufacturing burglar alarms needs an explanation of the operation of its burglar alarm system.
3. You wish to compare current stock prices for three companies that make integrated circuits.
4. For a consumer report, you need a comparison of three top-of-the-line washing machines according to cost, capacity, warranty, and special features.
5. A major publisher wants to show the sales coverage in the United States of its 125 sales representatives.

3. Using factual or statistical data from a recent report, construct a table. Then use the same data to construct a figure. Exchange with a fellow student. Analyze that person's figure and table to see which is more appropriate for the writer's purpose and content. Write a paragraph of evaluation.

4. Use the following information from the 1990 *Statistical Abstract of the U.S.* to create an effective graph for generalists, showing the most important information in five-year increments from 1965 to 1985. Consider which kind of graph will best explain this information to the reader.

Table 1. Municipal Solid Waste Generation, Recovery, and Disposal: 1960 to 1986
[In millions of tons, except as indicated. Covers post-consumer residential and commercial solid wastes which comprise the major portion of typical municipal collections. Excludes mining, agricultural and industrial processing, demolition and construction wastes, sewage sludge, and junked autos and obsolete equipment wastes. Based on material-flows estimating procedure and wet weight as generated]

Item and Material	1960	1965	1970	1975	1980	1981	1982	1983	1984	1985	1986
Gross waste generated	87.5	102.3	120.5	125.3	142.6	144.8	142.0	148.3	153.6	152.5	157.7
Per person per day (lb.)	2.65	2.88	3.22	3.18	3.43	3.45	3.35	3.47	3.56	3.49	3.58
Materials recovered	5.8	6.2	8.0	9.1	13.4	13.2	12.9	13.9	15.3	15.3	16.9
Per person per day (lb.)	.18	.17	.21	.23	.32	.31	.30	.32	.35	.35	.39
Processed for energy recovery	(NA)	.2	.4	.7	2.7	2.3	3.5	5.0	6.5	7.6	9.6
Per person per day (lb.)	(NA)	.01	.01	.02	.06	.05	.06	.12	.15	.17	.22
Net waste disposed of	81.7	95.9	112.1	115.5	126.5	129.3	125.6	129.5	131.8	129.7	131.2
Per person per day (lb.)	2.48	2.70	3.00	2.93	3.04	3.08	2.96	3.03	3.05	2.97	2.96
Percent distribution of net discards: [1]											
Paper and paperboard	30.0	33.5	32.4	29.6	32.5	33.1	32.1	34.1	35.7	35.5	35.6
Glass	7.8	8.8	11.1	11.4	11.0	10.9	10.7	9.9	9.3	8.9	8.4
Metals	12.8	11.1	12.0	11.5	10.1	9.8	9.7	9.6	9.3	9.0	8.9
Plastics	.5	1.5	2.7	3.8	5.9	5.9	6.5	6.8	6.9	7.1	7.3
Rubber and leather	2.1	2.3	2.7	3.2	3.2	3.1	2.9	2.5	2.4	2.5	2.8
Textiles	2.1	2.0	1.8	1.9	2.0	2.6	2.2	2.1	2.0	2.0	2.0
Wood	3.7	3.6	3.6	3.8	3.8	3.3	3.9	3.9	3.7	3.9	4.1
Food wastes	14.9	12.9	11.4	11.5	9.2	9.2	9.3	8.9	8.8	9.0	8.9
Yard wastes	24.5	22.5	20.6	21.7	20.5	20.3	20.9	20.4	20.1	20.4	20.1
Other wastes	1.6	1.7	1.6	1.7	1.7	1.7	1.8	1.8	1.7	1.8	1.8

NA Not available.   [1] Net discards after materials recovery and before energy recovery.

Source: Franklin Associates, Ltd. Prairie Village, KS, *Characterization of Municipal Solid Waste in the United States, 1960 to 2000*, 1988. Prepared for the U.S. Environmental Protection Agency.

Source: U.S. Statistical Abstract, 1990.

WRITING ASSIGNMENT

Prepare a minimum of one table and one figure to include in your formal report, proposal, or manual. Be sure that titles, labels, and keys are prepared correctly. Indicate the sources of your graphics or information. Place the graphics immediately following their first mention in the text. If you are working on a project as part of a team, decide as a group what graphics to include, and assign one person to create the graphics and write the text references.

REFERENCES

Michaelson, Herbert. 1990. *How to write and publish engineering papers and reports*. 3rd ed. Phoenix: Oryx.

Pacific Gas and Electric Company. 1989. *PG&E Progress*. San Francisco.

Price, Jonathan. 1984. *How to write a computer manual*. Menlo Park, CA: Benjamin/ Cummings.

Chapter 12

Designing the Document

When you write a workplace document, your primary concern is to communicate information and ideas to the reader as quickly and directly as possible. One way you do that is to use *content cues* within the text—meaningful titles, overviews, parallelism, transitions, and the like. Content cues are explained in Chapter 6 as ways of making information accessible to the reader.

In addition to content cues, you can provide *format cues* to direct the reader to important information and to make reading easier. Jan V. White, an expert in document design, says, "Page makeup is commonly called design. It is therefore thought to be an art-form intended to create *beauty*. It is nothing of the sort. Design is a *lubricant for ideas*" (1990, 6).

In the days when most documents were typed on typewriters, writers had very few choices in how a page could be designed. The text could be single-, double-, or triple-spaced, and the margins could be wide or narrow. Type variations were limited to those provided by the typewriter, and the only highlights were underlining and capitalization. Only if a professional printer was involved did design become a major factor, and then the choices were usually made by the printer or a graphics designer.

Today, however, nearly all writers have access to word processing, page-layout programs, and high-quality printers. That means writers themselves can, and often do, design the pages for the finished product, and that means *you*, as a writer, need to know the basics of using format cues for good document design.

How important is the "look" or design of a document? Think of how you feel when you open a report or textbook of solid, tightly packed words—page after gray page with no visual breaks. In contrast, how do you respond to a page with a picture or a drawing, with a list, with subheadings in boldface or color type? The more visually interesting the page, the more likely you are to respond positively—even before you start to read. Readers will respond the same way to your pages, and you always want to make your readers respond positively.

Anne Rosenthal, technical communications manager at a biotechnology instrument manufacturer, says, "Document appearance is important: it sells the documents both inside and outside the company." What Rosenthal calls "selling" may be as simple as motivating readers to actually begin reading: because technical content is often new and difficult, you need to pull readers into the text and make their reading job easy.

As White explains it, "One of the myths in the field of publication making is that readers are readers. In actuality, they start out as viewers. They scan, they hunt and peck, searching for the valuable nuggets of information. Reluctant to work, saturated by media, and a bit lazy, they literally need to be lured into reading. They need to be purposefully led through the information" (1990, 17).

Document design is your way of motivating readers to start reading and of helping them to find the information they seek.

12.1 Defining Key Terms

Within the fields of publishing, printing, and graphic design, a number of terms are used somewhat interchangeably. In this book, the term *document design* is the general term: it means the process of determining the way the report, manual, letter, brochure, or other document looks. A document that is well designed reinforces the internal organization of the contents by the way that organization is displayed on each page. In other words, the internal form of the document is made apparent through the *format*, or the general appearance. Moving from the document as a whole to individual pages, we use the term *page design* to describe the look of each individual page. Another term for page design is *page layout*: the way the printed document looks based on spacing and specifications for type.

In this chapter, you will learn the basics of good document design; for details you might want to consult a desktop-publishing or printer's guide such as *The Pleasures of Design: A Practical Guide to Layout and Typography in Desktop Publishing* by the Linotype Company (1988) or *Great Pages: A Common-Sense Approach to Effective Desktop Design* by Jan V. White (1990).

12.2 Planning the Whole Document

In planning the design of a whole document—whether it is a fanfold brochure, a 30-page manual, or a long report—you will want to consider four basic factors: the purpose, the reader, your publications tools, and the conventions you must follow. Addressing these factors will give you the specifics for a plan before you begin the formatting task.

Determine the Purpose

You've been addressing the issue of purpose at every step of the writing process; now once again, you need to ask "What am I trying to accomplish with this document?" Are you informing? Persuading? Instructing? Describing? Defining? Providing reference information? Related to *your* purpose is the reader's purpose: Why is the reader going to read what you have written? Is the reader seeking information for decision making? Learning a new skill? Trying to understand a new process? Installing a machine? And how motivated is that reader? It's highly motivated readers, for example, who read documents such as novels and creative nonfiction such as memoirs. These readers read at leisure, and what they read is intended to enlighten and entertain. What's more, the documents are read from beginning to end; thus, they require only a few format or design cues. With an attractive type and clearly marked paragraphs, most readers will navigate through the prose with little difficulty.

Technical and other workplace documents, however, are usually read under time pressure by professionals who want information or instructions. While they may be highly motivated to find what they want, they are not reading by choice. They seldom read a document from beginning to end; instead, they scan, looking for cues to the information they seek. To make their job easier, they need both the verbal or content cues explained in Chapter 6 and the visual or format cues discussed in this chapter.

Evaluate Readers and Their Needs

Beside knowing what readers want, you need to know under what conditions they may be reading. For example, if they are in the field, a thick three-ring binder stuffed with data may be too awkward to use. If they must read while performing a task, they may need a spiral binding so the document will lie flat. If the light will be poor, type sizes may need to be larger, and the document may require more white space to delineate items clearly. If the document must be updated frequently, it may need to be in a looseleaf binder, so new pages can be inserted.

As you consider reading conditions, also consider reader identity: *Who* readers are influences the design as well as the content. Older (and younger) readers may require a type size larger than the average. Operators may require more (and different) graphics than specialists or technicians. Specialists may accept denser documents than will managers. Generalists may need color to captivate their interest.

Assess Your Publishing Tools

As you make decisions about the way a document should look, remember that you are limited by your available tools and budget. It does no good to plan for color if you don't have color capability or the money to buy it. It's a waste of energy to plan for photographs if you can't import halftones into the document. Assess your current hardware and software capabilities, and make your decisions based on what you know you can do.

The document-designing stage is also when you decide whether to do the entire document yourself or turn it over to a professional graphics designer or printer. This decision is influenced by your tools and your expertise. As a student, you will probably do most of the designing yourself (though even students sometimes pay for résumé design), but as a professional, you may choose to design some documents but not others.

In these days of widespread and relatively inexpensive desktop-publishing tools, your chances are good of having at least:

1. word processing hardware and software that will allow you to
   - delete, insert, and rearrange text
   - handle equations and special symbols
   - prepare a table of contents, list of figures, and index
   - choose among type sizes and typefaces
   - use style features like bold and italics
   - center words and phrases
   - justify margins
   - set styles and sizes for various levels of headings

2. a page-layout program that lets you
   - design the pages, including width of columns, size of margins, and amount of white space
   - import graphics and enlarge or reduce them

- add style features such as boxes, rules (thin lines), and screens to direct readers' attention

3. a laser or other high-resolution printer that allows you to produce camera-ready or final copy

Consider Established Conventions

Conventions are traditional ways of doing things. Thus, in Part 3 of this book, you will learn conventional ways of designing typical workplace documents such as letters, résumés, instructions and reports. Over time, these documents have developed conventional page designs that work well to convey information. As you plan specific documents, you need to know how—conventionally—those documents have been designed; then, if you choose to change the design, you will make changes for specific purposes.

Other conventions may be dictated by your workplace. Many companies, for example, have already decided how their documents must look, and they provide a style manual or electronic style sheets (*templates*) for such things as the title page, table of contents, graphics captions, and list design (shape of bullets, type of numbers, and so on). The Tandem *Corporate Style Guide* (1989), for example, supplies sample pages of each major element in a manual and directs readers to use a specific computer template, macro, or program to prepare each element. Even if your company does not specify design for each document, you may want to establish style conventions to ensure that all similar documents have a similar look. Likewise, as a student, you may want to develop a consistent design for your documents.

The best way to plan the design of a whole document is to think about the four factors discussed here and then—if you have choices—to design and produce a few sample pages. Once you see how they look, you can format the whole document.

12.3 Designing Pages for Easy Reading

Each page is a picture composed of print and white space. The amount and relationship of that print and white space should be influenced by your knowledge of how readers read and what catches reader interest. If you think of a page as a picture, you can take eight steps in designing that will make the readers' job easier:

1. Use margins to frame the text and make the page open and inviting.
2. Choose a line length readers find comfortable.
3. Use line spacing (leading) that separates without distracting.
4. Consider unjustified (ragged right) margins.
5. Use white space to separate, to direct the reader's eye, and to establish hierarchy.
6. Set up vertical lists to isolate related items from the remaining text.
7. Insert boxes, rules, and screens to highlight and separate information.
8. Use color for highlighting.

Margins

Margins are the white spaces around text that frame the printed material, defining it as distinct text. Narrow margins make the page seem crowded and hard to read, while wide margins open the page and invite readers in. Actual margin size is determined by the finished size of the document: 8½ by 11 pages often use 1-inch margins at the top, bottom, and sides and 1½ inches at the bottom. However, company manuals that use large amounts of white space elsewhere on the page may have narrower margins, inserting headers and footers (see Chapter 6) into what would normally be the margin, and effectively reducing the margin to ½ or ¾ inch.

Line Length

Research indicates that for comfortable reading a line of type should be from 42 to 70 characters long—or approximately 10 to 12 words (Linotype 1988). That means you may want to consider two or even three columns of type instead of one, or you may want wide margins. Figure 12-1 shows how differences in line length affect the appearance of the whole page.

Line Spacing

Line spacing (*leading* in printers' jargon) is the amount of white space that separates lines of type. If the line is single-spaced, most type is set with one or two points of extra leading for easy reading. (A *point* is ½ of an inch, and type size is designated by the number of points.) For example, 10-point type may be set on a 12-point line. Different leadings have different uses, and the choice is based on function and intended reader. Lines of type are easier to read if they have more leading, but too much leading spreads the text out disproportionately.

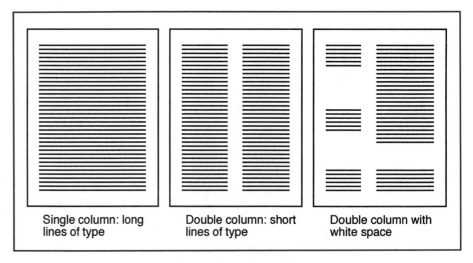

Single column: long
lines of type

Double column: short
lines of type

Double column with
white space

Figure 12-1 Line Length Variations in Page Design

A good way to separate single-spaced paragraphs is to insert a line or even a line and a half of white space between them. You can also double-space lines of type, add one and a half spaces, or even triple-space. Drafts are often double-spaced for easy reading and editing, but finished text is usually single-spaced. As a student, however, you may be required to double-space text so your instructor has room for comments.

Justification

As a reader, you are accustomed to blocks of type that are neatly squared off on both the right and left sides (right and left *justified*). However, in order to justify at the right margin, adjustments must be made to each line of type.

1. In a short line, tiny spaces must be inserted between words and even between letters in a word.
2. In a long line, the spaces must be reduced, thus crowding letters and words together.
3. Words at the end of a line must be hyphenated.

Many computer programs are unable to make these line adjustments in small enough increments. Thus, in justified right margins, lines may look either stretched or cramped, or many words may have to be hyphenated.

One way to avoid this problem is to leave the right margin *unjustified*—or *ragged*, as shown in Figure 12-2. Researchers in readability have also found that text with ragged right margins is easier to read because readers do not lose their place in the text as often (White 1990).

If text is right justified, page designers often indent paragraphs by three to five spaces to set them off. If text is unjustified (ragged right), they often do not indent, starting the paragraph at the left margin. Then, paragraphs are separated by a line or half line of white space (Linotype 1988).

Figure 12-2 Design Differences between Justified and Ragged Right Text

White Space

You use white space every time you insert a blank line between paragraphs or use margins to frame the text. But you should also think of white space as the frame around a list or heading. White space helps define the boundary between one idea and the next, helps relieve the monotony of gray print, and helps communicate by resting the eye. You can think of white space as a highlighter, and you should plan for it on the page as you design the document. Don't be afraid to use white space. Even a sizable chunk of it is permissible before a major heading. In a document packed with information—like a résumé—white space is especially important. Use it to lead the reader's eye to the points you want to emphasize.

Indentation by white space can also show hierarchy, or the relationship among points and subpoints. If you are consistent throughout the document, your reader will be aware of the hierarchy of ideas simply by the way white space sets off the information.

Lists

You can take advantage of both white space and the "chunking" effect if you use vertical lists to show the reader how bits of information relate to one another.

Numbered lists are effective to show sequence (first, second, third, and so on). Numbers are also effective to show hierarchy, as in a formal outline, or to emphasize the number of items to be considered. A good general rule for more than three subpoints of equal importance is to put them in a list. The physical arrangement of the list itself provides a visual clue about the relationship of items, and the white space around the list focuses the reader's attention.

If you do not want to show sequence or hierarchy, use *bullets* to set off items in the list. Bullets are usually solid or open circles—they can also be squares or diamonds. Bullets are available in most word-processing programs and are effective for highlighting items without indicating a specific order. When you design a list, set the numbers or bullets apart from the text so they are instantly visible.

Punctuate lists this way: If the items in the list are fragments, begin each item with a lowercase letter and use no punctuation at the end; if the items in the list are sentences or a mix of sentences and fragments, begin each item with a capital letter and end it with a period. List items should also be parallel in structure (see Chapter 8). Figure 12-3 shows two effective ways to design lists.

Boxes, Rules, and Screens

Boxes are thin lines that make a box surrounding a block of type or a figure. They are used to draw attention to the material within the box and to separate it from the rest of the text. Usually, boxed copy is related to the text but can be read and understood by itself. When writing for generalists, you can use boxes to set off interesting or explanatory material not needed for the main text. Such boxed material is sometimes called a *sidebar*. For technical readers, boxes are often used in instructions to set off warnings, cautions, and notes.

In the process of designing a document, follow these general steps:

1. Plan the whole document, considering purpose, reader, tools, and conventions.

2. Design individual pages, making them consistent with each other yet designed as individual entities.

3. Choose type sizes, fonts, and styles for ease of reading and for focusing attention while scanning.

4. Review the design to ensure that it follows good design principles.

Elements of page design include:

- margins
- line length
- line spacing
- justification
- white space
- lists
- boxes, rules, and screens
- color

Figure 12-3 Lists Designed for Easy Reading

Rules are horizontal lines used to separate or link text or to add emphasis. A rule across a single column of type isolates the text below it; a rule across two columns of type emphasizes the relationship of the two columns. Thus, especially with two- or three-column type, rules can help guide the reader through the page. Rules can also be combined with headings for emphasis, but the visual impact is very strong, so they should be used with care. Too many rules will confuse readers, so use them only to achieve specific desired effects.

Screens are areas of a page defined by pale dots over which type is printed. The screen has the same effect as a box; it draws attention to the material and sets it off from the rest of the text. Many page-layout programs have screening capability, but as with boxes and rules, you should use this feature only occasionally. Graphic design expert Jan White gives this advice about screens: "Use very pale screens to distinguish areas on the page; 20 percent is usually as dark as it should be" (1990, 20). Figure 12-4 shows screens printed in gray and in color.

Color

Color can be used to highlight, to link separate elements, or to reveal the structure of the whole document. While you are a student, you may not have access to color, but color will become increasingly important in documents as color monitors and printers become more available and less expensive.

"A document's structure is an important factor in visual communication. Studies have shown that dense text is intimidating to readers because it is difficult to read. Because there are no visual breaks in the text, readers cannot navigate through the page easily. Textual elements such as fonts, type styles, margins, and leading—when combined with structural elements such as headings, subheadings, graphics, and lists—provide visual breaks and landmarks for readers as they make their way through a document" (Van Brigle 1991,7).

"A document's structure is an important factor in visual communication. Studies have shown that dense text is intimidating to readers because it is difficult to read. Because there are no visual breaks in the text, readers cannot navigate through the page easily. Textual elements such as fonts, type styles, margins, and leading—when combined with structural elements such as headings, subheadings, graphics, and lists—provide visual breaks and landmarks for readers as they make their way through a document" (Van Brigle 1991,7).

Figure 12-4 Type Set on Screens for Emphasis and Interest

White says that one or two colors plus black are remembered best; many colors can be confusing (1990). Keep the following suggestions in mind for the effective use of color:

- Use the same color for the same elements within a document: first-level headings, rules, icons, and so on.
- Set type larger and bolder when it will be run in color. The same point-size type will appear to be smaller in color than in black.
- Use color for short elements—headings, annotations, features—but not for the main text. Large blocks of text in color are hard to read.

12.4 Choosing Type Sizes, Fonts, and Styles

Whether you are designing the document yourself or negotiating with a printer, you need some basics about type—appropriate sizes, fonts, and styles. Nearly 2,000 typefaces are available to professional printers (Linotype 1988), and computer printers may offer from 6 to 20. However, for an inexperienced document designer, too many choices may be distracting. In fact, most documents can be designed effectively with only one or two typefaces, such as the widely used Times Roman or Helvetica. The following paragraphs provide basic information on type; for more details, you can consult desktop-publishing guides or your printer manual.

Type Size

For the text itself, experts recommend a character size that is from 8 to 12 points, with 10 point being the easiest to read. A *point* is a measurement of type, with 72 points to the inch. Headings should be set in larger type, often 4 points larger than the point size of the text. Figure 12-5 shows sample paragraphs set in 10-point and 12-point type, with headings set in 14-point and 16-point bold type.

Type Font

A *font* is the set of characters that make up a complete typeface, and a *family* of typefaces is a range of weights and distortions of the same basic design. Many word-processing programs come with a variety of fonts from which to choose. Type fonts include both those traditionally used in the printing industry and more contemporary fonts. Proven favorites include Times Roman, Palatino, and New Century Schoolbook,

14 pt. Times Roman Bold

Writers Need Familiarity with Several DTP Systems

10 pt. Times Roman

In 1991, more than 1,500 DTP systems were available (double the number available in 1989), and writers must understand how to use as many of them as possible. A writer who only knows, for example, Ventura Publisher may not be hired if another candidate knows a half dozen or more systems; versatility is the key to getting and keeping a job. The more systems a technical writer can use, the more valuable that writer becomes to an employer (Gmelin 1991).

16 pt. Times Roman Bold

Writers Need Familiarity with Several DTP Systems

12 pt. Times Roman

In 1991, more than 1,500 DTP systems were available (double the number available in 1989), and writers must understand how to use as many of them as possible. A writer who only knows, for example, Ventura Publisher may not be hired if another candidate knows a half dozen or more systems; versatility is the key to getting and keeping a job. The more systems a technical writer can use, the more valuable that writer becomes to an employer (Gmelin 1991).

Figure 12-5 Type Size Variations

plus Helvetica, Chicago, Geneva, and typical typewriter fonts like Courier. The choice of font greatly influences the appearance of each page and the document as a whole; in addition, some fonts in the same type size take more space than others. Before you choose a font for your document, you should print out a page or two to see how the printed lines look on the page and how the spacing works. Figure 12-6 shows examples of both traditional and contemporary fonts with the names of each font printed in that particular font.

In general, Americans find type fonts called *serif* easier to read for the text because they have grown up reading this type of font. Serif fonts have little cross strokes on each letter that tend to join letters together as the eye moves over them. *Sans serif* type fonts are "cleaner" without those extensions, and they are often used for headings and subheadings or for visuals or short documents like brochures. Figure 12-7 shows the difference between sans serif and serif fonts with the letter *R*.

Most of the common type designs available on the computer are equally easy to read; you should print out a sample of text and heading choices to find a pleasing combination. In general, you should stay within one type family or limit variations to two families. Avoid fancy fonts like shadows and script except for very small examples because they tend to be distracting and hard to read.

Type Style

The term *type style* refers to those variations in type used for emphasis: boldface, all capitals, and italics. With a typewriter, underlining is often used in the same way.

The general rule is that style variations are best used to draw attention to individual words and phrases. When they are used in large blocks or too frequently on a single page, they lose their effect. Bold (darker) type and color type can be used effectively for headings and subheadings. All capital letters can be used for short headings but are hard to read in large chunks. Italics are effective when used within sentences to draw attention to individual words, and convention calls for italics for titles of books and periodicals. Italics show up less well than bold variations and, like

Times Roman	Helvetica
New York	**Chicago**
New Century Schoolbook	Geneva
Courier (typewriter)	Monaco

Figure 12-6 Common Type Fonts

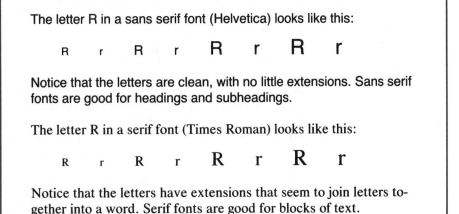

The letter R in a sans serif font (Helvetica) looks like this:

R r R r **R** r **R** r

Notice that the letters are clean, with no little extensions. Sans serif fonts are good for headings and subheadings.

The letter R in a serif font (Times Roman) looks like this:

R r R r **R** r **R** r

Notice that the letters have extensions that seem to join letters together into a word. Serif fonts are good for blocks of text.

Figure 12-7 Sans Serif and Serif Fonts

text in all capital letters, are hard to read in large chunks. Figure 12-8 (Van Briggle 1991, 8) shows the same section of text with three different style effects. You can also scan the pages of this book to see how varied type styles are used to highlight information and provide variety.

Headings

Most companies have a style guide that governs the kind and placement of headings and subheadings to show their position in the outline hierarchy. If you are planning the layout for a company document, you will want to consult that style guide. If you must choose the kind and placement of headings yourself, here is a typical arrangement:

SECTION 1
SECTION HEADING

1.1 FIRST-LEVEL HEADING
1.1.1 Second-Level Heading
1.1.1.1 Third-level heading. Text continues on the same line.

Note that the section heading is all capitalized, bold, and centered on the page. The text begins about three lines below the section heading. First-level headings are numbered, all capitalized, bold, stand-alone, and flush to the left margin. The heading number consists of two parts separated by a decimal point. Second-level headings stand alone flush to the left margin, have capital and lowercase letters, and are bold. Articles, prepositions, and conjunctions are not capitalized. Third-level headings are

Bold heading

Italics used to emphasize words

All capitals used for heading

Bold used to emphasize words

All capitals and underline used for heading

Underline used to emphasize words

Product- and User-centered Documentation
   Product-centered documents focus on a product's technical information, and such documents are organized according to the features of the product. *User-centered* documents, on the other hand, focus on the tasks a user must perform to operate a product successfully, and these documents are organized according to such tasks. In other words, user-centered documents are arranged according to the user's needs. For example, a tutorial arranged according to the steps a user must follow to learn a program is a user-centered document.

PRODUCT- AND USER-CENTERED DOCUMENTATION
   Product-centered documents focus on a product's technical information, and such documents are organized according to the features of the product. **User-centered** documents, on the other hand, focus on the tasks a user must perform to operate a product successfully, and these documents are organized according to such tasks. In other words, user-centered documents are arranged according to the user's needs. For example, a tutorial arranged according to the steps a user must follow to learn a program is a user-centered document.

PRODUCT- AND USER-CENTERED DOCUMENTATION
   Product-centered documents focus on a product's technical information, and such documents are organized according to the features of the product. User-centered documents, on the other hand, focus on the tasks a user must perform to operate a product successfully, and these documents are organized according to such tasks. In other words, user-centered documents are arranged according to the users' needs. For example, a tutorial arranged according to the steps a user must follow to learn a program is a user-centered document.

Figure 12-8 How Type Style Changes the Look of Text

run in with the text, use initial capitals, are bold, italic, or underscored, and end with a period.

Technical documents may use a decimal numbering system (as in the example above) because readers can then see at a glance where a numbered heading fits in the whole document. However, too many decimal places can be intimidating to generalist readers, so some documents use a modified system, combining numbers with heading size to indicate the hierarchy. This book, for example, uses only one decimal place (this is section 12.4) and uses type size to show subdivisions.

Unless your company style demands many decimal places to show levels of subordination, consider using paragraphs or bullets beyond the third or fourth level of subordination. The fourth level means four layers down from your major point; beyond that you begin to deal with very fine discrimination. For example,

1. _____
   1.1 _____
      1.1.1 _____
         • _____
         • _____
         • _____

Remember that the text must be comprehensible without the headings. In other words, the headings are not a part of the text but are really an embedded outline. Remember too that maintaining parallelism within each order of the headings will help the reader see relationships.

Because headings are signposts for readers, the chosen type size and font should help the readers understand the hierarchy. With carefully differentiated headings and subheadings, you may not need to use numbers at all. All capitals, especially in bold, make effective major headings *if* the headings are short. If the headings are long, all caps take too much room and are hard to read. Bold type is also good for subheadings, and if set bold, subheadings need not be four points larger than the text: They will stand out if increased by two points. Run-in subheadings (where the text continues on the same line) should be the same size as the text.

Designers often use sans serif fonts for headings and serif fonts for text, thus providing significant contrast between the two. Whatever option you choose, remember to be consistent throughout the document and to limit the number of fonts.

12.5 Ensuring Effective Design

Just as you must review content to ensure that it says what you want, so you must review design to ensure that it helps readers find information and invites them into the content. Four guidelines can help when you review.

1. Keep the design simple. Even though you have access to many fonts and sizes of type, style options like bold and italics, and highlighting devices like boxes and rules, use the minimum that will enhance the content.

2. Make the design consistent. Readers rely on repetition of design devices from section to section and page to page to help them know where they are.

3. Remember how readers move through a page. Readers of English read from top to bottom and left to right of each page. Therefore, it makes sense to put beginnings at the top and endings at the bottom; to put "before" on the left and "after" on the right. Because readers start looking at the top of the page, important information and headings should go there if possible. Headings should not be left by themselves at the bottom of a page; they should have at least two lines of text after them. Likewise, graphics are better placed near the top of the page than at the bottom.

4. Edit for format as well as for content. Because format involves so many details, it's easy to forget what you intended to do as you move through the document. Therefore, when you have finished formatting, you should systematically go through the whole document checking headings, type sizes and styles, rules, boxes, screens, and other design features for accuracy and consistency.

Good design does not occur by chance. It needs to be planned, reviewed, and revised if necessary. But good design can enhance any document, no matter how long or short, simple or complex. Senior technical writer Antonia Van Becker says, "I spend a lot of time adjusting leading or re-editing a passage to make room for complete paragraphs or file output examples on each page." Probation officer Judy Hopkins says, "My advice to students is simple: Understand the format. It will showcase your investigation and opinions. The format or scaffold, if it is well designed, can support as much weight as you need to put on it."

Checklist for Designing the Document

1. In designing the whole document, have I considered the following?
   __ the document's purpose and the reader's purpose for reading it
   __ conditions under which the reader is reading
   __ the reader's type and identity
   __ available hardware and software tools and the budget
   __ conventions dictated by the document type or company style
2. As I design each page, have I considered each of the following items?
   __ margins that will adequately frame the text
   __ line length that is easy to read (42 to 70 characters)
   __ line spacing that makes text more readable
   __ unjustified (ragged right) margins unless I have a sophisticated word processor and desire a formal appearance
   __ white space to frame lists, set off headings, or highlight figures and tables
   __ lists (numbered or bulleted) to group and set off related items
   __ boxes, rules, or screens to highlight or emphasize
   __ color (if available) to highlight, link, or reveal the document structure
3. In choosing type size, font, and style, have I considered the following?
   __ a type size between 8 and 12 points for text, or one that is equally easy to read
   __ a font that reads well and is consistent with the type of document
   __ serif fonts for large blocks of type; sans serif fonts for special uses or special documents
   __ style variations (bold, italics, capitals) in small doses and for special effects
   __ headings and subheadings that are either numbered or that show by type size and style where they appear in the hierarchy
4. Before I submit or release my document, have I reviewed it for each of the following items?
   __ simple design
   __ consistency of design elements
   __ readability

EXERCISES

1. Find and bring to class an example of each of the following: a chapter of a text in your field, a professional or company report, a journal or magazine article. Analyze the design elements of each and be prepared to discuss what works and what doesn't work.

2. Work in collaboration with two classmates to analyze the documents you found for exercise 1. Present a group report on the best and worst design elements of each.

3. The following information is adapted from a section of the *UL Standard for Safety for Microwave Cooking Appliances* (1981). Except for paragraphing, the visual markers in the original have been removed. Read the material carefully, seeking to understand the intent of the writer. Then determine what design features you would include and insert them. Be prepared to discuss in class.

40.1   The door and door interlock system, in conjunction with the complete appliance, shall be capable of completing 100,000 cycles of operation (opening and closing the door) for a household microwave cooking appliance and 200,000 cycles for a commercial microwave cooking appliance. During and at the conclusion of the test the performance of the system shall meet the following criteria: There shall be no electrical or mechanical malfunction that could result in the likelihood of fire, electric shock, personal injury, or excessive radiation emission. Only parts that fail safe and prevent completion of the test may be replaced or repaired. There shall be no loosening or shifting of adjustments or parts that could result in the likelihood of fire, electric shock, personal injury, or excessive radiation emission. Radiation emission shall not exceed 5 mW/cm² at any point 5 cm or more from the external surface of the enclosure.

40.2   Compliance with the requirement of paragraph 40.1 is to be determined by subjecting 2 complete samples of the appliance to the test described in paragraphs 40.3–40.5.

40.3   The appliance is to be connected to a voltage source as described under the input test section (paragraph 35.2). Any timers incorporated in the appliance are to be by-passed to facilitate uninterrupted operation of the appliance. The door interlock switches shall be operational during the test except that external controls (such as externally operated switches wired in parallel with the interlock switches) may be provided in order to by-pass each door interlock while microwave radiation emission measurements are being taken. The external controls shall not affect nor take part in the operation of the appliance in any other way.

40.4   The microwave cooking appliance door is to be cycled by an automatic device attached to the door handle or in the operating area of the door handle. The door closure force is to be predetermined so as to simulate intended operation. In no case shall the cycling mechanism apply a closure force greater than 20 lbf (89 N) unless agreed to by the manufacturer. The door is to be swung open from the closed position to any angle within 135 degrees to 180 degrees, or to within 10 degrees of its maximum travel, whichever is the smaller angle. A sliding door shall be opened to within 1 inch (25 mm) of its full travel. The cycling rate is not specified, but there shall be sufficient on time to permit generation of microwave energy that may be delayed by warm-up time of the magnetron, a time-delay circuit, and so forth.

40.5   The appliance shall be operated in accordance with the following sequence: An initial microwave radiation emission measurement shall be made as described in Section 34. A fresh coating of corn oil is then to be applied to all door sealing surfaces and any microwave absorbing load (such as a dry brick load) shall be placed and maintained in the oven cavity. The appliance is to be energized and the door operated for 10,000 cycles. Following the 10,000 cycles of operations, the cycling load is to be removed. Microwave radiation emission measurements are to be made as described in paragraphs 34.2–34.8. A microwave cooking appliance incorporating a resistance element in the cavity, a self-cleaning feature, or if intended for use in conjunction with another heating appliance shall, in addition, be operated as described in the temperature test until temperatures stabilize. Following the operation, microwave radiation emission measurements are to be repeated. After the microwave radiation emission measurement is made, the door sealing surfaces are to be cleaned of the corn oil with a soft cloth. The above procedure is to be repeated until a total of 100,000 cycles of operation are accumulated on a household appliance, and 200,000 cycles of operation on a commercial appliance.

4. The following process description, "Combine Harvester and Thresher" (1967), is intended for a general reader—not a farmer or a mechanical engineer. Its purpose is to explain how the combine harvester and thresher works. Follow these steps:

1. Read the whole process description.
2. Write down the single major point.
3. List the major subpoints.
4. Group minor points under the proper subpoints.
5. Rewrite and include the design elements you would use.
6. Indicate any drawings or figures that would be helpful.

You may find this exercise easier to do if it is a collaborative activity with two or three classmates.

The combine harvester is a combination of a grain harvesting machine and a threshing machine. The grain is cut, threshed and cleaned in one operation. The machine may be self-propelled or towed by a tractor.

The wheatstalks are cut by an oscillating knife while the revolving reel pushes them back towards the knife and auger (feed screw). Grain flattened by wind or rain is raised by the spring prongs on the reel. The cut wheat is conveyed into the machine by the auger and reaches the threshing cylinder which rubs the grain out of the heads against a "concave." The grains of wheat, together with the chaff and short fragments of straw fall through the interstices between the bars of the concave and into the cleaning "shoe." Some of the grain is carried along with the straw, is stopped by check flaps, and is shaken out of it on shaking screens on the straw rack of the machine. The straw drops out of the back of the machine and is left in a windrow for later baling, or is baled directly by a baling attachment or press, or is scattered over the ground by a fan-like straw spreader. The grain shaken out of the straw is also delivered to the cleaning shoe. In the shoe the grain is separated from the chaff and cleaned by sieves and a blast of air. The chaff and fragments of straw are thrown out from the back of the machine. The grains of wheat fall through the sieves and into the cleangrain auger (screw conveyor) which conveys them to an elevator and on into the storage tank or into bags. Any heads of wheat which fail to go through the sieves and are thrown backwards by the air blast fall short in comparison with the lighter chaff and drop into a return auger which, via an elevator, returns them to the threshing cylinder. Correct adjustment of the air blast—by throttling the intake of the fan and by altering the setting of baffles—is important in determining the degree of cleaning of the grain and the magnitude of the grain losses that occur.

REFERENCES

Combine harvester and thresher. 1967. *The way things work*. New York: Simon and Schuster, 432–433.

Felker, David et al. 1981. *Guidelines for document designers*. Washington: American Institutes for Research.

Gmelin, Ann. 1991. *Desktop publishing: Evolution, trends, and influence on technical writing*. Unpublished document. San Jose State University.

Linotype. 1988. *The pleasures of design: A practical guide to layout and typography in desktop publishing*. Hauppauge, NY: Linotype Co.

Tandem Computers. 1989. *Corporate style guide*. Number 17864. Cupertino, CA: Tandem Computers, Inc.

Underwriters Laboratories. 1981. *Standard for safety: Microwave cooking*. Number UL923. Northbrook, IL: Underwriters Laboratories.

Van Briggle, Gayle. 1991. *Document design: Increasing document readability*. Unpublished document. San Jose State University.

White, Jan V. 1990. *Great pages: A common-sense approach to effective desktop design*. El Segundo, CA: Serif Publishing Company.

Chapter 13

Documenting Your Sources

As a college student, you have probably written reports that require library research, references to or quotations from that research, and a bibliography of source materials. Many students think that the only reason for such required quotation and bibliographic documentation is to prove to the instructor that they know how to use the library; they resent the emphasis on details of spacing and punctuation in bibliographic entries, and they are sometimes tempted to "invent" missing information to complete an entry in order to save another trip to the library.

However, professionals who write know that documenting sources is more than an academic exercise. Carol Gerich, senior editor at a national research laboratory, says, "We take the citing of references very seriously. Our laboratory publishes a document titled *The Scientific Report: A Guide for Authors* that states, 'References to other works are frequently useful in technical reports. They enable you to call attention to supporting evidence for your statements in the briefest possible way.' Our authors document their sources, and our editors always check the accuracy of that documentation. If the references are incomplete, someone has to search for the missing information."

Documenting sources is, therefore, more than showing you can use the library and follow details of spacing and punctuation. It is also an ethical responsibility to your reader. Whether you are writing as a student or as a professional, you always need to ask: What is my relationship to the reader? What does the reader want from me? On the job, you will be the expert, knowing more about the subject than the reader. Therefore, your reader looks to you for accurate information, clear instructions, and reliable recommendations. In a way, you've been working for the reader while gathering information from sources like reports, letters, memos, articles, brochures, books, videos, interviews, experiments, direct measurements, and hands-on experience. When you write, you synthesize all the information into something new, and if you work carefully, the reader will accept what you say. But you also need to tell the reader the sources of your information for three important reasons: (1) to establish your credibility by citing support for your statements and conclusions, (2) to enhance your credibility by acknowledging those whose work influenced your own—honestly stating what is yours and what comes from someone else, and (3) to inform readers where they can find more or related information.

13.1 Definitions of Documentation Terms

In this book, the term *documentation* applies to the process of identifying your sources, but a similar often-used term is *citation*. To document is to furnish with evidence; the words *document* and *documentation* have as their root the Latin word meaning "teach." Scientific and technical papers seldom quote another source directly, usually summarizing or paraphrasing the key ideas. The sources of these summaries or paraphrases are indicated in two ways: (1) *within* the text, usually at the first natural sentence break following mention of the information, and (2) at the *end* of the text in a references, works cited, or references cited section. This final list contains complete information about the source that will allow readers to find it on their own.

The ways of documenting the source—both within the text and at the end—vary depending on the chosen system. The three major systems are explained in sections 13.4, 13.5, and 13.6.

Documentation is primarily found in journal articles, books, and formal technical reports. Even manuals, procedures, and descriptions can require documentation, however, if they contain material from some other source. It's important to acknowledge sources of information. For one thing, plagiarizing (using someone else's words or ideas as your own) is a form of stealing, and the penalties can be severe. If you plagiarize on the job, you could destroy your career and jeopardize the organization for which you work. If you plagiarize as a student, you could fail a course or even be expelled. In addition, you will be more believable if you admit that all your impressive information did not simply spring out of your own brain. In both science and technology, you need to convince readers that you know what's already been done and said on a subject; in fact, this increases the likelihood that the reader will be persuaded to accept your conclusions and recommendations.

Note that the term *documentation* is also used in industry to identify a series of publications that provide instruction for users of particular products. *Documentation* can also mean written records of what was done in a procedure or experiment. These specialized uses of the term are different from the one used in this chapter.

13.2   What to Document and Where to Find the Information

In order to serve your reader's needs, you need to know both what should be documented and where you can find that information.

What to Document

As you write, it pays to keep track of what you should document, because it's easier to note the sources correctly in your first draft than it is to go back and find the information later. Either write the source on the page itself or keep a separate index card for each source, keying it to the draft. All of the following kinds of information should be documented.

Direct Quotations

Though technical and scientific writers seldom quote an authority directly, if you do, you must identify the source of the quotation. Quoting directly means using a source's exact words, punctuation, and spelling. Such quotations are enclosed in quotation marks.

Paraphrases or Summaries

To *paraphrase* is to restate someone else's words in your own. To *summarize* is to briefly state the main points. Students often fail to document a paraphrase or summary since they are putting the information into their own words. Nevertheless, if the information or the idea is someone else's, you should document it. For example, suppose your information source (Shaw 1987, 96) says:

> It is now beyond question that even in rural areas acid deposition (both wet and dry) almost always stems from the activity of human beings: primarily the combustion of fuel for power, industry, and transportation.

You condense and rewrite the explanation this way:

> **Scientists are now sure that air pollution both from acid rain and from acid particles (called acid deposition) nearly always comes from human activity, mostly the combustion of fuel.**

Even though almost all the words are your own, the major point (as well as the authority for scientists being "sure") comes from your source, so the information should be documented.

Controversial Statements or Positions

It's important to give the source of any position or statement that is arguable. Your readers may want to examine the sources themselves in order to understand the controversy in more detail or to check on a related point.

Primary Source Material

If your information comes from personal—as opposed to published—sources, readers need to know the type of source, the identity of the informant, and the date. Sources can range from data sheets to personal letters to interviews to E-mail. Names of informants lend credibility and may allow the reader to find more information by contacting the source. Dates establish the immediacy of the information. In documenting dates, it's advisable to write them in military style (9 April 1993) instead of using a number for the month; that way there is no confusion about which number is the month and which number is the day.

Where to Find the Information

The best place to find the information to include in your list of sources is the original printed source—the title page of a book or manual, the table of contents of a journal or magazine. For a book, you always need to record the following:

- author and/or editor
- title and subtitle (if there is one)
- other pertinent information, such as volume number, edition, translator
- city of publication
- publisher
- date of publication

For an article, you need to record the following:

- author and/or editor
- article title and subtitle (if there is one)
- journal, manual, or newspaper title
- volume number and issue number (if there is one)
- date
- inclusive pages of the article

If more than one city of publication is listed, note only the first. If the city of publication is not well known, list the state or country as well (Menlo Park, CA). If you list a company's manual, give its order number or document number and the date. For unpublished information, you need the following:

- informant's name (and sometimes title, department, or company)
- type of source material (specification sheet, brochure, fax, letter)
- date

13.3   How to Choose a Documentation System

Unfortunately, no one documentation system is universally accepted by writers even in the physical sciences, let alone the social sciences and humanities. Thus, you must choose the documentation system most likely to be acceptable to the specific readers for whom you are writing. Here's how to make that choice:

- If you are working, ask to see either the company style guide or a previous paper that has an accepted documentation system, and follow the suggested system.
- If you are submitting an article to a journal, you can (1) look in the journal for the "Guidelines for Authors," (2) write and ask for such guidelines, or (3) study current issues of the journal to see what documentation system is used and then follow the examples.
- If you are a student, ask your writing teacher or professors in your major field what documentation system they recommend and where it is explained, and follow that system exactly.

You have choices among several systems, but once you've chosen a system, follow it exactly. Don't try to choose a bit from here and a bit from there. As *The Chicago Manual of Style* points out (1982, 440), a consistent style must be followed "because inconsistency in bibliographical details confuses readers and suggests careless research methods."

If your major field follows the documentation system of a particular style guide, you may want to buy that style guide or borrow it from a library. Be sure to use the most recent edition. Some of the common style guides in science and technology are the following:

- American Chemical Society. *Handbook for Authors of Papers in American Chemical Society Publications*.

- American Institute of Physics. *Style Manual for Guidance in the Preparation of Papers.*
- American Mathematical Society, *A Manual for Authors of Mathematical Papers.*
- American Medical Association. *Style Book/Editorial Manual.*
- Council of Biology Editors. *CBE Style Manual: A Guide for Authors, Editors, and Publishers in the Biological Sciences.*

Four other widely known style guides are:

- American Psychological Association. *Publication Manual of the American Psychological Association.*
- Modern Language Association. *MLA Handbook for Writers of Research Papers.*
- University of Chicago Press. *The Chicago Manual of Style.*
- United States Government Printing Office. *Style Manual.*

The three common documentation systems are the *author-date* system, the *author-page* system, and the *number* system. The author-date system identifies the source in the text by the author's last name followed by the year of publication (Davis 1991). The author-page system is similar but gives the author's name and the page number (Davis 342). The number system simply numbers the source and puts that number in parentheses within the text (6). For most systems, the list of sources is alphabetized by the author's last name. In the number system, the numbers are sometimes assigned in order as the citations appear, and the sources are listed in that order.

The author-date system is widely used in the biological, physical, and social sciences. The author-page system is used primarily in the humanities and is the system supported by the Modern Language Association (MLA). The number system is often used in the sciences because it minimizes distraction for the reader and keeps printing costs down in journals by eliminating the extra words within the text. This book uses the author-date system.

13.4 Author-Date System

The author-date system is used in fields where writers usually reference whole documents or sections of long works. Details are explained in *The Chicago Manual of Style*. In the author-date system, you indicate the source in the text of your document at a natural break or at the end of the sentence following the first mention of the information. For example:

> Limitations of adaptive optics include (1) loss of some of the telescope's gathered light in the optical systems used to correct the image, and (2) the complicated corrections needed to correct the starlight in the wide swath of turbulent air observed with a big mirror (Powell 1991).

The complete reference is then listed at the end of the document in a list titled *References*. The source cited in the example would be listed in this way:

> Powell, Corey S. 1991. Mirroring the cosmos. *Scientific American*, November, 112–23.

Because all the references are alphabetized by author's last name, that is listed first, followed by the first name or initials. The year of publication is given next because that helps identify the source. The article title is given in full, followed by the name of the periodical, the date, and the inclusive page numbers.

Several minor variations of this order can be used, but the important thing is to be consistent. Generally, *within the text* you include in parentheses the author's last name and the date with no punctuation. You can also add a comma followed by the page number, and if you are quoting directly, you must include the page number. In the list of references, you include information in the following order—each major bit of information separated by a period.

Books

> Author's Last Name, Author's First Name or Initials. Year. *Title of book*. City: Publisher.

Reports and Journal, Magazine, and Newspaper Articles

> Author's Last Name, Author's First Name or Initials. Year. Title of article. *Title of Journal or Magazine*. volume number (or date): inclusive page numbers.

Notice that all lines of the entry other than the first are indented two spaces. The date, if given in addition to the volume number, appears in parentheses. In titles of articles and books, only the first word and any proper names of the title and any subtitle are capitalized. In titles of journals, all major words are capitalized. Titles of books, reports, journals, and magazines are either underlined or italicized. Specific examples follow based on *The Chicago Manual of Style*.

Book with a Single Author

> Horton, William K. 1990. *Designing and writing online documentation: Help files to hypertext*. New York: Wiley.

Work with Two or More Authors

> MacNeill, Jim, Pieter Winsemius, and Taizo Yakushiji. 1991. *Beyond interdependence: The meshing of the world's economy and the earth's ecology*. New York: Oxford Univ. Press.

Two Works by the Same Author

Note that the works are arranged chronologically by date of publication. For the second book, the author's name is replaced by three hyphens or a three-em dash.

> Peters, Robert Henry. 1983. *The ecological implications of body size.* New York: Cambridge Univ. Press.
> ———. 1991. *A critique for ecology.* New York: Cambridge Univ. Press.

Two Works by the Same Author in the Same Year

Note that the works are arranged alphabetically by the first major word of the title— lowercase letters *a* and *b* differentiate one work from another.

> Collins, Randall. 1988a. *Sociology of marriage and the family: Gender, love, and property.* Chicago: Nelson-Hall.
> ———. 1988b. *Theoretical sociology.* San Diego: Harcourt Brace Jovanovich.

Journal or Magazine Article

> Michaelson, Herbert. 1990. How an author can avoid the pitfalls of practical ethics. *IEEE Transactions on Professional Communication* 33 (June): 58–61.
> Jewell, Jack L., James P. Harbison, and Axel Scherer. 1991. Microlasers. *Scientific American,* November, 86–94.

Report

> U.S. Congress, Office of Technology Assessment. 1991. *Rural America at the crossroads: Networking for the future.* OTA-TCT-471. Washington: U.S. Government Printing Office.

Manual

> Hewlett-Packard. 1991. *Writing style guide.* Palo Alto, CA: Corporate Learning Products.

Television Program

> Pearl Harbor: Two hours that changed the world. 1991. Host David Brinkley. ABC. 5 December.

Personal Communication

References to personal communications are best given only in the text itself. If they are included in the references, they are listed this way:

> Lundquist-Frank, Kristin. 1991. Letter to author. 21 October.
> Stewart, Amy. 1992. Interview with author. Austin, TX. 3 March.

To summarize, if you plan to use the author-date system, you should (1) identify the source within the text by putting the author's last name and the year of publication in parentheses, and (2) arrange the list of references alphabetically, following the examples in this section *exactly*, including punctuation and spacing. For most documents you will write as a student, this summary of the author-date system will give you enough detail. However, if you need more specifics, consult *The Chicago Manual of Style*.

13.5 Author-Page System

The author-page system is used in fields where writers need to quote or cite specific information; it sends the reader directly to the page on which the information is located. This system is explained in detail in the *MLA Handbook for Writers of Research Papers*, and explanations from that handbook are summarized here.

Like the author-date system, the author-page system places references within the text in parentheses, in this case the author's last name and the page number. Do not separate them with commas. For example:

> You can divide information by chunking it in short paragraphs, a dozen printed lines or so. You can then test if the paragraph is a logical unit by putting a temporary title above it. Now, is that what the paragraph is really about? If so, then you might want to leave the title there (Bell 87).

The reference is to page 87. The source is a chapter by Paula Bell in a book edited by Thomas T. Barker. In the section called *Works Cited*, this source would be listed in full, alphabetized by the author's last name in the following way:

> Bell, Paula. "Cognitive Writing: A New Approach to Organizing Technical Material." *Perspectives on Software Documentation*. Ed. Thomas T. Barker. Amityville, NY: Baywood, 1991.

Note that the general form is similar to the author-date system, but in this style the publication date is moved to the end of the entry, the chapter title is in quotation marks, and key words in the chapter and book titles are capitalized. Here are the general forms to follow for entries in the list of works cited.

Books

> Author's Last Name, Author's First Name. *Title of Book*. City: Publisher, date.

Reports and Journal, Magazine, and Newspaper Articles

> Author's Last Name, Author's First Name. "Title of Article." *Title of Journal or Magazine*. Volume number (year): inclusive page numbers.

Notice that each major entry is followed by a period. Titles of books, reports, magazines, and newspapers are underlined or italicized. Lines after the first are indented five spaces.

In compiling the list of works cited, alphabetize entries by the author's last name. If the author is not identified in your source, alphabetize by the first major word of the title. Sample entries for various kinds of sources follow.

Book with a Single Author

> Horton, William K. *Designing and Writing Online Documentation: Help Files to Hypertext.* New York: Wiley, 1990.

Work with Two or More Authors

Note that only the first author is listed last name first; the others are listed first name first.

> MacNeill, Jim, Pieter Winsemius, and Taizo Yakushiji. *Beyond Interdependence: The Meshing of the World's Economy and the Earth's Ecology.* New York: Oxford UP, 1991.

Two Works by the Same Author

Note that a line made with three hyphens or a three-em dash replaces the name for the second entry.

> Peters, Robert Henry. *A Critique for Ecology.* New York: Cambridge UP, 1991.
> ———. *The Ecological Implications of Body Size.* New York: Cambridge UP, 1983.

Journal or Magazine Article

> Jewell, Jack L., James P. Harbison, and Axel Scherer. "Microlasers." *Scientific American* Nov. 1991: 86–94.

Article or Book with No Listed Author

Alphabetize by the first major word of the title.

> *Valley of the Drums: Bullitt County, Kentucky.* Washington: U.S. Environmental Protection Agency, 1980.

Television Program

> *Pearl Harbor: Two Hours That Changed the World.* Host David Brinkley. ABC. WPVI, Philadelphia. 5 Dec. 1991.

Data from Computer Search

If the information is from a computer online data search, add the vendor's name and system identification number.

> Sun, Qiangnan, ed. "Computer Aided Techniques in Manufacturing, Engineering, and Management." *Computers in Industry* 8. 2–3, (Apr. 1987): 111-269. DIALOG file 8, item 1766278 EI 8706055851.

Personal Communication

If the information comes from a letter, list the writer's name first, then the writer's title and company (if appropriate), "Letter to the author," then the date in military style, with the month abbreviated.

> Green, Frank. Manager of Engineering Documentation, XYZ Company. Letter to the author. 14 Apr. 1992.

Do the same thing for a personal interview, identifying it as such.

> Smith, May. Director of Marketing, RRR Company. Personal interview. 3 Mar. 1992.

In summary, if you plan to use the author-page system, you should (1) identify the source within the text by putting the author's last name and the page number in parentheses, and (2) compile the works cited section alphabetically, following the examples *exactly*, including punctuation and spacing. For most documents you will write, the examples given here will provide enough information; however, if you need more detail, consult the *MLA Handbook for Writers of Research Papers.*

13.6 Number System

The number system of documentation is often used in the sciences. A simple reference number is put in parentheses after the reference in the text, like this (12). Sometimes the relevant page reference is also included, separated from the reference number by a comma, like this (12, 243). For example, here is a text entry:

> One problem with satellite imagery is that areas with low amounts of vegetation may not show up on remote sensing images, yet these areas may be vulnerable to Desert Locust infestation (3).

The citation at the end of the report would read:

> 3. U.S. Congress. A plague of locusts. Washington: OTA-F-450; 1990.

The list of references (called *References Cited*) can be done in two ways: In the first method, the reference number used in the text refers to the position in a list alphabetized by the author's last name. This method is sometimes called the alphabet-number system. Here, the citation numbers will not be sequential in the text. In the second method, called the citation-order system, the reference numbers are given sequentially in the text, and the references themselves are listed in that order in the references cited. The number system is explained in detail in the Council of Biology Editors *CBE Style Manual*, which is the basis for the following summary.

The general form for entries in the references cited is similar to the other systems discussed in this chapter, with these exceptions:

1. Titles of books and articles are not capitalized (except for the first word), nor are book and journal titles underlined or italicized.
2. No quotation marks appear around article titles.
3. The date appears at the end of the entry, preceded by a semicolon.
4. The system is generally streamlined, often using initials instead of an author's first name, omitting subtitles, and abbreviating journal titles and publishers' names.

Here are the general forms to follow.

Books

1. Author's Last Name, Author's Initials. Title of book. City: Publisher; year.

Reports and Journal, Magazine, and Newspaper Articles

2. Author's Last Name, Author's Initials. Title of article. Title of journal. Volume number: inclusive pages of article; year.

Remember that this form will vary slightly from one journal to another, so be sure to consult a specific journal or style book in your field. The following examples show how to write entries for the list of references cited.

Book with a Single Author

1. Horton, W. Designing and writing online documentation. New York: Wiley; 1990.

Work with Two or More Authors

2. MacNeill, J.; Winsemius, P.; Yakushiji, T. Beyond interdependence. New York: Oxford; 1991.

Journal or Magazine Article

3. Jewell, J.; Harbison, J.; Scherer, A. Microlasers. Sci. Amer. (Nov.) 86–94; 1991.

Technical Report

4. U.S. Environmental Protection Agency. Valley of the drums. Washington: EPA-430/9-80-014; 1980.

Unpublished Letter

>   5. Kuhn, H. [Letter to F. Whitehouse, Lawrence Livermore Laboratory]. 13 Feb. 1992.

Interview

>   6. Purcell, E. Interview with author. St. Louis, MO. 7 Nov. 1992.

To sum up, if you plan to use the number system, you should (1) identify the source within the text with a number in parentheses, and (2) compile the References Cited, either alphabetically with the numbers keyed to the source or sequentially by the order of the citations. In either case, follow the samples in the style guide *exactly*, including punctuation. If you need more detail, consult the *CBE Style Manual* or *The Chicago Manual of Style*. The two style manuals differ slightly in details.

13.7 Annotation of Reference Lists

When you get ready to document your sources, consider the advantages of annotating each source. In an annotated reference list, you follow the bibliographic information with a very brief summary of the contents of the article or book. Annotations seldom run more than 200 words and are often shorter, but they provide valuable information for the reader because they go well beyond the scanty identification provided by a title. Figure 13-1 shows several entries from an annotated bibliography by a student (Van Briggle 1991). This student followed the author-date system of documentation.

Checklist for Documentation

1. What specific material in my manuscript should be documented? Do I have any of the following?
    ___ direct quotations
    ___ paraphrases, summaries, or ideas from my sources
    ___ controversial statements or positions
    ___ primary source material
2. Which documentation system is commonly used in my field or recommended by my instructor?
    ___ author-date
    ___ author-page
    ___ number

REFERENCES

> Duin, Ann Hill. 1989. Factors that influence how readers learn from text: Guidelines for structuring technical documents. *Technical Communication* 36 (2): 97–101.

This article provides a good overview of the following elements of document design: audience's prior knowledge, organization, reader's goals and future tasks, study strategies (rereading, outlining, mnemonics), monitoring comprehension, and textual effects (structure, coherence, unity and audience appropriateness). The author ends with a suggested framework for designing documents, which includes analysis of audience, purpose, and effect.

> Holland, Melissa, Veda Charrow, and William Wright. 1988. How can technical writers write effectively for several audiences at once? *Solving problems in technical writing.* New York: Oxford Univ. Press.

The authors define reading ability, goals of reading, and different reading tasks. They recommend techniques for treating multiple audiences, and they conclude with case studies that demonstrate effective use of the recommended techniques.

> Huckin, Thomas. 1983. A cognitive approach to readability. *New essays in scientific communication: Research, theory, and practice.* Farmingdale, NY: Boynton.

The author proposes a new definition of readability and describes "relevant research in cognitive psychology." He also discusses theory, textual features, reader variables (reading styles), and readability guidelines, as they are related to or derived from cognitive psychology.

Figure 13-1 Annotated List of References

EXERCISE

For practice in using the style of documentation preferred in your major field or one recommended by your instructor, organize, correct, and rewrite the following citations:

> Flexer, Carol, Denise Wray, and JoAnne Ireland. 1989. Preferential Seating Is Not Enough: Issues in Classroom Management of Hearing-Impaired Students. *Language Speech, and Hearing Services in Schools.* (January): 11–21.

Hawkins, PhD, David B. and James F. Fluck MA. 1982 A Flexible Auditorium Amplification System. *ASHA*. Vol. 24. No. 4: 263–267.

Laws of the 93rd Congress—First Session. "Rehabilitation Act of 1973." *United States Code: Congressional and Administrative News*, Vol. 1, 1974: 408–409.

Phillips, Gene. Interview with Artemisa Valle. San Jose California, 27 March 1989.

Vaughn, Gwenyth R., PhD. 1986. "Assistive Listening Devices and Systems (ALDS) Come of Age." *Hearing Instruments*, Vol. 37, No. 7: 12–14.

 WRITING ASSIGNMENT

Consult your writing instructor or a professor in your major field to find out what documentation system to choose for your term project. Then follow that system exactly to document your sources. If you are working on a collaborative project, decide on the documentation system to use before writing the draft. Assign each writer to provide full information for his or her written material. Also assign one person in the group to prepare the in-text references and the final list of references; assign another to review that documentation.

REFERENCES

CBE Style Manual Committee. 1983. *CBE style manual: A guide for authors, editors, and publishers in the biological sciences*. 5th ed. revised. Bethesda, MD: Council of Biology Editors, Inc.

The Chicago Manual of Style. 1982. 13th ed. Chicago: Univ. of Chicago Press.

Gibaldi, Joseph and Walter S. Achtert. 1988. *MLA handbook for writers of research papers*. 3rd ed. New York: Modern Language Assn.

Shaw, Robert W. 1987. Air pollution by particles. *Scientific American*, August, 96.

Van Briggle, Gayle. 1991. *Document design: Increasing document readability*. Unpublished report. San Jose State University.

Chapter 14

Writing for International and Multicultural Audiences

"Thinking globally and writing for an international audience may well be a communicator's greatest challenge in the nineties," according to writing consultant Gloria Sciuto (1990). Sciuto works in the Boston area, but her words are echoed by technical writers across the United States. In Illinois, university extension coordinator Karen Zotz writes about nutritional and family concerns "in a language the public can understand" (1991). Sometimes, she says, that information must be translated into other languages, even though her clients live in the United States. In California, Roger Connolly directs a group of writers at the American branch of a Japanese computer company. His group prepares English versions of original Japanese documents both for use in English and for translation into other languages. "Writing for non-native speakers of English," he says, "requires that we choose words and sentence structures very carefully" (1991).

In Chapter 1, you learned the importance of focusing on your reader at the planning step. You analyzed readers by type: generalists, managers, operators, technicians, and specialists. You also looked at where in a company readers worked because you learned that people in different jobs have different needs. In the 1990s, however, most of you will need to add another concern to your reader focus: the needs of multicultural and international audiences.

How important is it to think of other cultures and other languages? Here is just one example. In the first quarter of 1991, Apple Computer, Inc., reported that over 50 percent of its revenues came from the international market. Apple documentation has been translated into 35 different languages, and many documents are routinely translated into from 15 to 20 languages (Allen 1991). In addition, large numbers of international users read Apple documentation in English, but often, English is their second language, and their language proficiency varies.

Therefore, whether your employer supplies goods (paring knives, machine tools, bridges) or services (education, social assistance, banking, insurance), the chances are increasing that you will be writing for readers in and out of the United States whose cultural and linguistic background is different from your own.

14.1 Understanding the Global Marketplace

What is the global marketplace? Many large American companies report that half their annual sales are now outside the United States, and nearly every company sells products or services to clients within the United States from a variety of language groups and cultures. Thus, the global marketplace requires that we think of both multilingual and multicultural communications.

Amana Refrigeration, Inc., for example, sells its major appliances in Portugal; according to an Amana spokesperson, "The distributor there says they can sell our products more readily if we stop assuming that everybody speaks English" (Deutsch 1991). Even those who do speak English have varying language ability. A 1990 survey by the Hong Kong Chamber of Commerce, for example, found that "new, young employees in 70 percent of the companies contacted encountered English language difficulties (both spoken and written). Companies responding to the questionnaire included those in banking, shipbuilding, construction, trading, the service sector, and the legal and accounting fields. The total work force of these companies was about 85,000" (Winters 1992). And language aside, "Our cultures manifest themselves in our information needs and our styles of communication. In other words, our cultures define our expectations as to how information should be organized, what should be included in its content, and how it should be expressed" (Hein 1991).

Unclear and unorganized writing in English or other languages and insensitivity to other cultures can restrict company sales, lose customers, and even cause legal problems. According to Berlitz Translation Services, "Languages define and reflect the cultural boundaries of the different people with whom you do business. Crossing these boundaries requires a thorough knowledge and understanding of the cultural differences of your readers" (n.d.). In other words, to meet all your readers' needs, you need to understand the global marketplace.

As a global communicator, you need to consider at least four potential audiences:

1. readers living in the United States who understand American culture but who learned English as a second language
2. international readers living in countries where English is the first language, but it is not American English (for example, Britain, Australia, Canada, parts of Africa and India)
3. readers living outside the United States who read English as a second language and do not know American culture or understand American idioms
4. readers outside the United States who use translations into their own languages

This chapter will show you how to write in English so you can better meet the needs of all four reader groups.

14.2 Defining Key Terms

In reading this chapter, there are a number of terms you may need defined. *Culture* means "ways of living": those customs, attitudes, and beliefs built up by a group of people and transmitted from one generation to another. *Languages* are "ways of communicating," with various groups of people having their own languages. *Translators,* who know both the *source* language and the *target* language, determine "what the author is trying to communicate in the source language and then render this accurately into the target language" (Berlitz n.d.). Translation is also increasingly accomplished by computer—a process called *machine translation.*

Translation is thus an important step in communicating with multicultural and international audiences, but beginning in 1989, American companies began to move beyond translation to localization and globalization.

The term *localization* describes the process of adapting a piece of writing or a product to both culture and language, thus making it seem local. Graham Vanentine of Hewlett-Packard Switzerland differentiates the terms this way: "Translation is purely the conversion of words. Localization is the process of giving your product the look and feel of a locally built product" (qtd. in Allen 1991). Through localization, all the sounds and colors, all the religious, social, and political references of the target language are adapted, and the accent of the source language is lost (Cyphers 1991). For example, Fisher Controls, a subsidiary of Monsanto Company, already has customer training guides in French, Spanish, and Portuguese. Now it is tailoring those guides to match the way Spanish and Portuguese are spoken in Colombia and Brazil. "We want to accommodate any culture that uses our equipment," says Fisher marketing manager Joseph M. Morris (Deutsch 1991).

The concept of *globalization* is broader: it involves thinking of yourself as an employee of a worldwide company and aiming to fully communicate with the other employees, the customers, and with anyone else interested in your company's product or service. This broadest concept of communication requires (1) recognition that not everyone speaks and reads English (or does it well) and (2) awareness of the ways in which other cultural groups structure documents and carry on business. Even within the United States, writers must be global communicators—meeting the needs of a pluralistic society composed of groups with very different languages, cultures, backgrounds, and beliefs.

14.3 Considering Cultural Differences

Awareness is the key to dealing successfully with people from other cultures, because many things taken for granted in the United States are done differently in other parts of the world. This section discusses some of the general areas that require your awareness; sections 14.4 and 14.5 deal specifically with writing, design, and graphics concerns.

Geography, Environment, and Culture

In setting your writing on an international stage, you need to avoid stereotypes. Many of us characterize other countries with convenient visualizations that are actually stereotypes: French cathedrals, Chinese sampans, Islamic mosques, African grass huts, Brazilian jungles. At the same time, we have difficulty seeing ourselves cast in stereotype. Are you comfortable seeing an Old West cattle ranch as a symbol of the United States? Or a big-city slum? Or Hollywood?

In the business world, most readers—in all countries—will live in an urban setting much like that of any large U.S. city, will drive the same kinds of vehicles, and will work in offices with similar equipment. However, those readers will not have the same phone service or mail service, and the pace and structure of their business day may be different. To communicate effectively, you need both to accept the similarities and account for the differences.

Therefore, in your writing it's important to avoid geographical, environmental, or social references like these:

- General locations like the "South" or the "East Coast." In the United States, these words have specific geographical and historical meanings; elsewhere, they could mean nothing or could be ludicrous in context.

- Specific terminology like the "Twin Cities," the "Sun Belt," the "Rocky Mountain area," or the "Tennessee Valley," which have little meaning to readers in Brisbane, Munich, or Osaka.
- Seasons. It's not cold everywhere in January; fall is not always in October; there may, in fact, be no fall season.
- Holidays. Not all countries celebrate Christmas; Halloween is unique to the United States.
- The work week. Not every country observes an eight-hour day or a "weekend" of Saturday and Sunday.

Politics and History

Unless your writing is directly concerned with political or historical information, you probably should avoid such references. The breakup of the Soviet Union and the unification of Germany are just two examples of political change that could make your writing obsolete and inaccurate. Readers in countries that have been invaded, colonized, or repressed in their past will resent images of that repression. Therefore, images like rickshaws, queues, and pith helmets, or symbols like the hammer and sickle, the swastika, or the rising sun are inappropriate in your writing.

Numbers, Dates, Addresses, and Colors

While numbers may seem to be totally objective, not all cultures use the same measurement systems. Here, it's the United States that is out of step: internationally, the metric system is widely used (as it is in American science and engineering). But if you're writing for generalist readers, you may find yourself writing about miles per hour and gallons of gasoline; about yards of cloth or cups of sugar; about foot pounds of torque or BTUs of energy; of degrees Fahrenheit or inches of mercury. International and multicultural readers may need those terms changed into those they can more readily comprehend. (Can you visualize how far it is to the next town if I tell you it is 42 kilometers? Can you estimate the length of a rope that is 2.5 meters?)

When numbers are used in dates and addresses in typical U.S. usage, they can be very confusing to international readers. In the United States, we often display the date as month, day, and year, but in Germany and most of Europe, it's day, month, and year. Other countries (Japan, for example) may use year, month, day. You can avoid this problem by always writing out the name of the month instead of giving it a number equivalent. Some company style guides also recommend writing the date military style, *9 September 1993*. Addresses in other countries are also organized very differently; it's important that you follow the local custom exactly, including punctuation. Countries in the former Soviet Union reverse the usual American order, beginning the address with the country and progressively becoming more specific. Many countries insert commas between house numbers and street names or numbers and between other items in an address. In currency, some countries reverse the positions of commas and decimal points; for example, U.S. custom calls for 2,303.14, but in many countries that number would read 2.303,14.

Research tells us that color in text improves readability and helps readers access information; color also adds interest. But some colors carry symbolic overtones. Green, for example, is considered a "good" color by Muslims; red and gold indicate celebration in India and China. But blue and white are associated with mourning in China, and black is associated with darkness and evil in India. To the Malays of Singapore, yellow and purple should be reserved for royalty. As a writer and as a document designer, you must be aware of potential advantages and possible problems with color use.

Gender Roles

In recent years, most writers in the United States have made serious efforts to show equality between men and women in the workplace. Text and graphics usually show approximately equal numbers of men and women performing specific tasks; women as well as men appear in managerial roles and are shown leading discussions and directing projects. However, in some cultures and some geographic areas, gender roles are changing more slowly, and the workplace is occupied primarily by men. You need to learn about the areas and cultures where your writing will be read and, without compromising your own ethical stance, acknowledge the makeup of the workforce in that area or culture.

Business and Social Customs; Body Language

Some cultures thrive on detail, punctuality, and protocol, while others emphasize generality, relaxed time concepts, and informality. And those characteristics do not necessarily group as I have listed them. Americans are sometimes considered rude and abrupt by readers in other cultures because of the very qualities I call admirable in this book—stressing organization, getting right to the point, repeating the company name, summarizing. Americans are often informal, calling business associates by their first names, yet proceeding to business matters without establishing social relationships. Asians, by contrast, may plan for a business lunch or dinner but will avoid any business talk during the meal. They use the meal itself for social interaction and save business for the end of the meal.

You cannot be expected to know all the customs of each culture or language group for which you write, but the more you know, the more effective your speaking and writing will be. It may be especially valuable to learn about body language—the personal "space" people find comfortable, hand signals and handshakes, touching, table manners. While you may not actually write about these subjects, the more you know about other cultures, the less likely you are to offend.

Sources of Information

When you are working, your marketing department or localization group can help you in writing letters and memos and in monitoring your tone. While you are a student, you can learn awareness by getting to know international students at your college or university and learning about a variety of customs. In addition, both government agencies and private concerns provide written information about other cultures and countries. The United States Department of State publishes two valuable sources of information: *Background Notes* and *Post Reports*.

Background Notes are "short authoritative pamphlets about various countries, territories, and international organizations. . . . Each *Background Notes* features information either about a country's people, land, history, government, political conditions, economy, and foreign relations, or about an international organization's makeup, history, programs, and purposes" (1992). More than 170 countries are described, and information is updated every three to four years. A reading list is included. Printed documents are available in the government documents section of libraries that are depositories or can be ordered from the Superintendent of Documents, Government Printing Office, Washington, DC 20420.

Background Notes are also available on CD-ROM from the National Trade Data Bank. Access to the information is easy, and complete documents can be printed out. Information is constantly updated. The National Trade Data Bank is also an excellent source of other information for international marketing.

Post Reports are prepared primarily for official United States government employees and their families but provide a wide variety of information about individual countries for interested readers. Information includes geography and climate; food, clothing, and religious activities; currency, banking, and weights and measures; local holidays and recommended reading. *Post Reports* are updated frequently and are available in depository libraries and from the U.S. Government Printing Office.

Some states also publish useful information about ethnic and cultural groups. The California Department of Education, for example, has a series of pamphlets on the cultures of the people from Laos, Vietnam, and Cambodia. Intercultural Press, an independent publisher, publishes a variety of books on cross-cultural communications for business people.

14.4 Developing Writing Strategies

Good writing in English will yield good translations and help second-language readers understand your text. Therefore, if you follow the suggestions in Chapter 8 about diction and syntax choices, you will help your international readers and your potential translators. You should pay special attention to three writing areas: word choice, sentence construction, and specialized usage.

Word Choice

Diction, or word choice, is even more important to international and multicultural readers than to native speakers. The single most important consideration is consistent terminology. You have may learned the value of variety in writing, and you may habitually use synonyms for important terms, simply to avoid monotony. For example, in this book I use the terms *readers* and *audience* interchangeably, yet the dictionary definitions of these words are not the same. In a procedural description, the words *participants, learners,* and *users* are often used at random, but the meanings are slightly different. In a report, you might easily use *show, display, exhibit,* and *illustrate* as synonyms. A translator or a second-language reader may assume that because you used different words, you meant something different with each one. Therefore, the translator may select inappropriate or even incorrect terms to represent your words in the target language. The second-language reader may have to go to the dictionary again and again to unlock meaning.

The easiest way to avoid confusing the reader is to choose *one* term, define it in a glossary, and use it consistently. Roger Connolly, who directs writers dealing every day with international writing, calls this "one concept, one word." Translators tell me that the first thing they translate in a document is the glossary (if there is one) or the index. Once the important terms are fixed in meaning, readers and translators can deal with the text itself.

In addition to limiting meaning by avoiding synonyms, you can help international readers by reducing what Canadian technical writer and translator Charles Velte calls "semantic breadth" (1990). Some common English words have an enormous range of meanings; the same word is used as a noun, a verb, or even an adjective. For example, you can say (Cyphers 1991):

> A belt *drives* the main shaft. (verb)
> The assembly includes a belt *drive*. (noun)
> Check the *drive* belt for slack. (adjective)

According to Velte, the word *run* has 99 meanings as a verb, 39 as a noun, and 4 as an adjective. As you can see, looking up the word *run* in a dictionary may not be of much help to a translator or a second-language reader. Replace *run* in your text with a more specific term, and you will make the translator's or reader's task easier.

Finally, help your readers deal with acronyms. Technical writing in English is often loaded with acronyms and abbreviations. Good writing practice calls for spelling out acronyms the first time they are used, and for international readers such identification is critical. For example, you should say *scanning electron microscope (SEM)* and *subject matter expert (SME)*. Acronyms can also change meaning from industry to industry, so you can't assume that even U.S. readers will recognize an acronym. It's also wise to replace abbreviations like *i.e., etc., cont.* with their word equivalents: *that is, and so forth, continued.*

Sentence Construction

For international communication, many company style guides recommend keeping sentences to fewer than 25 words and simple in construction, without complex clauses or long phrases and the consequent layers of meaning that are difficult to untangle.

Experts in international communication also recommend using the active voice, which has a clear doer and a clear recipient of the action (see section 8.3 for explanations of the active and passive voice). It's easier for translators to change the active voice into passive voice than it is to go the other way. Also, it's easier for second-language readers to find the agent (doer) in an active voice sentence.

Finally, careful punctuation and spelling are essential. Missing or misused commas can easily cause confusion, especially around long prepositional phrases or with restrictive or nonrestrictive clauses. Misspelled words may keep second-language readers from finding the meanings in their bilingual dictionaries.

Dependent clauses that use the subordinating conjunctions *while, as, if,* and *when* cause particular problems for international readers. For example:

> The program did not execute while the document printed.
> While the document printed, the program did not execute.

Do these two sentences mean exactly the same thing? The problem is that *while* can mean either *although* or *at the same time.* In the sample sentences, native readers might infer that in the first sentence *while* meant simultaneous action, but in the second sentence *while* meant *although.* But since the words are the same, international readers could easily be confused, and translators would have to make a decision. (It's not all that clear to English readers either.)

Specialized Usage

You learned in section 8.2 that *jargon* is the highly specialized vocabulary of a particular trade or profession. I called this "good jargon" because properly used, specialized language that is understood by readers saves time and makes communication easier. But jargon can baffle both translators and second-language readers, because jargon terms are often adopted from culturally specific metaphors where they have different dictionary meanings. For example, in the computer industry, the following terms are common:

abort	download	spool (verb)
crash (verb)	downtime	motherboard
boot (verb)	online	mouse

While these terms have specific meanings for computer users, they are difficult to translate. Therefore, whatever your profession, you should either avoid such specialized terms or define them in your glossary.

Interestingly, some American words have become the standards in other countries, especially in aeronautics and data processing. According to Berlitz, "the word 'display,' for example, is now 'standard' worldwide, as are 'hardware,' 'software,' and 'computer.' The words 'marketing' and 'know-how' are now international and current, even in the [former] Soviet Union. In fact, in Japan and certain European countries, it is considered fashionable to include as many English words as possible, adding tone to copy" (n.d.).

In addition to jargon, English expressions that cause problems are *idioms* and *Americanisms.* Idioms are expressions that are peculiar to a particular language, and the words that make up the expression, when translated, do not mean the same as the expression. For example, we might say someone *kicked the bucket* when she died, or a pilot *bought the farm* when he crashed. Literally, of course, such phrases mean something totally different. In day-to-day interaction, we often use idioms to "spice up" our writing: expressions like *put up with, we've been there, fly-by-night operation,* and *strike a bargain* add interest to our writing, but can be difficult for non-native readers to understand.

Americanisms are expressions that are peculiar to American English, and they often use sports terminology. Terms like *slam dunk, touchdown,* and *grand slam* have little meaning to people unfamiliar with American sports. Other Americanisms may come from television, the comics, and popular music.

Metaphors, expressions that attempt to clarify by comparing two unlike things, are like idioms in that they may not be accessible to other cultures. If you say, "He went on a wild goose chase," or "Don't be a wet blanket," your writing may be incomprehensible. Localization experts in the target country or culture may be able to substitute local metaphors that approximate those you have used, but you can't count on that kind of dual-culture expertise.

14.5 Planning Page Design and Graphics

You learned in Chapter 12 the importance of designing the pages of your documents for ease of reading, accessibility, and interest. In Chapter 11 you learned about the effective use of graphics. You can adapt many of these principles in international and cross-cultural communication because page design and graphics are, if anything, even more important when you write for these readers. However, that international audience also has special needs that you can plan for as you write and design the English version.

Page Design

Problems begin with paper size. The standard American paper size of 8.5 x 11 inches is not the paper size used in most countries. The so-called A4 standard converted to inches measures approximately 8.25 x 11.75 inches, which means that your side margins may not be wide enough and your top and bottom margins may be too wide. Paper size causes problems in faxing documents and in using computer printers as well as in designing pages. The key is to recognize the differences and to make good use of the common "live" area between the two formats.

An even greater problem, though, is *text expansion and contraction.* In translation, depending on the target language, the text can expand from 15 to 200 percent; in some languages, text can contract. Table 14-1 shows some examples of text expansion and contraction. Text expands because words in the target language are longer or because more than one word is needed for a single English word. For example, the word *go* increases to five letters in German with *gehen.* In Spanish, *on* and *off* become *conectado* and *esconectado.* Text contracts when the target language uses shorter words, or when a character can represent a whole word, as in Chinese.

Table 14-1 Text Changes in Selected Languages

English to Spanish or Italian	= + 15 to 20 percent
English to French	= + 20 to 25 percent
English to German	= + 30 to 50 percent
English to Indonesian	= + 150 to 200 percent
English to Swedish or Danish	= − 20 to 30 percent
English to Japanese	= − 30 to 50 percent

In long documents like manuals and reports, you can solve space problems by using *modular* design: that is, giving each major point or operation its own segment of text, and numbering the sections separately (instead of numbering pages). Modular design also allows for generous use of white space; that way text can expand or contract without affecting the basic organization.

Graphics

When you travel by air, you may have noticed that the airplane evacuation placard in your seat pocket uses line drawings with practically no words to tell you what to do in an emergency. That "wordless" information is understandable by everybody whatever their language or their reading ability. Thus, such placards are good examples of effective graphics for international communication.

Here are some guidelines for effective graphics use:

- Use graphics that an international audience can understand. For example, a power plug and receptacle should be generic, not detailed enough to look like any particular type.
- Use one figure or table to illustrate each concept; do not try to incorporate several concepts.
- Do not superimpose callouts (word identifications of parts of graphics) on the graphic itself. Leave plenty of white space around the graphic to allow for text expansion.
- Use action graphics: arrows showing direction, hands performing actions, or people in action poses.
- Do not use human hands in symbolic (as opposed to action) poses. Many hand positions that have one meaning in U.S. culture have quite another meaning in other cultures.
- In photographs or drawings of people, use enough representatives of various cultures to show variety. Remember the guidelines on gender roles and historical stereotypes given in section 14.3.

These guidelines will help make your document design and graphics in English also appropriate for international readers. And all the authorities agree that extensive use of graphics will help readers from other languages and cultures.

The need for effective international communication is not something to concern us in 5 or 10 years; the need is current and daily becoming more important. As one of my students wrote in a report on her internship experience:

> Because this is a worldwide company, the task of translating documents into other languages is a daily reality. If a document is not clearly written, it must be revised before it can be translated; therefore, the publication department works diligently to produce consistent documentation to simplify the translation (Black 1991).

International and multicultural communication also extends beyond clear writing and translation. According to Donna Hamlin, a consultant to companies that carry on business internationally:

> If we want to go global, we have to get global. We have to think differently—about how we manage mixed cultural teams, organize work, and develop relationships. . . . Global thinkers build these cultural sensitivities into their plans from the beginning (Hamlin Harkins 1991).

Checklist for International and Multicultural Writing

1. Who are my potential readers?
   ___ U.S. residents who learned English as a second language
   ___ international readers who know an English other than American
   ___ international residents who read English as a second language
   ___ readers who will use a translation
2. Have I shown cultural awareness in my writing in the following ways?
   ___ omitting cultural stereotypes
   ___ avoiding American cultural references like names, seasons, or holidays
   ___ eliminating inappropriate political or historical references
   ___ using numbers, dates, and addresses in ways that will be clear to all readers
   ___ choosing colors with regard to cultural traditions
   ___ portraying gender roles in ways that acknowledge the makeup of the workforce in that area or culture
   ___ taking into account the business and social customs of other cultures and other peoples' sensitivity to body language
3. In choosing words, have I followed these guidelines?
   ___ chosen a single word for a single concept or task and used that word consistently
   ___ defined my terms in the text or in a glossary
   ___ narrowed the use of words that have a range of meanings to avoid confusing my readers
   ___ identified acronyms and avoided abbreviations
4. In constructing sentences, have I done the following?
   ___ kept sentences to fewer than 25 words
   ___ avoided ambiguous subordinating conjunctions (or clearly differentiated their use)
   ___ used the active voice whenever possible
   ___ punctuated with care
   ___ double-checked my spelling

5. In choosing written expressions, have I followed these guidelines?
___ avoided jargon or clearly defined it
___ omitted idioms and Americanisms
___ checked metaphors to ensure reader understanding
6. In planning visual elements, have I taken these actions?
___ taken varying page size into account
___ planned for text expansion or contraction in translation
___ used modular design
___ designed graphics with generic characteristics
___ used a single graphic for each concept
___ left white space around graphics for text expansion of callouts
___ used action graphics whenever possible
___ avoided human hands in symbolic positions
___ used photographs and drawings of people that show a variety of ethnic groups

 EXERCISES

1. Interview a student on your campus who is from a different cultural group or whose first language is not English. Ask questions about customs, holidays, business activities, and writing style and tone. Write a short report on the information following the guidelines in Chapter 15 or prepare a short speech after reading Chapter 24.

2. Search magazines, reports, and newspapers for examples of language use that would not translate or localize well. Look for idioms, Americanisms, jargon, and metaphor. Copy examples or clip and bring to class for discussion.

3. Search magazines, newspapers, reports, and textbooks for at least three graphics examples that would work well for an international reader. Be prepared to explain why they would be effective or what could be changed to make them better.

4. Working collaboratively with three or four other students, look up at least five of the following words in a dictionary. See how many meanings each word has as a verb, noun, or other part of speech. Determine how you could define individual words to limit their meanings in your field of study or to substitute synonyms that would be more exact. If this is a class exercise, each group of students can do five or more words and explain their decisions to the rest of the class.

enter	run	drive	close	view	dash
head	screen	abort	heap	work	stop
end	see	read	stand	balance	drill
tax	time	start	exit	can	raise
use	field	alert	bias	pay	get

5. Locate a bilingual dictionary (German-English/English-German or Spanish-English/English-Spanish) and turn to the first page for the letter *B*. Compare that page with the comparable page in a pocket English dictionary (or a college edition of an

English dictionary). If your foreign reader relied exclusively upon that particular bilingual dictionary in order to read a memo you had written, what words in your English dictionary would you want to avoid using? Make a list of those words and be prepared to discuss your findings in class.

REFERENCES

Allen, Jeanette. 1991. *Localizing ESD documents.* Cupertino, CA: Apple Computer, Inc.

Background notes. 1992. Office of Public Communication. U.S. Department of State. Washington: Government Printing Office.

Berlitz tips: Writing copy for better translations. n.d. New York: Berlitz Translation Services.

Black, Maggie. 1991. Internship report. San Jose State University.

Connolly, Roger. 1991. Writing for international markets. Speech at San Jose State University.

Cyphers, Tony. 1991. *Writing for an international audience.* Unpublished document. San Jose State University.

Deutsch, Claudia. 1991. Demand for translators is on the rise. *San Jose Mercury News.* PC1, April 28.

Hamlin Harkins. 1991. *hh reports* 7.2 (June). San Jose, CA: Hamlin Harkins Ltd.

Hein, Robert G. 1991. Culture and communication. *Technical Communication.* 38.1 (February) 125.

Post reports. 1992. U.S. Department of State. Washington: Government Printing Office.

Sciuto, Gloria. 1990. Suitable for all viewing audiences: Tips for preparing international documentation. *Proceedings* 37th ITCC, RT 145.

Velte, Charles. 1990. Is bad breadth preventing effective translation of your text? *Proceedings* 37th ITCC, RT 5.

Winters, Elaine. 1992. Exactly how ready for the workplace? *Technical Communication.* 39.2 (May) 266.

Zotz, Karen. 1991. Letter to author. 11 November.

Part Three

The Forms of Technical Writing

Chapter 15

Memorandums and Informal Reports

Much of the day-to-day work of business and industry is carried on by means of memorandums and informal reports, so you will find yourself writing—and reading—hundreds of them in a single year. Fourteen of the 20 professionals I surveyed before writing this textbook said that memos and informal reports were the types of writing they did most often. These professionals included an electronic design engineer, a medical writer, a corporate attorney, a civil engineer, a marketing assistant, and a consumer affairs coordinator. Whatever your intended profession, then, you will need to understand the forms and organization required for effective memos and informal reports. Even with electronic mail, the basic forms persist; technology has simply changed the way they are transmitted. In many ways, a memorandum is the easiest kind of technical document to write because it follows a standard form, has specific identified readers, is short and direct, and follows a clear organization. Informal reports often follow memorandum form.

15.1 Definitions of Memorandum and Informal Report

A *memorandum* is a short document directed to a reader or readers *within* the same organization as the writer. The plural term is either memorandums or memoranda, and the word is often shortened to memo. Memos generally contain fewer than four typed, single-spaced pages, and they carry on routine business, such as giving or requesting information.

An *informal report* is a written document of fewer than 10 pages and is often directed to a decision-maker. If intended for internal use, it may follow memo form or be attached to a transmittal memo. If directed outside the organization (to a customer, for example), it may be in the form of a letter or attached to a transmittal letter. The form may vary depending on the purpose. In general, informal reports can (1) provide factual information about tests, events, meetings, experiments in the laboratory or the field, surveys, interviews, the progress or status of a project, or (2) analyze or evaluate information in order to market an idea or product or to recommend action. Such reports might evaluate courses of action or products or discuss the impact of an event or procedure. Both types are shortened and simplified versions of the formal reports called recommendation, feasibility, research, and project reports that are discussed in detail in Chapters 22 and 23.

15.2 Memorandum Form and Organization

Memos begin with four items of identification: the date, the reader's name, the writer's name, and the subject. Some memos also include a reference line that directs the reader to a previous document or subject. Figure 15-1 shows several standard arrangements; when you are working, you should follow your company's accepted form. Many companies have computer templates for memo formats that will automatically set up the page for you.

TELFORD ENGINEERING   **MEMO**

Date: 14 July 19___
To: Roxanne Miller
From: Tom Lipscott   *T.L.*
Subject: Report on the ME 250 Conference

INTEROFFICE

To: C. Wu

From: A. Jarvis   *A. Jarvis*

Subject: Change of SE-3 Task Force Meeting Date

Date: September 9, 19___

Blandings Associates

To: Distribution List   From: R. S. Angell
                                                  R.S.A.
Subject: Due Dates for RQ   Date: 7 March 19___
                   Proposal

Reference: RQ203.1

Figure 15-1 Formats for Identifying Items in Memos

It's important to provide the full calendar date (10 April 19—) for later reference. How formal you are with the reader's name depends on four things:

1. who the reader is (whether a co-worker or a supervisor, for example)
2. how well you know the reader
3. what kind of information is in the memo
4. how many people will read what you write

Because they are official pieces of business, most memos use either first and last names or first initials and last names. Include job titles as well as names in a memo that will become a permanent record (one that reports information or results) or a memo going to a person you do not know.

> To: J. Lockwood
>   Manager, Final Test
>
> From: S. Chu
>   G-15 Module Designer

Direct the memo to those people who must *do* something with the information. Others, who will simply read it to be informed, should be "copied." That is, their names should be listed at the end, preceded by *cc* for "carbon copy" or "courtesy copy."

> cc: C. Lorimar
>   H. Sanchez
>   T. Watt

As the writer, be sure to sign or initial your memo either immediately after your typed name or at the end of the document. Follow company style in this matter. Your signature or initials indicate that you have read and approved the memo; with that signature you bear responsibility for what it says.

The subject line is like a report title. It should tell the memo's purpose and indicate its content with key words. For example:

> Subject: Schedule of Meetings for the A2 Task Force
>
> Subject: Report on Design Agreements with the Madison Facility
>
> Subject: Biweekly Status Report on Analysis of Surge Protectors

Some memos also include a reference line that directs the reader to a previous memo, telephone conversation, or report. See Figure 15-1 for an example of a reference line, and consult Chapter 6 for more information on titles and reference lines.

The memo itself should be typed single-spaced with double-spacing between paragraphs. Most memos follow the traditional organization of introduction, body, and conclusion, which is explained in Chapter 5. Keep memo paragraphs reasonably short and deal with only one item of information per paragraph. In longer memos, use headings to identify major topics. But keep a memo generally under four pages; if your document will be longer, consider making it an informal or formal report and attaching a transmittal memo to it. Whatever the length, be sure to include all relevant information and remember that once it leaves your hands, any memo can become a permanent record. Figure 15-2 shows a typical memo informing recipients of important due dates. Like many organizations, Teltech uses a computer template that reminds writers of the items to include. This memo is written to 12 individuals on a standard distribution list; sometimes individual names would be listed at the top instead of the general term "distribution." The writer signs or initials the memo. Courtesy copies are sent to others who need the information—in this case, the manager of another group.

15.3 Types of Informal Reports and Their Organization

Short informal reports vary depending on their purpose and the type of report. For convenience, reports can be classified into types by their characteristics, as Table 15-1 shows. In practice, reports often combine types and do not have the neat distinctions you see in this table. Four types of informal reports will be discussed in this section, because these four typify the reports you might write and because each one has different important parts. These four are

1. investigation reports
2. trip and conference reports
3. work activity reports
4. information reports

Investigation Reports

A report form with which most students are familiar is the *laboratory report*, which is a kind of *investigation report*. Laboratory reports are not usually organized sequentially, but they do include required items, which are identified with subheadings. The introduction for a lab report will typically include the following:

- purpose
- problem or objective
- scope
- equipment or materials
- procedure or methods

The central portion, or body, reports on the findings or presents the data acquired. This section is often called "Results." It's tempting in this section to follow a sequential or narrative organization, reporting results in the order they were discovered, but, in fact, a sequential method probably wastes the reader's time. Instead, you should select significant data and present it in whatever hierarchical organization best fits the purpose. (Review Chapter 4 for details on types of organization.) You may also want to include tables or graphs of some kind.

Teltech
Interoffice Memorandum

To   Distribution   File No. FC2.10.1
Subject F200 Submission   Date   July 17, 19__
                                   From   H. R. Lewison *H. L.*
                                   Of   Dept. FCO
                                   At   Downtown Central
                                             City
Copies to B. O. Smally   ext 6-7010

As we all know, the F200 package will be submitted to Transcom on
Thursday, August 1, 19__. In order to complete this package on that
date, we must all meet the following intermediate dates:

July 26 Submit one blueprint to the downtown office of each
         drawing to be included in the F200 package.
July 29 Attend the drawing review session at the downtown
         office at 1:30 p.m.
July 30 Make any corrections to the drawings.
July 31 Submit all drawings (one full-size vellum each) to the
         downtown office.

Contact the downtown office if you have questions or to coordinate
changes. Please review the preliminary design review comments to
ensure that they are incorporated as appropriate. All quantities,
drawings, and specifications must be checked to ensure that they are
consistent with each other.

Distribution:

G. Anderson   C. Yang   D. Feriera
B. Fieffer   O. Patterson   B. Lagemeyer
C. Krisfeld   A. Shaval   K. Mak
S. Reiner   A. Short   F. Williamson
               I. Tahin

Figure 15-2 Typical Information Memo

The conclusion of a lab report is usually called the "Discussion." In this section,
you explain

- What principles or generalizations the results support.
- How these results agree (or disagree) with earlier published or unpublished
  results.
- What the results mean and how the conclusions are supported.

Table 15-1 Types and Characteristics of Informal Reports

Type	Purpose	Reader	Examples	Method of Organization	Important Parts
Occurrence	Reporting results	Manager Decision-makers Peers	Accidents, incidents Equipment breakdown Tests Natural Occurrences	Hierarchical	Introduction Conclusions and Recommendations Method Facts Discussion
Trip and Conference	Documenting observations and activities Sharing new information	Manager Peers	Trip summary Highlights of conference	Hierarchical	Purpose and date Primary tasks or personal activities Questions raised Information gained Recommendations
Work Activity	Reporting progress or status on a project	Manager Team leader	Periodic reports Status reports Progress reports	Hierarchical Sequential	Accomplishments Work in progress Work scheduled Anticipated problems General information
Meeting Minutes	Recording discussion and agreements	Managers Those who attended and those who did not	Minutes	Hierarchical and sequential	Identifying information Committee reports Old business New business (or)Action items and parliamentary points Time of adjournment
Investigation	Reporting results and observations	Managers Peers	Field reports Laboratory reports	Hierarchical	Introduction Methods and materials Results Discussion
Information	Sharing information with a new audience Marketing	Generalists Specialists Peers	News releases Marketing Interviews	Hierarchical	Introduction Key information Ramifications and uses
Archival	Making a permanent record	Self Managers Peers	Report to file	Hierarchical	Introduction Background Key decisions Implications

Field reports (in which you report on collected data from some location) are a second kind of investigation report and follow a similar introduction-body-conclusion organization but have subheadings relating directly to the material you are presenting. The introduction of a field report includes the purpose, the problem, and the methods used to collect information; the body details the findings; and the conclusion discusses what the findings mean. A field report might also include a recommendation based on that conclusion. In Figure 15-3, the writer reports on a field inspection of automatic swimming pool chlorinators. Notice the specific detail in the report and the support provided for the conclusions and recommendations. Because this report is short and internal, it is written in memo form.

AUTO-CHEM, INC.

Date: September 15, 19__
To: F.C. Powers
           Quality Control Manager
From: S. Hioki *S. Hioki*
           Customer Service Engineer
Subject: Inspection of C6432T Automatic Pool Chlorinators

INTRODUCTION

Reason for field study

On September 14, 19__, I inspected the C6432T Pool Chlorinators at four locations in Addison County to determine the reasons for customer complaints of residual chlorine levels below the legal minimum of 1.0 ppm. These four chlorinators were all less than one year old and installed in 20,000 to 30,000 gallon public pools in either apartment buildings or motels. The locations were:

Background

1. Highland Motel
   342 Green Avenue
   Centerville
2. Conway Motel
   1442 Olney Drive
   Harkness
3. Crestview Apartments
   836 River Parkway
   Centerville
4. Olympia Apartments
   1135 Fourth Avenue
   Rockland

Figure 15-3 Example of a Field Report

Methods

I collected information on these chlorinators in four ways:

1. Checked the chlorine level in the pool water.
2. Visually inspected the chlorinators and the method of installation.
3. Disassembled the chlorinator and checked the components.
4. Interviewed the person responsible for pool maintenance.

FINDINGS

1. Chlorine levels at all four locations were below 1.0 ppm.
2. The pump at location 3 was improperly installed, leading to a reduced feed of chemical crystals and a consequent level of chlorine of 0.80 ppm. In addition, chlorine was stored in a plastic garbage can instead of the 30 gallon crock with a pump mount we recommend.

Factual information

3. At locations 1, 2, and 4, all the equipment was properly installed, but chlorine levels were 0.92, 0.85, and 0.87, which had led to warning citations by the county pool inspector.
4. Disassembly of the chlorinators at locations 1, 2, and 4 showed that the metering mechanism was clogged with chemical crystals in all three.
5. When I interviewed pool maintenance personnel, I found:
   a. The instructions for acid wash of the metering mechanism at location 1 were missing, and the pool cleaner had never performed the required acid washes.
   b. Pool cleaners at locations 2 and 4 said they performed acid washes approximately twice a month.

Conclusions

CONCLUSIONS AND RECOMMENDATIONS

Clogging of the metering mechanism by chemical crystals is caused primarily by improper or insufficient acid washing. Although our instructions recommend weekly acid washes, maintenance personnel performed this procedure only two times a month. Most maintenance personnel are apartment or motel managers rather than trained pool technicians, so we should either simplify the acid washing procedure or reduce the number of acid washes needed.

Recommendations

I recommend the following actions:

1. Rewrite the instructions to clearly specify times for acid washes based on hours of chlorinator operation.
2. Simplify the acid wash instructions by numbering the steps and adding an illustration of the metering mechanism.
3. Redesign the metering mechanism—perhaps with a squeeze-feed roller tube—to eliminate the need for acid washes.

Figure 15-3 Continued

Trip and Conference Reports

If you are sent to another location for a project or meeting, or if you attend a conference, you will usually write a brief report on the significant activities you participated in. Although it is tempting to write this report sequentially—telling what happened first, next, and so on—you need, instead, to keep the readers' needs in mind. Readers want to know what happened that was important, what information you gained, and the conclusions you reached. A good general organization includes the following:

1. the purpose
2. a summary of primary tasks or personal activities
3. a discussion of questions raised or information gained
4. conclusions and recommendations

Figure 15-4 shows a conference report written in memo form. The writer is a graphic designer at a computer company; his report stresses both the activities he participated in and the information that is significant to his company.

Hi-Tech Associates
Interoffice Memo

Date: 21 October 19__
To: Technical Publications Department
From: R. Geimer, Graphics Designer *R. H.*
Subject: Seybold Computer Publishing Exposition

Purpose and background

Purpose: The Seybold Computer Publishing Exposition ran for four days, from October 1-5, 19__, at the San Jose Convention Center. C. Gates and I attended to investigate new products that might enhance our publications effort. I also attended minicourses on the use of high resolution printing languages and the integration of incompatible computers on a network.

Relevant information is summarized. Presentation is not chronological.

Summary of Activities:

1. Apple Computer released a new scanner and two new printers. According to Apple, these new devices will increase the use of photographs in laserprinted materials due to their ease of use and increased ability to handle grayscales and halftones. The new printers, the LaserWriters IIf and IIg, are equipped with FinePrint and PhotoGrade, new technologies that allow the printer to control the size and shape of the toner dots to achieve maximum quality out of a 300 dpi (dots per inch) printer.

Figure 15-4 Conference Report in Memo Form

2. Sun Microsystems demonstrated NewsPrint, which turns a Sun workstation into a network printing interpreter. With NewsPrint, a Sun workstation can take files in any printing format and translate them into the optimum language of the printer being used. To demonstrate this, they had various Apple, Hewlett-Packard, and Sun computers all printing PostScript documents to non-PostScript printers. As an analogy, imagine that you are a police artist. A French tourist has just been mugged, and you have to somehow communicate with the tourist in order to draw a sketch of the mugger. You will need an interpreter to bridge the language barrier. In the computer world, which abounds with printing languages, a Sun workstation equipped with NewsPrint serves as this interpreter.

3. NeXT Computers had a hands-on workshop where users created a newsletter on their Color NeXT Stations. We used Adobe Illustrator, Word Perfect, and NeXT Mail to create the newsletter, and printed it on a Canon color copier specifically adapted to function as a network printer.

4. Many vendors announced their latest software releases. Quark demonstrated Quark XPress version 3.1, which now includes style and color palettes. Adobe gave presentations on Illustrator 3.0 and Photoshop 2.0, programs that are rapidly becoming industry standard for affordable desktop graphics. On the high end, Silicon Graphics gave a fascinating demonstration of their workstations, showing computer graphics clips from "Terminator II" and various commercials that were all created using their technology.

Value of the conference

Recommendations

Conclusions: For our purposes at Hi-Tech, the conference was most valuable to keep us informed about latest advances in desktop publishing. As we continue to update our publications hardware, we will want to have an inhouse demonstration of the new Apple scanner. The DTP task force may also want to do a comparative study of scanners currently on the market so we can use more photographs in our manuals. At the department meeting on Nov. 2, I will summarize information from the two minicourses I attended; if you have questions before that date, please call me at extension 5-2124.

Figure 15-4 Continued

Work Activity Reports

Most people who work in technical occupations will regularly report on their work to a supervisor or client by writing either status or progress reports. A *status report* is a written account of accomplishments on the job during a designated period of time. In business and industry, status reports are written at regular calendar intervals—weekly, biweekly, or monthly. A *progress report* is a written account of accomplishments at stated intervals in a project—after completion of a test phase, for example. Since both types of reports are often for internal use, they follow memo format. Progress reports may also go to clients or customers, and then could be in letter form or informal report form and attached to a transmittal letter. The form and organization for these reports can vary widely, and your company or instructor may have specific requirements. If not, the form in Figure 15-5 works well and can be adapted to a variety of occupations. Even though the sections of the status report are not labeled introduction, body, and conclusion, the form does contain those elements. *Tasks Completed* serves as an introduction or overview, while *Tasks In Progress* makes up the body, and *Tasks Scheduled* or *General Information* serves as a conclusion.

Sometimes you will have to include a problem section in your status report. This could go in any section, but is more likely to be included in section 1.2. Divisions of the problem might be as follows:

- Old Problems: Resolved
- Old Problems: Unresolved
- New Problems

Date:
To:
From:
Subject:
Reporting Period: (From _____ to _____)

1.0 Status/Progress/Accomplishments (one of these)

    1.1 Tasks Completed
        [Be very specific about what you have done, including the date.] Example: The clutch assembly test unit was completed on 17 April 19__. Eight units completed destructive torque tests on 1 May 19__. K. P. Smith wrote the report on the results.

    1.2 Tasks in Progress
        [Again be specific. Give completion dates if appropriate.] Example: Eight clutch assemblies are currently in "Normal Use Life Test." This test is on schedule and targeted for completion on 30 May 19__.

    1.3 Tasks Scheduled
        [This would include new assignments or tasks you have been assigned that you have not yet started.]

2.0 General Information
    [You can include here (or in a section 1.4) your log of time spent on various tasks, such as setting up test equipment and calibrating the tester. You can also include anticipated problems or potential sources of information.]

Figure 15-5 Organization of a Status Report

Another format is the Problem Report. Here is a possible outline:

- Statement of the problem
- Impact of the problem (cost, schedule, procurement, safety)
- Activity or action plan
- Person responsible

Do not underestimate the importance of status and progress reports: managers often use them to assess your accomplishments for salary and promotion review, and clients use them to evaluate the effectiveness of your work. In a status or progress report, you are, in effect, writing a self-evaluation. But their importance is more than personal. As consultant and college professor Christine Barabas says:

> Traditionally, progress reports used to be little more than time cards, reports of work done and work to be accomplished. The more industrious the employee appeared to be, the better the report and presumably the greater the "progress." Today, however, managers need to know more than what someone is doing or will be doing. They need to know the value of the work. They need more than facts; they need assessments, speculations, and judgments to help guide them in their decision making (1990, 132).

Figure 15-6 shows a variation of this status report organization written by a software writer in industry. This format is standard at that company, and the categories are already provided (Rhodes 1991).

Date:	August 2, 1991
Writer:	Jennifer Rhodes *J. R.*
Status for Week Ending:	August 2, 1991
Current Assignment:	MV MANAGER for DBCTL
Release Number:	N/A
Begin Date:	July 15, 1991
Projected End Date:	August 27, 1991

Planned Milestones or Tasks Accomplished

- I continued to incorporate primary edit comments to the User Guide and Reference, and began incorporating primary edit comments to the Customization Guide.
- I used file aid to make data set changes from Linda to the UG&R.
- I worked on the cover letter and review form for the technical review of the CG.
- I worked on the UG&R preface using the CG preface as a base.
- I printed out the UG&R manual part by part to check my editing before its verification edit.

Figure 15-6 Status Report

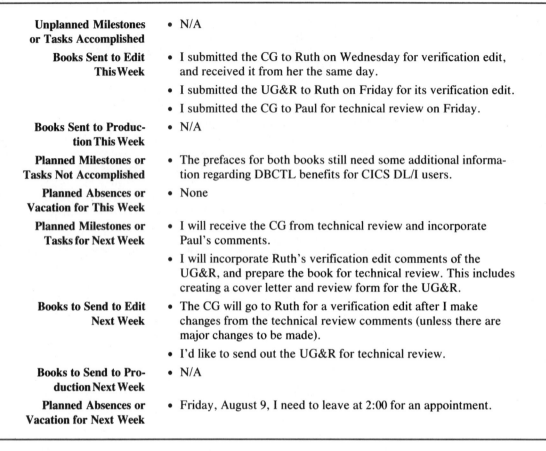

Unplanned Milestones or Tasks Accomplished	• N/A
Books Sent to Edit This Week	• I submitted the CG to Ruth on Wednesday for verification edit, and received it from her the same day.
	• I submitted the UG&R to Ruth on Friday for its verification edit.
	• I submitted the CG to Paul for technical review on Friday.
Books Sent to Production This Week	• N/A
Planned Milestones or Tasks Not Accomplished	• The prefaces for both books still need some additional information regarding DBCTL benefits for CICS DL/I users.
Planned Absences or Vacation for This Week	• None
Planned Milestones or Tasks for Next Week	• I will receive the CG from technical review and incorporate Paul's comments.
	• I will incorporate Ruth's verification edit comments of the UG&R, and prepare the book for technical review. This includes creating a cover letter and review form for the UG&R.
Books to Send to Edit Next Week	• The CG will go to Ruth for a verification edit after I make changes from the technical review comments (unless there are major changes to be made).
	• I'd like to send out the UG&R for technical review.
Books to Send to Production Next Week	• N/A
Planned Absences or Vacation for Next Week	• Friday, August 9, I need to leave at 2:00 for an appointment.

Figure 15-6 Continued

Information Reports

Short reports that present information can go to internal or external readers. You might, for example, write to management reporting on a failure-tracking system to identify areas that can be improved in the company's design, manufacturing, and test processes. You might write an information report to a client company assessing the toxicology of the metals and solvents they anticipate using in a new laboratory so they can take proper precautions.

If your report goes outside your organization, you might write to a specific client or to generalists in the form of a news release. One way experts share their information with general readers is to disseminate it through radio, television, or print media. In Figure 15-7, for example, consumer affairs writer Kristin Lundquist Frank writes a news release to share information about new food labeling laws and their effect on industry (Lundquist Frank 1991).

FOR IMMEDIATE RELEASE: August 27, 1991
Contact: Kristin Lundquist Frank
       Communications/Consumer Affairs Coordinator
       1-(800) 944-MILK (6455)

NEW FOOD LABELING LAW WILL FUEL FOOD PRICES

Introduction and background

A recently enacted federal food labeling law will cause major overhauls in food processing procedures that will increase food prices, according to speakers at a recent writers' conference sponsored by Dairy Council of Wisconsin.

Under the new law, Congress required the Food and Drug Administration to create guidelines for individual ingredients in processed foods as well as descriptive phrases and health claims commonly used to advertise foods, according to Douglas R. Engebretson, vice-president of quality assurance and regulatory affairs for Land O' Lakes. The nutrition information shown on the label will also change almost completely.

Engebretson, whose Minneapolis-based company processes and packages dairy products, said the new law creates many gray areas for manufacturers. For example, gum is commonly used as a stabilizer in ice cream. Since it contains fiber and that content is now required to be listed, it raises questions as to the extent of a manufacturer's responsibility to declare that ice cream contains fiber.

"There is an incredible amount of information that is invisible to the consumer that will have to be recognized and administered by those of us involved," said Engebretson.

Consumers are also likely to be confused over the new label information, according to Amy Barr, director of the Good Housekeeping Institute. And while consumers want nutrition claims to be used in advertising and are very interested in food and nutrition, they don't understand many key issues.

"There's a lot of confusion out there," said Barr. "And what we're also seeking is that our technical and detailed nutrition information does not appeal. People just want the facts, ma'am, and unfortunately in food and nutrition, there is no black and white.

Body includes factual information and commentary

"Our consumer readers think they understand, but they latch onto certain words and that's about it. Many of those are misconceptions."

The labeling law requires new information about fat, cholesterol, fiber and carbohydrate content, leaving room for consumer confusion when interpreting the new data, according to Emerita Alcantara, director of manufacturer relations for Dairy Council of Wisconsin.

"One of the most important things we can do is to help consumers interpret what I call the new 30 percent rule," said Alcantara. "Consumers have heard that they should limit their fat intake to no more

Figure 15-7   Informal Report as a News Release
Reprinted with permission—(Kristin Frank) Dairy Council of Wisconsin.

than 30 percent of total calories. It is important to remember that this recommendation is for the entire diet, not for individual foods."

Barr continued that she sees evidence of confusion over the "30 percent rule" every day.

"Consumers are going to start eliminating foods out of their diets if their fat content is 32 percent, 34 percent, because they don't understand that whole diet concept," said Barr. "It's a real problem, and we're already seeing it at the magazine."

She said consumers are afraid of fat, and their fear is paradoxical. "People are giving up whole milk so they can eat more ice cream. Or they're giving up ice cream for ice milk, but they're not losing weight, and they're not eating less fat, they're just eating more."

Implementing the necessary label changes is a headache for the manufacturer, according to Engebretson.

Report conclusions emphasize the association viewpoint

"Every packaged item may easily have up to three layers of labeling," said Engebretson. "We're looking at maybe $1,000 in every case, just to do the analysis, per component, and whatever the plate changes are for the packaging, and the print runs, to say nothing of the existing packaging that will have to be destroyed.

"Even in the rather simplistic business that Land O' Lakes is in, butter and cheese mostly, we have over 3,000 end items, which means over 9,000 layers of packaging that have some kind of content information on it," said Engebretson.

In addition to packaging, food processors will have to reevaluate formulas, procedures and databases to see whether their manufacturing processes comply with the law.

"The industry is being warned to build this into their budgets, because like it or not, those costs will be incurred, and the consumer will have to pay for them," said Engebretson.

Dairy Council of Wisconsin is a non-profit nutrition communication and nutrition marketing association serving Wisconsin, northern Illinois and northwest Indiana.

Figure 15-7 Continued

15.4 The Process of Writing Memos and Informal Reports

Planning

In planning a *memo*, first determine its purpose. Ask yourself why you intend to write this information instead of telephoning. Usually you'll choose to write if (1) you can't reach the person by phone, (2) the information must go to several people at the same time, or (3) you want to provide a permanent, written record. Here are common purposes for writing memos:

- to request information or action
- to respond to a question or request
- to give information: to announce, explain, describe, or recommend
- to formalize policies or confirm agreement reached in a phone conversation or meeting
- to boost morale or compliment people for work well done
- to document action taken. This might be a "memo to file": that is, written to yourself, for the record, and to help you remember what was done. It also could be meeting minutes.
- to propose new procedures or ideas. Short, internal proposals are often written in memo form. See Chapter 21 for details on how to write proposals.

In planning an *informal report,* your first step is also to clarify the purpose. If your report will be more than 10 pages, consider using the formal report form instead. The additional front matter (such as abstract, table of contents, and list of figures) and the additional back matter (such as list of references and appendixes) will help clarify complex matters for readers of a long report. Chapter 22 tells how to create front and back matter for readers of a long report.

Your second consideration in planning is the reader. Most memos go to a specific reader, and you need to identify that reader by background and job. Then you'll know how much detail to include and whether you can use technical terms without definitions. If you have only one reader (who is not your supervisor), you can write informally, but the more readers you have, the more formal the memo or report should be. Your relationship to the reader will also affect the tone, which you should keep in mind as you write. Even though you are writing a short document, it pays to think it through by using the planning sheet in section 1.6.

The memo shown in Figure 15-8 offers to provide a service (Hopkins 1991). Notice that the reference line is in the heading and is explained in the memo's first line, and that the memo is directed to specific individuals. The writer uses abbreviations that are part of the jargon of the profession—VOMP stands for Victim Offender Mediation Program and P.O. for parole officer.

Gathering Information

You might already have the information for a memo before you begin to write, but if you are writing a short report, you might have to do additional research, interview experts, or conduct additional experiments. Concentrate on accuracy, remembering that you are providing a record of your work.

Take special care in constructing documents like status reports, which report directly on the work you have been doing. Some employees are very casual in writing status reports, not realizing that at evaluation time managers frequently study a series of status reports to determine how well the employee has been performing. In effect, you are evaluating your own performance when you submit a status report, so it pays to

Victim/Offender Mediation Program

In Santa Clara County

Friends Outside
551 Stockton Avenue
San Jose, California 95126
(408) 295-6033
FAX (408) 295-2907

Probation Department, Juvenile Division
840 Guadalupe Parkway
San Jose, California 95110
(408) 299-3078
FAX (408) 971-8471

To: John Sutro, Central
    Harold MacLean, North
    Bob Creamer, South
    Ray Ortiz, East
    Mike Berry, West
    Marty Brewer, Court Unit

From: Judy Hopkins, VOMP, x3078 *J. H.*

Re: VOMP Services

Date: 9/11/91

Reference and background

My last memo of 2/1/91 suggested two <u>additional services</u> besides face-to-face mediation VOMP might provide to line deputies: (1) <u>receiving and disbursing money when payments are involved, and (2) assisting victims organize their restitution claims.</u>

Purpose of the memo

Many of your staff have made requests for these services, and I think we have been able to make contributions that both relieve the supervision probation officers of many details and enhance the service the department provides. In the course of doing these additional services, ANOTHER option has developed. The purpose of this memo is to explain another enhancement we can provide.

Factual information

3. <u>Preparing documentation and recommendations for contested restitution hearings.</u> I have handled several referrals where I spoke with both the offender/family and the victim to review records and documentation, checked into insurance coverage limits, priced certain replacement items, and then gave the P.O. an update of my findings and a suggestion as to what might be fair restitution. Sometimes this has resolved the issue, and both parties agree to the sum, and we set up a contract/payment. Sometimes, however, they don't agree, and the P.O. has used my investigation as the basis for a contested restitution hearing. As you are aware, these hearings require (a) documentation from the victim, (b) the statement/position/circumstances of the offender, and (c) a recommendation for a dollar amount, if money is involved. It occurs to me there might be a benefit to the supervision officer if s/he is not the person who does this; namely, that the information come from a neutral source.

Action request

Please share my suggestions with the people in your unit, and thank them for their continuing support of VOMP.

A Juvenile Restitution Program sponsored by the County of Santa Clara Probation Department and Friends Outside in Santa Clara County, a United Way Agency

Figure 15-8 Memo Offering a Service

Reprinted by permission of Judy Hopkins, Project Director, Victim Offender Mediation Program, Santa Clara County, CA.

have all the information at hand and to organize it well. The same is true of nearly all memos you write. Someone is likely to look at them when your performance is evaluated.

Organizing

Clear organization of a memo or report will help the reader understand the content. As in all writing, you need to decide on your one major point, choose supporting subpoints, and arrange those subpoints according to a method of organization that will work for that document. Most memos and informal reports have a three-part organization:

1. Statement of the purpose and subject. This section may require background or a summary of previous actions.
2. Details and discussion, organized in a way that fits the purpose and subject.
3. A conclusion that tells what you want the reader to do (action). Memos that report or provide information do not always require a conclusion.

If your document contains bad news, however, you may need to modify this three-part organization. Most readers will be offended by a blunt opening statement containing bad news. In this case, you need an opening that begins positively or with the background of the problem. In the middle section, once the reader is prepared, you can present the bad news. In the conclusion, you indicate action but try to close on a positive note. See section 16.5 for more on this method of organization.

Be aware that good organizing skills can enhance your job performance. Stacey Lippert, a senior editor in the marketing communications group of a major computer company, says this about what may seem to be a routine assignment—writing meeting minutes:

> In my experience, writing meeting minutes for a meeting that occurs weekly (like a task force meeting) is an excellent way to establish yourself in a company as being knowledge-able in a certain area. It's also a great way to achieve visibility! If minutes are written well (information is easy to access, important information comes first, action items are clearly outlined), people perceive the meetings as being successful because they can see all that was accomplished. Then they perceive you as a key information point because you disseminate the information, and they will go to you with questions because your name is the most visible.

Writing, Reviewing, and Editing

Review your organization before you write; review and edit your memo or report before you send it out. Make sure that you have answered the pertinent questions posed in Chapter 7 for reviewing documents. Sign or initial the final copy of a memo, and remember that your initials indicate your approval and acceptance of respon-sibility for the content and form. Informal though a memo or short report might be, its contents can be of vital importance to you personally and to the company for which you work.

Checklist for Writing Memos and Informal Reports

1. In writing this memo, have I included the following information?
   ___ date
   ___ reader's name
   ___ writer's name and signature
   ___ subject line
   ___ subheadings for major topics
   ___ cc list if appropriate
2. In writing this memo, have I taken the following actions?
   ___ clearly organized the information, with an introduction (or background), body (discussion), and conclusion (action)
   ___ included subheadings to guide the reader through long or complex material
   ___ stayed below the four-page limit
3. In writing this informal report, have I chosen the appropriate type?
   ___ occurrence (accidents, incidents, tests, natural occurrences)
   ___ trip or conference
   ___ work activity (progress or status)
   ___ meeting minutes
   ___ investigation (field or laboratory)
   ___ information (news release, marketing, interview)
   ___ archival (report "to file")
4. Will this informal report do the following?
   ___ have clear divisions that correspond to what the reader expects
   ___ respond to a reader's real needs
   ___ stay below the 10-page limit. (If not, consider the more formal reports discussed in Chapters 21–23.)

EXERCISES

1. After consulting Table 15-1, compose a list of five specific situations from your major field in which an informal report might be needed. Choose one of these situations and write a two-page report, following the guidelines in this chapter.

2. Attend a lecture, presentation, or conference in your major field of study. Write a short report on what you learned to your classmates following the guidelines in this chapter.

3. Choose a subject from your major field that is new, controversial, or would be of interest to a general reader. Write a short report in the form of a new release. You might want to target a specific audience like local high school students and write so that they can understand a technical subject.

WRITING ASSIGNMENTS

1. If you haven't already done so (see Chapter 1), write your instructor a short memo explaining the chosen topic for your term report. Include the reason for your choice.

2. As your group meets to plan its term project, assign rotating team members to record the minutes of the group meetings. Distribute minutes to each group member and to your instructor.

3. If you have not already done so as part of your term project, write a status report in memo form to your instructor. Include as much of the information as is pertinent from the status report outline given in this chapter. If you are working as a team, assign one person to write any required status reports. Three days before each status report is due, submit to that person a one-page explanation of the status of your part of the project. The writer can then assemble this data into one report.

REFERENCES

Barabas, Christine. 1990. *Technical writing in a corporate culture: A study of the nature of information.* Norwood, NJ: Ablex.

Geimer, Rick. 1991. Conference report.

Hopkins, Judy. 1991. Memo. Probation Dept., County of Santa Clara, CA.

Lundquist Frank, Kristin. 1991. *New food labeling law will fuel food prices.* Dairy Council of Wisconsin.

Rhodes, Jennifer. 1991. Boole & Babbage Status Report. 2 August.

Chapter 16

Business Letters

Even though you work in a technical field, you will write many business letters during your professional career. For example, civil engineer Gordon Anderson currently supervises the design of portions of a mass transit system for a major city. "In the civil engineering field," he says, "the majority of the work is for government agencies. Numerous letters, reports, and other documents are required, and the importance of technical writing cannot be overemphasized."

Much day-to-day business in engineering and other technical fields is conducted by telephone, but letters are important both because they can be read and referenced at a time convenient to the reader and because they provide a permanent record. Sometimes you will write a letter when you can't reach a person by telephone. Sometimes a letter is more appropriate than a phone call. For sensitive matters, you can plan a letter carefully to project the tone, style, and information you want to convey. For sales letters, you can write one letter and address it personally to many prospective clients. Letters, like internal company memos, are fairly easy to write because they follow a standard form, are directed to specific, identified readers, are short, and have a clear method of organization.

16.1 The Definition of a Business Letter

A *business letter* is a short document addressed personally to a reader or readers who are usually *outside* the writer's organization or place of business. You will write letters both as a private individual and as a representative of your company. Generally, the letters you write for your company will be typed on letterhead stationery, and most letters should take only one or two typed, single-spaced pages. On very formal occasions, letters may be written to people *within* a company, but usually internal correspondence is by memo. A short informal report can be written in letter form if it goes outside the company. However, if the material will cover four or more pages, it should be written as an informal or formal report and accompanied by a one-page letter, called a *cover letter* or *letter of transmittal*.

16.2 Letter Form and Parts

Most business letters follow a standard form and include the same basic parts. These parts identify the reader and the writer, give the date, and open and close the message in a traditional way. With standard parts, the reader can focus on the letter's content rather than its form. Entire books have been written on proper letter form and organization, so the material in this chapter is necessarily condensed, but all the basic information you need to write successful letters is here.

Form

Follow your company's standards or computer template for letter form, but if there is no template or you are writing on your own, you can choose one of the arrangements shown in Figure 16-1, 16-2, or 16-3. By following this general form, you will write correct, well-designed letters.

Figure 16-1 shows the *block* or flush left form preferred by most businesses and many students because it is easy to use: you simply line up all the parts of the letter at the left margin. Paragraphs are not indented but are set off by double-spacing.

Figure 16-2 shows the *semiblock* or *modified block* form, in which the heading (the writer's address and the date) and the complimentary close and signature are placed to the right of the center of the page. In this form, paragraphs can either begin at the left margin or be indented five spaces. If you choose the semiblock form, line up

CONWAY Research in Agriculture
300 S. 17th Ave.
Oklahoma City, OK 73111

Date — July 13, 19___

Inside address —
Mr. Frank Smithers
Photon Applications, Inc.
555 North Lake Street
Kansas City, MO 64118

Salutation, with a colon — Dear Mr. Smithers:

Single-space paragraphs

Double-space between paragraphs

Complimentary close, with a comma — Sincerely yours, _____

Signature — *Edwin Campbell*

Identification —
Dr. Edwin Campbell
Director of Research

Figure 16-1 Block or Flush Left Letter Form

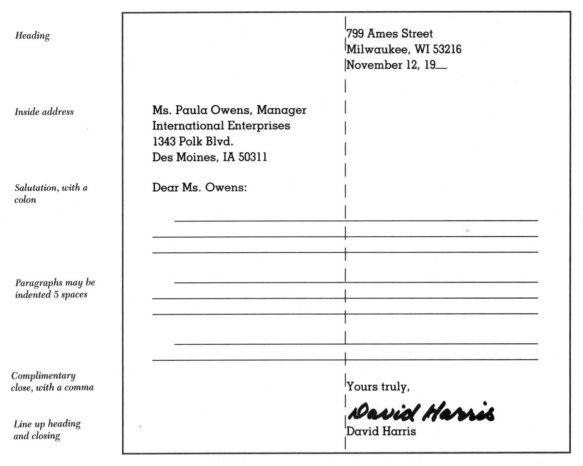

Heading

Inside address

Salutation, with a colon

Paragraphs may be indented 5 spaces

Complimentary close, with a comma

Line up heading and closing

799 Ames Street
Milwaukee, WI 53216
November 12, 19__

Ms. Paula Owens, Manager
International Enterprises
1343 Polk Blvd.
Des Moines, IA 50311

Dear Ms. Owens:

Yours truly,

David Harris

Figure 16-2 Semiblock or Modified Block Letter Form

the heading and complimentary close along an imaginary line just right of the center. Always think of a letter as a picture to be framed on the page, and balance margins and white space to provide an attractive frame. All letters should have margins of at least one inch on the top and sides and one and a half inches at the bottom. You may have to adjust margins after typing the letter in order to balance text and white space. Fortunately, word processing programs make such adjustments easy.

The third form, shown in Figure 16-3, is called *simplified* because it omits two traditional parts of the standard letter: the salutation and the complimentary close. A subject line substitutes for the salutation, and the signature and typed name appear at the bottom without a complimentary close. The simplified form is useful if you don't know how to address your reader, or if the letter will have many readers. On the other hand, to some readers this form may seem abrupt and somewhat rude, so use it only if the form is standard practice for your company.

Heading

4673 Westlake
Fort Worth, TX 76132

April 17, 19___

Inside address

DMC Machine Corporation
741 Washington
Salt Lake City, UT 84101

*Subject line
substitutes for
salutation*

Subject: Recall of the SMX Lathe

_____
_____
_____
_____
_____
_____
_____
_____

*No complimentary
close*

Walter Collins
Walter Collins

Figure 16-3 Simplified Letter Form

All letters should be typed on standard 8 ½ by 11 white or off-white bond paper. Do not use onionskin or erasable paper. The second page of a letter should be on plain paper even if you use letterhead stationery for the first page. Type the following information at the top of each additional page: the reader's name, the date, and the page number:

Smithers   page 2
July 13, 19___

Parts

A standard business letter has the following basic parts:

Heading

The heading includes the writer's street address, city, state and ZIP code, and the date. On company letterhead stationery, the company name and address are already printed, so you need type in only the date. Notice that the writer's name does *not* appear in the heading.

For your quick reference, here is a list of the postal abbreviations for the states. These abbreviations should be used in addresses, and increasingly, they are also used within the text of documents.

Alabama	AL	Montana	MT
Alaska	AK	Nebraska	NE
Arizona	AZ	Nevada	NV
Arkansas	AR	New Hampshire	NH
California	CA	New Jersey	NJ
Colorado	CO	New Mexico	NM
Connecticut	CT	New York	NY
Delaware	DE	North Carolina	NC
District of Columbia	DC	North Dakota	ND
Florida	FL	Ohio	OH
Georgia	GA	Oklahoma	OK
Hawaii	HI	Oregon	OR
Idaho	ID	Pennsylvania	PA
Illinois	IL	Puerto Rico	PR
Indiana	IN	Rhode Island	RI
Iowa	IA	South Carolina	SC
Kansas	KS	South Dakoka	SD
Kentucky	KY	Tennessee	TN
Louisiana	LA	Texas	TX
Maine	ME	Utah	UT
Maryland	MD	Vermont	VT
Massachusetts	MA	Virginia	VA
Michigan	MI	Washington	WA
Minnesota	MN	West Virginia	WV
Mississippi	MS	Wisconsin	WI
Missouri	MO	Wyoming	WY

Inside Address

Include in the inside address the reader's name (preceded by Mr., Ms., or other title such as Dr.), title if appropriate (Publications Manager, Vice President of Sales), department, company name, street address, city, state, and ZIP code. No comma is used between the state abbreviation and the ZIP code. Sometimes you must include a building number or internal mail stop. Duplicate the inside address on the envelope. (See section 3.3 for information about finding addresses of companies.)

Salutation

Address the reader as "Dear (person's name)" and end the salutation with a colon. If you know the person well, you can use his or her first name only. Otherwise, use either a title (Dear President Jones) or Mr. or Ms. If you know that a woman prefers Mrs. or Miss, use that. If you are uncomfortable with Mr. or Ms. in a salutation or do not know the reader's sex, use the reader's full name: "Dear Chris Harkness." Letter salutations can easily become sexist if you don't write to a specific, named individual, so if you can't determine the specific name, use an appropriate title such as "Dear Sales Manager." Avoid salutations like "Dear Sir" or "Gentlemen": these terms no longer can be used in a generic sense. The simplified form in Figure 16-3 omits the salutation and substitutes a subject line. Subject lines can also appear in letters that use a standard salutation.

Body

For easy reading, keep sentences and paragraphs fairly short in the body of the letter, and include only one idea per paragraph. Single-space within paragraphs and double-space between them. Section 16.3 tells what to include in the body of a letter. Try to keep a letter to one page, two at most.

Complimentary Close

Traditional closings are "Sincerely," "Sincerely yours," or "Yours truly." The complimentary close begins with a capital letter and ends with a comma; other words are not capitalized. If you know the reader well, you might chose a less formal closing like "Cordially" or "Regards." The simplified form omits the complimentary close.

Signature and Identification

Letters are always signed, but below the signature you should type your name. Many of us have only semilegible signatures, and you want to make sure that the reader knows how to spell your name. If you write as a representative of your organization, type your title below your name. If you are a woman and prefer to be addressed as Mrs. or Miss, indicate that with your typed name "(Mrs.) Carol Green." Sometimes, too, the company name appears in this identification; follow your company style in this matter.

Sincerely yours,
ELECTROVISION, INC.

Elizabeth Swanson

Elizabeth Swanson
Engineering Products

Optional Information Placed in the Lower Left Corner

- If someone else types the letter, the writer's initials appear first in capital letters, followed by the typist's initials in lowercase letters.

    LMR/st   or   LMR:st

- If you enclose additional material with the letter, you can indicate that by

    Enc.   or   Encl: résumé

For multiple enclosures:

    Enc. 2   or   Encl: Proposal
                          Background data sheet

- If you send copies of the letter to other people, it's a courtesy to let the primary reader know.

> cc: A. Wright
> T. Liu

Other Optional Information

- If you want to remind the reader of previous correspondence or a numbered contract or document, you can insert a reference line just before the salutation:

> Re: Order #E43487

Alternatively, you can include this information in the opening paragraph.
- You can also insert a subject line just before the salutation. (Remember that in the simplified form, the subject line replaces the salutation.)

> Subject: Cost Estimates of the Halifax Project

Again, this information can appear in the opening paragraph.

16.3 Types of Letters

All letters—no matter what the purpose or content—fall into three general categories: (1) letters that *ask*, (2) letters that *answer or provide information*, and (3) letters that intend to *persuade*. Table 16-1 lists many of the types of letters you may be called on to write. In letters that *ask*, you appeal to the reader to respond to a request; in letters that *answer or provide information*, you convey information, appreciation, or complaint; and in letters that intend to *persuade*, you attempt to sell readers on your position or your product.

Table 16-1 Common Types of Business Letters

Letters That Ask	Letters That Answer or Provide Information	Letters That Persuade
order	response to request, complaint, or invitation	sales
request	follow-up information	job application
complaint	announcement (good or bad)	invitation
collection	report	recommendation
credit	cover or transmittal	request
invitation	congratulations	complaint
	appreciation	proposal
	remittance	
	acceptance	
	resignation	

In addition to these specific purposes, letters build and maintain relationships in the business world. It's quite possible to learn to know a person well over the years simply through exchanging letters to carry on the routine business of your companies. If your letters are both professional and cordial, your correspondents will consider themselves your friends and will think well of your organization.

Sometimes types of letters overlap categories, but in general their purposes fall into one category or another. The following sections discuss in detail the content of one type of letter from each of the three categories in Table 16-1. In addition, Chapter 17 has examples and information about job-application letters, and Chapter 22 discusses cover or transmittal letters.

Letters That Ask

One subtype of a letter that asks is a request letter. In a request letter, you are asking your reader to do you a favor, so you should carefully plan both the content and organization. Your letter should be direct and short—one page if possible. Fill out a planning sheet (see section 1.6) to remind yourself about purpose, reader, topic, and tone. Keep your key words in mind as you draft the letter, and use as many key words as you can. Whatever you say, write the request to make it as easy as possible for your reader to respond.

Opening Paragraph

In one or two sentences, the opening paragraph must state your subject and purpose and establish a satisfactory tone. Rather than introduce yourself or give lengthy explanations, get to the point: be as specific as possible about what you want to learn. Here's an example of an effective opening paragraph by a professional:

> Dear Ms. Asher:
>
> I am writing to ask for your help in identifying the skills needed by students who want to become professional technical writers. Central University plans to set up a certificate program in technical writing, and we want that program to meet the needs of local industries. Please help us plan that program by answering the following questions.

Here is an example of a student's opening paragraph:

> Dear Mr. Bradley:
>
> Since I am considering the purchase of a high-quality four-track cassette recorder for my home studio, please send me information on your Studiomaster Studio 4.

Note the friendly, positive tone in both examples. Part of the success comes from the frank appeal for help and the shift at once from the *I* of the opening to the *you* of the rest of the sentence. One expert on business letters says to remember that the first paragraph belongs to the reader; your purpose is to introduce the reader to the topic. You can also write good openers if you use action verbs like *request, ask,* and *send.*

Middle Paragraphs

Middle paragraphs should explain the request and give the specifics of what you need. If you ask for more than three items or pieces of information, put them in a numbered list for easy reading. If appropriate, frame your request as a question or a series of questions to make your reader's job in responding easier. Here is an example of the middle paragraphs from a student letter that uses a list:

> As a student of technical writing at Green Hills State University, I am working on the investigation and evaluation of surge protectors. A large part of my report will be based on manufacturer-provided specifications. Those specifications I will be most concerned with are the following:
>
> 1. maximum spike dissipation
> 2. maximum spike voltage
> 3. maximum spike current
> 4. clamping spike voltage
> 5. surge current clamping ratio
> 6. clamping response time
> 7. noise attenuation
> 8. frequency range
> 9. filter network
>
> Even if you are able to supply me with only a few of the above specifications, it will be a great help.

Note how the middle paragraph asks for *very specific* information. It also explains the reason for the inquiry and tells how the information will be used.

Another student asks specific questions in the middle paragraph:

> I am writing to you because of the high ratings the Monark 872 received in the *Consumer Reports Buying Guide Issue*. Would you please answer these three questions about the Monark 872:
>
> 1. How is resistance applied to the flywheel?
> 2. What is a high-momentum flywheel, and what are some of its advantages?
> 3. Could you supply me with the name and address of a distributor or a local retailer of the Monark 872?

Don't word your request in such general terms that your reader doesn't know how to reply. If you say, "Please send me information on radial tires," you could be looking for the history, specifications, sales records, durability, or road test results of radial tires. You probably won't get the information you need, and you may be lucky to get any information at all.

Closing Paragraph

Use the last paragraph to motivate the reader to respond to your request. If you need the information by a certain date, give that specific date in this paragraph. Be sure to allow enough time to respond though; otherwise, you risk alienating your reader by sounding too demanding. It's better to give a specific date like "October 14" than to say "within three weeks," so your reader doesn't have to calculate the date. And be courteous enough to give the reason for your cutoff date.

If possible, the closing paragraph should show some benefit to the company or person addressed, perhaps by your interest in purchasing the product or by your offer to send a copy of the finished report. Also indicate your appreciation of an answer. You don't have to gush, but a simple "thank you" is appropriate. Remember that you want to end the letter without unprofessional begging but with a positive tone. Unless you are writing to a private individual, enclosing a stamped, self-addressed envelope is usually not necessary.

Here are closing paragraphs from student letters:

> I would much appreciate receiving brochures on General Electric Model #WWA8364V and Hotpoint model #WLW3700B by October 15 so that I can include the information in my report. I will be grateful for any assistance you are able to give me.

> My project is to be completed by November 14, 19__, and I would appreciate any help you could give me in answering these questions. I would be happy to send you a copy of my analysis when it is completed.

In the course of your work, you will often need to write letters that ask for information or for clarification of previous agreements or information. The letter in Figure 16-4 is an example of a request letter that forms part of the ongoing correspondence between the project director and the client. Notice that background information is given before the request. Notice also the friendly yet professional tone.

Letters That Answer or Provide Information

Another type of letter provides information, either initiated by the writer or responding to a request. The two situations may be rather different, even though both are informational letters.

An Informational Letter

A letter conveying factual information that is not intended to persuade can be straightforward and positive in tone. Because you are writing directly to a real human being, make the letter as personal as possible, using names and personal pronouns like *I* and *you* and short, active sentences. However, you should probably keep the tone semiformal rather than informal since you want your information to stand objectively on its own. You can again follow a three-part organization, as does the example in Figure 16-5.

Southern California
300 South Grand Avenue
Suite 3120
Los Angeles, CA 90071

May 29, 19___

Mr. Robert H. Dillar
City of Los Angeles Department of Airports
1 World Way
Los Angeles, CA 9009

Reference line

SUBJECT: Metro Green Line North Coast Segment
Guideway Clearance Through Parking Lot "C"
File GL 5.17.1

Dear Mr. Dillar:

Background

As discussed at our coordination meeting of May 16, 19___, the vertical clearance beneath our aerial guideway through Parking Lot "C" could be reduced to lessen the impact to the landing lights for Runways 24 R and L. At our meeting, a minimum clearance of 11'-0" was men-

Action request

tioned. Please advise if this is acceptable for Parking Lot "C" or if another clearance (such as 14'-6") is required for bus and emergency vehicle access.

We will study the vertical alignment in this area and keep you apprised of the results of our work. Thank you for your continued cooperation.

Sincerely,

Benjamin Olson

Benjamin Olson
Project Manager

cc: Joe Smith, Transcal II
Dave Simpson, RCC

Figure 16-4 A Letter of Request

WQC | Water Quality Control
Indianapolis, Indiana 46206

March 30, 19___

H.J. Rogers
Interstate Water Monitoring Agency
8800 River Parkway
Central City, Ohio 45801

Dear Hal:

Purpose

Your agency will be interested in the success we had during the last month with a carbon absorption system for emergency treatment of contaminated runoff water. This surface runoff came from improperly stored drums of industrial chemical waste and threatened both Watson Creek and the nearby water wells of private dwellings.

Our sampling of water in the drainage basin indicated the presence of 142 different chemical compounds, including organic compounds like benzene, toluene, ketones, xylene, styrene, and polychlorinated byphenyls (PCBs).

Details

We constructed three devices to meet the immediate problem:
1. trenches to intercept the surface runoff
2. a catchment basin to collect the runoff from the trenches
3. a carbon absorption unit to treat the contaminated water in the catchment basin.

The carbon absorption unit was constructed from two watertight trash dumpsters ($20' \times 10' \times 5'$) placed on a bed of crushed rock. The contaminated water was pumped into the bottom of the first dumpster and released through a perforated pipe into 8000 lbs. of crushed limestone with an aeration ring at the bottom to remove volatile organic compounds. The overflow from this dumpster went into the second dumpster and flowed through two separate cells of activated carbon. The treated water was then released into Watson Creek.

Closing—what use is to be made of information

James Chemical Laboratories conducted chemical analyses of the released water at four different locations and found a 99 percent reduction of organic compounds after treatment. We were pleased with the quick results we attained with this carbon absorption system on an emergency basis.

Hope you find this information useful.

Sincerely,

Tess

Tess H. Wade
Water Quality Control

Figure 16-5 An Informational Letter

A Response Letter

As you work in your profession, you will frequently need to respond to requests of different kinds. Response letters differ from informational letters because the reader looks to you to fulfill some specific need. Your letter will be judged by that reader in three ways:

1. *Content.* Does the response answer the questions and give the information needed?
2. *Tone.* Is the tone both friendly and businesslike? Is the response gracious rather than grudging? Will the reader react positively to what is said?
3. *Form.* Does the letter follow standard business form? Is it correct and clear?

In Figure 16-6, a customer relations supervisor responds to a student's request for technical information for a college report (Clark 1985). Notice how the opening paragraph both acknowledges the letter of request and sets a friendly tone. The writer can only partly fulfill the request for information, but the first paragraph in the body tells what he can do, emphasizing the attempt to be helpful. In the second body paragraph, instead of simply saying "We don't have that information," the writer suggests an alternate source. Finally, the writer closes with an action request of his own—and the positive "Good luck" as a final word.

A different kind of response letter is shown in Figure 16-7. Because the content of this letter is so positive ("You got the job!"), the writer probably did not need to worry about tone. Yet this letter is carefully crafted to be welcoming and to center on the reader and the reader's needs. Note the careful mix of *we* and *you* and yet the concentration on the specifics of the contract (Bell 1985). The student's response was "Isn't this a *nice letter*!"

Response letters sometimes must deny a request or disappoint the reader in some way. Under the circumstances, tone is a special consideration. Strive for a tone that is friendly but factual, and tactful but objective.

Letters Intended to Persuade

Whether you are selling a product, an idea, or a course of action, persuasion requires special sensitivity. You must consider the reader's needs and the reader's position relative to the writer. You must carefully plan the tone. You may want to review sections 1.2 on reader evaluation and 1.4 on tone before you tackle the draft of a persuasive letter.

Like most letters, a persuasive letter has three parts: an introduction that explains the purpose of the letter, the body, which elaborates on the services, products, or ideas and their superiority, and a conclusion that calls for some kind of action. Figure 16-8 is an example of a sales letter sent, along with detailed information, to prospective new customers. Using a computer mail-merge program, this letter can be personalized to specific individuals.

U.S. SUZUKI MOTOR CORPORATION

October 4, 19__

Michael A. Murray
199 So. 15th Street
San Jose, CA   95112

Dear Mr. Murray,

Reference

Thank you for your interest in Suzuki motorcycles. We will be more than happy to assist you with technical information for your comparison report.

Response to request

Please find enclosed brochures, spec sheets and photocopies from the Owner's Manual that should provide you with the majority of the information you have requested.

Other sources

Unfortunately, U.S. Suzuki does not publish some of the information you have asked for. Horse power, quarter mile performance, top speed, and fuel consumption are the specific items that we will be unable to provide you with. I would recommend, however, that you search through the motorcycle magazines in the library for their test reports on these models as they should provide the balance of the information you need.

Action request

If it would not be too great an inconvenience, we would be pleased if you would send us a copy of your results when it has been completed.

Good luck on your report!

Sincerely,

U.S. SUZUKI MOTOR CORPORATION

Mike Clark
Customer Relations Regional Supervisor

MC/kks

enclosures

Figure 16-6 A Response Letter

□ Silicon Compilers

2045 Hamilton Ave. • San Jose, CA 95125
408/371-2900

May 5, 19___

Marian Cochran
1253 S. 7th, #42
San Jose, CA 95112

Dear Marian:

Purpose and details

It is my pleasure to offer you the technical publications summer internship, with a salary of $8.00 per hour. This position will begin on May 12 with a completion date of July 15, 19___. May 12 through May 23, you will be working on a part time basis. From May 27 through July 15, 19___ you will be full time, working approximately 40 hours per week.

Action request

We look forward to your joining us at Silicon Compilers in what we hope will be a mutually advantageous internship. Please sign and return the enclosed copy of this letter as your acceptance of our offer. If you have any questions please call me at (408) 371-2900.

Sincerely,

Paula Bell

Paula Bell
Manager Technical Publications

I accept the position at Silicon Compilers Inc.

_____   _____

Candidate's Signature   Date

Figure 16-7 Response Letter to a Job Applicant

Space for inside address

Purpose

I have enclosed some information on our product line and a listing of current prices on our add-on peripherals.

Details intended to persuade

Over the past six years, PDP Systems has grown to become one of the largest distributors of computer peripherals in the country. PDP is presently an authorized distributor of Cyrix math co-processors, GTK video cards, and Kingston proprietary modules. Some of PDP's more popular products include genuine MS DOS and MS Windows 3.0 and hard disk drives from Seagate, Conner, and Western Digital.

In addition, PDP Systems works directly with Hyundai, Toshiba, Oki, and Samsung, as well as a number of other manufacturers. PDP's established relationships with the leaders in the memory industry assure the quality of our products and the competitiveness of our pricing.

As your distributor, PDP can offer you the following:

- Same-day shipping
- 24-hour RMA policy
- Strict quality control procedures
- Technical support staff
- One-year warranty on all products (five-year warranty on math co-processors)
- 15-day evaluation on all video cards

Action request

After reviewing our literature, please feel free to give me a call for more information and up-to-date pricing. I look forward to hearing from you soon.

Sincerely,

Space for personal signature

Sales Representative

2140 Bering Drive ● San Jose, California 95131 ● 408.944.0301 ● Fax 408.944.0811

Figure 16-8 A Letter Intended to Persuade

16.4 Electronic Communication

Business communication has been changing dramatically in the last 20 years, and those changes will continue in the next decades. Most of us now take word processing and sophisticated desktop publishing hardware like laser printers somewhat for granted.

Though writers are still responsible for content and accuracy, letter and memo writing is easier and faster with word processing. Text can be centered, displayed, changed, and stored, and letters can be customized without retyping.

Electronic mail, or *E-mail,* is an extension of word processing; letters and memos, instead of being printed and sent by mail, are transmitted by satellite or telephone lines to the addressee's computer. When that person logs on to the computer, the message is available. The reader can read, store, or print the message or send it to other people. The advantages of E-mail include instant transmission all over the world and the saving of money and paper. Disadvantages include the proliferation of messages (busy readers may not want to face 25 messages the first thing in the morning), readers who do not respond to a message (so you don't know if it was actually received), and the potential lack of a written document that records agreements and accomplishments. Note also that you may never know who else your reader may forward your message to.

One way to take advantage of electronic communication's speed and still have a written document is to fax it instead of using E-mail. A *fax machine* (facsimile maker) electronically scans a printed document and transmits the data to another fax machine, which prints a facsimile of the original document. Graphics as well as text can be faxed, and because of its advantages, faxing is often used for communication with those *outside* of one's company, while E-mail is used for communication *inside* one's company.

Even with electronic communication, though, the basic letter and memo forms are still in use, and you need to master the conventions so you can use them to your best advantage.

16.5 The Process of Writing Letters

Planning

In planning a business letter, first decide your purpose: what you want this letter to accomplish. After consulting the list in Table 16-1 and reviewing the material in Chapter 1, fill out a planning sheet like that shown in section 1.6.

Always try to write a letter to a specific, named individual, even if you have to call the company to find out who that individual is. Try to learn as much as you can about the primary reader, categorizing him or her by type, job title, and background (see section 1.2). If you should send copies to secondary readers, plan the letter to meet their needs as well—for example, by avoiding technical jargon if secondary readers are generalists or managers.

Since letters are personal, direct communications from you to the reader, your tone is very important. You can easily offend a reader by being rude, condescending, curt, or servile. Determine your position relative to your reader, asking yourself if you are writing *up* to a superior, *down* to a subordinate, or *horizontally* to a peer. Also determine whether the content of the letter is good news or bad news; the content will dictate both your tone and the method of organization you should adopt.

For example, here is part of a request letter that has a tone problem. The letter is written to volunteer CPR (cardio-pulmonary resuscitation) instructors asking for their help in teaching CPR to 5,000 people.

> . . . We need 250 CPR instructors to teach at the seven sites listed below. . . . We need instructors to sign up for four-hour shifts. We will be starting at 7:00 a.m. both days and will finish at 5:00 p.m. on Saturday and 3:30–4:00 p.m. on Sunday. You may sign up for as many shifts as you wish. We will try to give you your choice of site, but it may not be possible.
>
> All instructors MUST attend an orientation, as the classes won't be taught in the traditional modular method, and all instructors MUST be able to show proof of retraining. Only three orientations will be held, and your attendance at one is MANDATORY. . . .

Since this letter is written to experienced instructors, the readers are at least peers of the writer, but notice the brusque and rather rude repetition of the words *need, must,* and *mandatory,* giving the impression that the readers are subordinates (and not very bright ones either.)

If the writer wants to attract volunteers, the tone needs to be friendly and persuasive, recognizing how much these instructors will contribute. It could be done like this:

> . . . The only way we can train 5,000 people in this lifesaving skill is with your help: we need 250 CPR instructors to teach at the seven sites listed below. . . .
>
> Can you help by signing up for one or more four-hour shifts on Saturday and Sunday, October 18 and 19? We will be starting at 7:00 a.m. both days. . . . We will do our best to give you your choice of sites.
>
> These classes won't be taught in the traditional modular method, so retraining sessions are planned for the following three dates: October 2, 11, and 14. After retraining, you'll have an opportunity for a skill checkout with an instructor-trainer, and you'll be issued a new instructor card.
>
> Call (215) 333-2614 for details and to reserve a spot in a retraining session. You've already contributed much as a volunteer instructor, and we hope you can help us again.

Gathering Information

You may have at hand all the information you need before you start your letter. If not, spend some time verifying details. Be specific, whether you are asking for something, giving explanations and information, or trying to persuade the reader. Letter-writing costs are high because of the time involved in writing them, and you can avoid second letters of clarification if you are accurate and complete the first time.

Organizing

Good or Neutral News and Positive Persuasion

If your letter conveys good news, urges the reader to undertake positive action, or simply presents objective information, you should use this three-part organization:

1. Tell the reader the main point of the message.
2. Substantiate that main point with details.
3. Close with the action you want the reader to perform, or repeat the main point you want to make.

This direct approach works well with busy readers. They want to know at once the purpose and main point of your letter, and they will not object to a straightforward presentation.

Figure 16-9 shows how this basic organization works in a transmittal letter that is attached to an informal report. The first paragraph begins with the reason for the letter. Middle paragraphs deal with two major points: (1) a general assessment of the safety of the materials if they are properly used, and (2) questions the writer needs answered. The closing paragraph requests action and emphasizes the cooperative tone.

Bad News and Negative Persuasion

If your letter must present bad news, persuade a reader *not* to do something (or to do something he or she doesn't want to do), you need to change the method of organization. Your letter may still have the three basic parts, but they must be sandwiched between expressions of goodwill. You shouldn't begin with a direct statement of the purpose and main point. Sometimes the news is too harsh ("You didn't get the job"), and at other times the reader is hostile and needs to be met on some common ground. With bad news, you need to prepare the reader before you state the main point. Here is the organization to follow:

1. Set up a common ground with the reader or make a positive comment.
2. Give background information.
3. Tell the reader the main point of the message.
4. Support the main point.
5. State the action you want the reader to perform.
6. Close with a positive statement or one that conveys goodwill.

What types of letters will contain bad news and negative persuasion? Most often, these will be letters of complaint or replies to complaint letters (especially when the complaint must be denied). Sometimes you must write a letter about your inability to meet a schedule or about problems in production or personnel. As you advance in your chosen career, you may have to write negative letters to job applicants. In all these cases, you need to combine a roundabout method of organization with careful

TOX Inc.
1432 River Road
Austin, TX 78734

April 17, 19___

Paul Sanchez, Manager
Laboratory Operations
RPD Associates
8414 Wayfield Avenue
Waco, TX 76711

Dear Mr. Sanchez:

Purpose

An interim assessment of the metals and solvents you anticipate using in your new UHV lab is attached to this letter.

Details

In general, I do not anticipate any problems with these substances if the incompatibility and reactivity warnings are followed. I recommend that proper protective equipment should be used when handling any of the substances. For example, weighing of compounds should occur under a chemical exhaust hood, and safety goggles, gloves, and lab coats should be worn by laboratory workers.

Questions

To help me complete the assessment, please provide answers to the following two questions:

1. What isotope of Uranium are you intending to use? The literature indicates that there are three artificial isotopes that are not radioactive.
2. Do you anticipate cleaning this system at any time? If so, how do you intend to do this cleaning? My concern is that if cleaning of the system is to be done by your department, your employees could be exposed to serious health hazards.

Action request

If there is an aspect I have not addressed, or if you would like additional information on any of the compounds, please contact me at the above address or by calling 512-464-1323 extension 6541.

Sincerely yours,

Lisa Sisack

Lisa Sisack
Chemical Analysis

Enclosure

Figure 16-9 Informational Transmittal Letter

consideration of tone and style. Negative letters most often call for formal or semiformal style and for a tone that is positive or neutral, but always courteous. Remember the image of yourself and your company that you want to project. Figure 16-10 is a rejection letter to a job applicant—a difficult letter both to write and to receive. Notice that the bad news is cushioned by expressions of goodwill and that the tone is courteous and encouraging.

Writing, Reviewing, and Editing

Once you have determined an appropriate method of organization, write a draft of your letter. Make sure that you write the letter in terms of what's important to the reader. This is sometimes called the "you" attitude. For example:

Not   We will ship your order on August 17, 19__.

But   You can expect delivery of the complete order by August 20, 19__.

Not   We have sent the application to our engineering managers, and we will notify you of any action within two weeks.

But   Your application is currently being reviewed by engineering managers, and you will hear from us within the next two weeks.

When your draft is written, review the letter by checking the content against your planning sheet. Edit for technical accuracy, for punctuation, and for correct spelling of names as well as other words. The acceptance of your message often depends as much on the correctness of your letter as it does on the content, so you must take extra care with the details. Follow standard forms for salutations and closings, but avoid tired clichés like "In re your letter of February 12" or "Pursuant to the matter of." Instead, write in direct, uncomplicated sentences. See Chapter 8 for help on avoiding clichés. Have a classmate or co-worker read the draft. Then type the letter, following one of the three standard forms. Make sure to allow adequate margins, framing the letter on the page. Proofread the final copy for accuracy before you sign it. Since a letter is a communication from you and your company directed personally to a reader, your reputation as a professional rides on each letter you write.

Expression of goodwill

Rejection

Expression of goodwill

Dear Mr. Willis:

   Thank you for your interest in Carbide Engineering. We enjoyed meeting you at the plant on July 12 and were impressed with your background in CAE workstations.

   While you were one of the three finalists for the engineering position, we are sorry to tell you that the position has been filled from within the company. This in no way reflects on your credentials for the job, which are outstanding. We will keep your résumé in our active file should another opening occur.

   We wish you success in your career goals.

Figure 16-10 Bad-News Letter

Checklist for Writing Letters

1. Which standard form shall I choose?
   ___ flush left
   ___ modified block
   ___ simplified
2. Have I included all the standard parts and the necessary subparts?
   ___ heading
   ___ inside address
   ___ salutation
   ___ reference or subject line (if appropriate)
   ___ body
   ___ complimentary close
   ___ signature and identification
   ___ typist identification (if needed)
   ___ enclosure (if needed)
   ___ cc (if appropriate)
3. What type of letter do I need to write?
   ___ a letter that asks
   ___ a letter that answers or provides information
   ___ a letter than intends to persuade
4. What is the content?
   ___ good news
   ___ bad news
   ___ neutral news
5. What is my position relative to the reader, and how should I write to that reader?
   ___ up to superiors
   ___ down to subordinates
   ___ horizontally to peers
6. Have I carefully reviewed and edited to eliminate errors?

EXERCISES

1. Study letters A, B, and C to determine what they do well and what problems, if any, they contain. Look at the form, content, and tone. Be prepared to discuss your responses in class.

A.

September 24, 19___

Sanyo Electric Inc.
200 Riser Road
Little Ferry, NY

Attn: Microwave Oven Customer Service

Dear Sirs:

I am a university student doing a comparative study on microwave ovens for a semester project. I would appreciate it if you could send me some information regarding your microwave ovens. I want to know about the advantages of your brand, the warranty, safety, cooking features, and cost.

This information will enable me to compare your ovens with other brands and to compile a report recommending the best oven to meet the criteria mentioned above. This study is due the first week in November, and I would appreciate it if you could send me some information by the third week in October.

I would like to thank you in advance for your cooperation. The information will be beneficial to me now as well as in the future when I need to purchase a microwave oven.

Sincerely,

Carol

Carol A. Rekidder
60 Lake Drive
Kearney, NB

B.

September 17, 19___

Condor Computer Corp.
PO Box 8318
Ann Arbor, MI 48017

Subject: Condor 3 Database Software

I am researching and reporting on various database software on the market today. This report will be the foundation of several purchase decisions.

Could you please send me the following information on your Condor 3 software package:

1. Available technical reference material
2. System compatibility data

Could you also send any other information that is available on this subject.

Thank you,

Gregory T. Hart

Gregory T. Hart

C.

2330 Walnut Street
Springfield, MA
February 25, 19__

Apple Computer
20525 Mariani Avenue
Cupertino, CA 94014

Dear Sir or Madam:

I need some information about Apple computers. This information will be used in a report comparing different brands of personal computers.

This report is part of the technical education curriculum at the university. I will be comparing different brands of computers by price, expandability, software expansion, color or graphics capability, and service contracts offered.

I hope to bring out the advantages of the Apple over the other computers. With the help of the needed information, I shall be able to make consumers aware of the features of the Apple.

I would really appreciate your cooperation in this matter. It will help me if the information is sent right away.

Sincerely yours,

Lee Anson Judge

2. Analyze letter D for tone, organization, and total effect. Determine how you would respond to such a letter, and write a short paragraph explaining your response. Submit to your instructor and be prepared to discuss your response in class.

3. Choose a subject from your major field that you understand well. Write a brief informational letter to your writing professor explaining that subject in an objective manner and using language that your professor can follow even if he or she is not a specialist in that field.

4. Write a letter to the chairperson of your major department discussing some problem or concern you have noticed. Make the letter persuasive, giving the background of your concern and the reasons why you advocate a specific action.

5. If you are currently working, write a letter to your boss or manager suggesting a change in procedure or practice that will improve workplace conditions or enable you to better serve clients. Be objective in presenting the information, but try to persuade your superior to understand the reasons for making the change.

D.

Julie Munk
1431 Saratoga Ave. Apt. #101
Orlando, Fl 32825

Dear Ms. Munk:

You have an outstanding charge for the reason(s) indicated:
[ ] KEYS   [ ] DAMAGES   [X] FEES

Our records indicate that your balance is $19. If it is not paid immediately, holds will be placed on your records, grades, and transcripts. If these charges remain unpaid, your account will be sent to the tax offset program, where the charges will be deducted from any tax refund due you in the future.

If you have any questions regarding your charges or payment, please call or write to me at the number or address above. Any documentation related to your charges would be helpful.

Thank you for your immediate attention to this matter.

Cordially,

Ramona Clark
Ramona Clark
Accounting Technician

WRITING ASSIGNMENT

Write a letter of request for information for your term project. Follow the suggestions in section 3.3 to find addresses of specific companies, and obtain the names of individuals at those companies. Send the letters after review and revision. If this is a collaborative project, assign one or two people to write request letters. Before sending the letters, review them as a group for accurate content and correct form. After three weeks (or when your instructor suggests), send a second letter of request to those companies that have not responded.

Chapter 17

Résumés and Job-application Letters

The résumé and job-application letter you submit to a potential employer show your ability to organize and write effectively. Whatever your field of study may be, if you present yourself well in those two documents, you have an excellent chance of launching your career. Later, as that career advances, you will probably need to "market" yourself to other employers—again carefully crafting a résumé and application letter. For now and the rest of your career, this chapter may contain the most important information in this writing textbook.

Listen to an expert on the subject of résumés and job-application letters. Suzanne Birdwell is the western manager of human resources for a major telecommunications company. Her job responsibilities include all human resource activities: compensation and benefits, safety, training, affirmative action, employee relations. But a major responsibility is employment: As part of her job, Birdwell looks at hundreds of résumés for technical and professional positions every month.

When I asked her to comment on the job-application process for this book, Birdwell sent me more than a dozen annotated résumés and an equal number of job-application letters, all written within the past year. Here is a sampling of her comments on those documents: "letter reads well, résumé is well organized"; "letter is too long"; "I like the way he tried to balance his schooling with job experience"; "too formal in presentation of qualifications"; "too wordy—difficult to read"; "sloppy"; "doesn't look professional"; "tried to be cute—it's awful."

As you can see, when you market yourself and your abilities, the *content* of the documents and the *presentation* (or page design) are equally important. This chapter will give you suggestions for content and page design and offer examples that you can follow.

17.1 Definitions of Job-search Documents

The primary job-search document consists of two parts: a résumé and an accompanying letter of application or request. A *résumé* is a short and highly organized summary of your education, activities, and work experience. Its purpose is to tell the prospective employer of your skills and abilities. The word *résumé* comes from a French word meaning "sum up," and résumés are sometimes called fact or data sheets.

A *job-application letter* is a short business letter directed to a specific person at a company, which (1) tells of your interest in a specific job, (2) highlights applicable information from your résumé, and (3) requests an interview. The purpose of the letter is to obtain an interview. Job-application letters are sometimes called transmittal or cover letters, but while a standard cover or transmittal letter is subordinate to the main document, a job-application letter is at least as important as the résumé.

Other letters that are part of the job search include

- follow-up letters
- thank-you letters
- acceptance letters
- refusal letters or those withdrawing from consideration

Students sometimes ignore these additional letters, but writing them marks you as a professional, and can help secure the job you want, both now and in the future.

Because résumés and job-application letters differ significantly in form and organization, they will be discussed separately in this chapter. You should plan and write your résumé before you write a job-application letter. When you compose your résumé, tailor it to your own qualifications and the potential employer's needs; do *not* simply copy one of the forms given here.

17.2 Preparing for the Job Search

Planning

Before you write either a résumé or a job-application letter, you need to spend some time analyzing yourself, your abilities, and your short-term and long-term professional goals—and how you plan to achieve those goals. Begin by listing 10 achievements that have given you both personal satisfaction and a sense of accomplishment. These achievements can come from work, school, or personal experiences. Then, analyze these achievements to see where your interests and abilities lie. Next, write down each of the jobs you have had, and list three to five specific things you accomplished on each job. Finally, on a separate sheet of paper write down personal considerations, asking yourself questions like these:

- Where do I want to live? Am I willing to relocate? To travel on the job?
- Do I want to work for a big corporation? A small company?
- Can I work at night? On weekends? More than 40 hours a week?
- What's most important to me in a job?

Answering these questions will help you determine what kind of job you want and will help you construct your résumé and job-application letter.

Now is also a good time to begin a folder of information about yourself. Include your college transcripts, but add a list of your professors and the courses you took at all the schools you attended. Keep notes on each job you've had—your duties, your title, your supervisor's name and title, and the address of the company. Keep a list of the addresses where you have lived, and the time you lived there. Note organizations you've joined and the dates. In other words, keep track of all the things that happen to you. If you keep this folder current, you'll always have at hand the data needed for job applications, résumés, and security clearances. In the long run, such a folder can save you considerable time and effort.

Gathering Information

The next step in the job search is learning about prospective employers. Several months before you need a job, begin gathering information on companies. Attend job fairs at your college or in nearby cities and collect brochures from employers who attend. Find out what your college or university career planning and placement center can do for you; register for their training sessions and campus interviews. Research specific companies through company brochures and annual reports, or through articles in business and technical periodicals. Study the *Occupational Outlook Handbook* published by the U.S. Department of Labor to see where the jobs are expected to be. Begin to read newspaper ads, especially on Sundays when many companies advertise heavily, to learn what kinds of jobs are being advertised in your field. When you read, remember that only 10 percent of open positions are actually advertised, so other jobs are also available. In addition to investigating the sources listed above, you should tell your friends, relatives, and acquaintances that you're looking for a job.

Collecting Documents

Whatever type of job you hope to find, there's a good chance it will involve communication of some kind. As you prepare for the job search, then, it makes sense to collect documents that prove you know how to communicate. Such documents can include school assignments: a report from an engineering lab, a set of instructions written in your technical writing class, graphics designed for a feasibility study, an outline and slides for an oral report, an abstract or summary from the literature search for a science paper.

You can also include documents from work experience. Whatever your field, you should save samples of work you've created by yourself or as part of a team: letters, memos, reports, graphics, status reports, newsletters, instructions, procedures, descriptions. Even if your work experience does not directly relate to the job you want, employers are interested in proof of your ability to work with others, plan, organize, use word-processing tools, and write coherently.

Select three or four of your best documents, make good copies of them, and plan to bring them with you when you go for an interview. You may want to have a different selection for each company you interview, so it's helpful to have a variety.

In fact, if writing is an important part of the job you seek, you should not only collect these documents, you should organize them in a portfolio. Use a quality notebook with dividers and plastic sleeves, and include at least four or five different kinds of documents. If you've written a long manual, choose two or three pages to include, and write a paragraph explaining what this is, what your part in it was, and how it was created. At the beginning of the portfolio, place a table of contents. The whole portfolio should demonstrate you at your best—not only your writing skill, but your organizational, design, and presentation ability.

17.3 Résumé Form and Organization

Once you have planned your job search and gathered information about yourself and potential employers, you are ready to write a résumé. Most résumés contain the same basic components:

- personal data
- job or professional objective
- educational information
- work experience
- skills and abilities
- references

Sometimes, a summary of experience is also included. The order and arrangement of the components will differ depending on the kind of résumé you write. In this chapter, you will learn about the two major résumé types—chronological and analytical—and the advantages and disadvantages of each. You may want to try writing both to see which works better for you and the job for which you are applying.

Both chronological and analytical résumés should be short and succinct, preferably one page or at the most two. The typical résumé gets a first reading of *only 15–20 seconds;* therefore, if you ramble, the reader may never get to your key information. That very short reading time also means that the form must be easy to read. You need to arrange information carefully on the page: use white space as a frame and highlighter and choose boldface type, underlining, capital letters, and lists to direct the reader's eye. Always put the most important information at the top of the page. You need not write complete sentences in a résumé (in fact, you probably shouldn't), but you should emphasize with verbs what you can do or have done. Your résumé should be *absolutely free of errors or smudges;* that paper represents you to the reader, and spelling or typing errors will usually disqualify you at once. You don't need to pay for professional typesetting, but make sure that the copies made from your original are clear and on good quality white or off-white paper. You need not use fancy or colored type; make it your best professional effort without fancy touches. (See Chapter 12 for more on designing documents.)

Chronological Résumés

A *chronological résumé,* also called a traditional or "obituary" résumé, is relatively easy to write because it simply tells what you have done in the past. Figure 17-1 shows the chronological résumé of a graduating senior, with all the information on one page for easy reading.

Components of a Chronological Résumé

Each of the six major components is allotted a separate section on the page, with the most important or relevant information placed near the top of the page.

1. Personal Data Always put your name, address (including ZIP code), and phone number (including area code) at the top. If you have a temporary and permanent address, include both; it's important that a prospective employer be able to contact you easily. At one time, personal data also called for information like height, weight, age, marital status, children, health, hobbies, and interests. Today, however,

Olga Fernandez
4326 N. Elm Street
New London, IA 51003
(319) 296-3329

*The objective
provides focus*

**CAREER
OBJECTIVE**   An entry-level position in marketing, preferably in the areas of product development and research.

*Emphasis on
relevant education*

EDUCATION   UNIVERSITY OF IOWA, Iowa City, IA

Bachelor of Science in Marketing with a minor in English, expected December 1993. GPA: 3.5/4.0

Pertinent course work:

Marketing Management	Marketing Channels
Industrial Marketing	Marketing Research
Consumer Behavior	Business Logistics
Business Statistics	Programming in BASIC
Technical Writing	and FORTRAN

EXPERIENCE
1-91 to present   MARKETS, INC., Burlington, Iowa

*Emphasis on job
titles and specific
experience*

Marketing Research Interviewer

- Conduct surveys under diverse conditions and among varied consumer groups including college students, retail store and motion picture patrons, and housewives.
- Compile data from various studies to form basis for a national research project on consumer spending potential for microwave ovens.
- Write biweekly reports on research projects.

*Experience in
reverse chronological
order*

1-90 to 12-90   THE DAILY IOWAN, Student Newspaper, University of Iowa, Iowa City, IA

Advertising Manager

- Increased the number of regular advertisers from 9 to 21 over a period of two semesters.
- Eliminated paper's budget deficit.

ACTIVITIES   Member, American Marketing Association, University of Iowa; served as Vice-President

*White space sets off
items for easy reading*

Tutor, Statistical Analysis course (two terms)

Member, Future Farmers of America

REFERENCES   Available upon request.

Figure 17-1 Chronological Résumé of a Graduating Senior

most résumés omit that information. For one thing, it's illegal for employers to discriminate on the basis of sex or age, and for another, that information is often judged irrelevant to your job potential. You may choose to include some bits of personal information, but do so *only* if they will enhance your application and if they bear directly on the job you want.

Personal information beyond your name, address, and phone number should not go at the top of the résumé but at the bottom. Examples of personal information that might be appropriate include: (1) the fact that you are single if the job calls for extensive travel, (2) your fluency in a second language if the job involves international contact, (3) your status as a parent if the job calls for working with children or involves nurturing qualities. Evaluate personal information carefully to ensure that it will enhance your application. If it won't, leave it out.

2. Job or Professional Objective   Career advisers do not agree about whether or not you should clearly state the job you want on the résumé. If you make the objective too specific ("applications programmer"), you may eliminate yourself from consideration for related positions. If you make the objective too broad ("challenging position using my abilities in business"), the statement means nothing. Some experts thus recommend putting the objective only in your application letter. Remember, though, that the job objective provides a focus for the résumé. In many ways, it's like a thesis sentence—with the details of the résumé supporting the objective.

The best policy is to tailor each résumé to a specific situation. If you are responding to a job opening for which you qualify, stating the objective specifically is to your advantage. But if you're prospecting—to see what possibilities exist—you might want to be more general or list multiple objectives. In all cases, you should have researched a company *before* you write and submit a résumé, so you know what kinds of jobs are available there. You may want to create more than one résumé, each with a different objective and a slightly different emphasis. The added work and expense is worth it if you find the job you want. Using a word processor will make adapting your résumé to specific jobs much easier. If you are entering the job market, you may want to list both present and long-term objectives. For example:

Current Objective	To create clear, concise documentation and training aids for computer system and software end-users
Professional Goal	To supervise and/or manage a group producing technical documentation and training materials

3. Educational Information   If you are a graduating senior or a recent graduate, education should be near the top of the page because your education is probably more relevant to the job objective than is your work experience. In this section, you should list in *reverse chronological order* the schools you have attended, with city and state, inclusive dates of attendance (1989–1993), and degrees received or the date a degree is expected. Add your grade point average (GPA) or class standing if it is better than average. If your GPA is better in your major, indicate that. This section can also include such educational highlights as honors, awards, research studies, projects, particularly relevant course work, offices, and activities. Such activities are important to employers because they show your active involvement in college life. (While you're in college, it pays to do more than just attend class. It is never too late to join organizations that interest you.) Do not include high school activities on your résumé unless you are still in college or the high school work is unusual or relevant to the job.

4. Work Experience   List your work experience, again in *reverse chronological order,* leading off with your current or most recent job. Identify each job with a job title and tell what your primary duties were. Instead of writing complete sentences, use short phrases with active verbs and specific nouns. For example:

> *Trained* new employees on etching and masking procedures.
>
> *Wrote* documentation for the EZY interface.
>
> *Reported* monthly sales plan figures and variances from prior years to upper management.
>
> *Assigned* duties and projects for a staff of 7–10 salespersons.

Name the company, identify the city and state, and give the inclusive dates of employment. You may also want to name your supervisor. If you have had many part-time jobs unrelated to your career objective, you can group them ("summers of 1989–1991 fast-food cashier at McDonalds and Burger King"). Explain any large gaps in your employment history ("1992–1993 travel in Europe").

5. Skills and Abilities   Some chronological résumés add a section highlighting special skills. These could include fluency in a second language, knowledge of specific programming languages, skills such as welding or wirewrapping, or experience with systems like specific word processors or laboratory equipment. Determine what to include by testing the skill's relationship to your job objective, and omit this section unless you have pertinent skills to list.

6. References   Most employers want you to have references, but they do not expect to find them listed on the résumé. A simple line such as "References available on request" is sufficient. You may think that you need to include the references to fill up the page, but by themselves references don't mean anything to the employer. It's more important for you to find solid information from your education and work experience to fill this page.

Even though you don't list references, you *should* type a list of them, including name, job title, place of business, address and phone number; and you should bring several copies of that typed list with you to an interview. References should be people who know the quality of your work and can speak enthusiastically about it. Have three to five names, including at least one work manager, and for new graduates, one professor. Be sure to ask these individuals if they would be willing to give you a good reference. You don't want your sources to be surprised by a phone call asking about you and your work.

Arrange these six components of the résumé on the page so they are easy to read. Figure 17-2 shows a chronological résumé for an experienced technician seeking a job as a software engineer. Notice the emphasis on relevant job experience.

A Summary of Experience

Some résumés, especially those that cover more than one page, include a "summary of experience" or a "professional summary." Such a summary is placed immediately following the career objective; it usually is short—three to five lines— and it highlights the significant items that pertain to the career objective. For example, the beginning of a two-page résumé for a production management position could look like this:

Leonard J. Weilker
1810 Alabama Avenue
Ft. Collins, CO 80502
(303) 226-1234

Career Objective:	Management position in an operations, production control, or materials environment
Summary of Experience:	Over 15 years of successful operations production control experience supporting a wide variety of products. Experience includes 6 years of management.

If you are beginning your career, a summary of experience is probably not necessary unless the various jobs you have had during your college career contribute substantially to the career objective.

Advantages and Disadvantages of a Chronological Résumé

The advantages of a chronological résumé are (1) employers are familiar with this traditional format, (2) the chronological organization traces your past and tells the reader what you have done, and (3) the résumé is easy to compose. Disadvantages are (1) readers must figure out from your past what you are capable of doing because the résumé itself does not indicate your potential, (2) your experience may not support your job objective, and (3) the common traditional aspect may make your résumé look like all the others.

Analytical Résumés

An *analytical* or *skills* résumé emphasizes your qualifications and skills, especially as they relate to your stated job or professional objective. This kind of résumé requires you to think through your past and isolate the three to five skills or abilities that will make you a valuable employee in a particular career slot. Those skills and abilities can come from your education, paid or volunteer work experience, school or extracurricu-

David Godfrey
3609 Bailey Drive
Columbia, MO 65203
(314) 443-3250 or 442-7135

OBJECTIVE
To obtain a position as Software Engineer in a research and development environment.

EDUCATION
9/90 - 5/93

University of Missouri	Minor-Mathematics
Columbia, MO	Member of Upsilon Pi Epsilon
B.S. Computer Science	Honor Society
expected May 1993	Dean's List-G.P.A. 3.4/4.0

Honors and GPA show above-average scholarship

6/87 - 3/88

Suburban Technical School	Digital Computer Technology
Hempstead, NY	Received Diploma

Relevant Coursework		Relevant Job Experience	
Unix	C	DOS	Logic Analyzers
CMS	Pascal	Logic Design	VME Bus
MVS	JCL	PCAD/ORCAD	Ethernet LAN
VMS	BAL	Fiber Optics	Video Circuits
Rbase	ADA	Medical Imaging	68000

Specifics of coursework and job experience appeal to managers' needs

EMPLOYMENT
Computing Services University of Missouri, Columbia, MO 65203
9/90 to present

Senior Computer Technician. IBM/ZENITH authorized service center. Service computer equipment, install software, and provide technical assistance to users.

Work experience enhances degree in computer science

Fonar Corporation Melville, NY 11747.
10/87 - 7/90

Senior Electronics Technician. Involved in R&D projects based on an MRI scanner used in biological imaging. Project involvement included prototype stages: schematic capture, prototype assembly, debug, modifications, documentation. Installed the first model into an existing MRI scanner in the field at UCLA Medical Center. Appointed technical representative to Fonar Corporation for the Radiological Society of North America (RSNA) convention 1989.

Listed experience is specific

REFERENCES   Available upon request.

Figure 17-2 Chronological Résumé Emphasizing Work Experience

lar activities, hobbies, or research projects. An analytical résumé includes the same six components as a chronological résumé, but their emphasis and organization differ. In an analytical résumé, you want to emphasize your abilities and accomplishments in the order of importance, regardless of their chronological sequence. Study the analytical résumé in Figure 17-3 to see how the writer has grouped her skills and abilities to support the job she seeks.

Components of an Analytical Résumé
Here's how to treat the six sections in an analytical résumé.

1. *Personal Data*   As in a chronological résumé, the key data at the top should include your name, address with ZIP code, and phone number with area code. If you have a temporary and a permanent address, list both to make it easy for the employer to contact you. Other personal data may be located at the bottom, but your space is usually limited. One reason for including personal data is that it might give an interviewer a topic of mutual interest for discussion, but observe the cautions about personal data discussed in the section on chronological résumés.

2. *Job or Professional Objective*   Clearly stating your objective is crucial to an effective analytical résumé because it acts like a thesis sentence, with the résumé itself supporting the objective. If you are simultaneously applying for several jobs, you may have to tailor both the job objective and supporting résumé to each company. Objectives can be stated in many ways:

> A position in financial planning or management leading to a career in corporate finance
>
> Teaching woodworking in junior or senior high school, with an interest in coaching baseball or softball
>
> Technical editor
>
> Seeking a position in developing software systems. Flexible in area of assigned work but most interested in system programming, computer network systems, microprocessor applications, and graphics

3. *Skills and Abilities*   In an analytical résumé, this section is the most important, for it supports the job objective and sells you to the employer. At the left margin, you list three or four particular skills or qualifications, and you follow each skill with specific accomplishments that back it up. If you can show how you saved money with a new process, reorganized for more efficiency, and designed or implemented a successful operation, you will demonstrate your value. Skills, abilities, or qualifications can be listed in several ways:

- Names of job functions: manager, salesperson, designer, planner, coordinator, writer, programmer
- Adjectives describing the skill: supervisory, analytical, mechanical, organizational, research

Ann B. Dennis   Home (408) 258-8315
1637 Alta Vista Drive   Work (408) 904-0514
San Jose, CA 95127-1310

OBJECTIVE   *A full time position as a technical writer/editor.*

LEADERSHIP AND ORGANIZATIONAL SKILLS

Leadership
-- Performed duties of Executive Officer for a 120 soldier electronic communications/signal unit in the U.S. Army
-- Supervised 15 soldiers and civilians in the Personnel Administration Center responsible for job assignments and promotions

Organization
-- Managed unit property account in excess of $15 million
-- Coordinated and managed annual unit training budget of $100,000
-- Implemented budget guidelines and internal control regulations

WRITING AND EDITING EXPERIENCE

Writing
-- Co-wrote document for the Environmental Resource Center on the environmental and economic aspects of recyling
-- Compiled and analyzed technical data and prepared reports
-- Prepared annual personnel reviews and evaluations

Editing
-- Edited Electric Power Research Inst. *Battery Charger Manual*
-- Edited article for *The Annals of Dyslexia*
-- Edited unit standard operating procedures handbook for my unit

EMPLOYMENT HISTORY

Editing Consultant, April 1991 to present

Editor, Electrical Engineering, San Jose State Univ. and Electric Power Research Inst., San Jose, CA, May 1991 to October 1991

Administrative Assistant, Industrial and Systems Engineering, San Jose State Univ., San Jose, CA, August 1989 to present

Executive Officer, U.S. Army Information Systems Command, Berlin, Germany, First Lieutenant, June 1986 to July 1989

EDUCATION

Technical Writing Certificate, San Jose State Univ., Dec., 1991

BA in Instructional Technology, Calif. State Univ., Chico. May, 1986

Details from work experience support specific skills

Job titles and places of employment are summarized

Education is deemphasized

Figure 17-3   Analytical Résumé of a Person Changing Careers

- Applications of abilities: art, budgets, promotion, production, troubleshooting, human relations

The more specific you are with your backup information, the more believable you will be. For example, if you list a skill such as "budgeting," you need to say something like "Planned monthly budgets for Goldmark Department Store. Prepared daily and monthly cash flow records and reports. Improved cash flow reporting time by 50 percent." Then be prepared during your interview to back up any claims you make with specific evidence. What you are doing here is confidently driving home your accomplishments to the reader.

4. Work Experience   You may have already covered your work experience in the skills section, but many effective analytical résumés add a brief summary of employment in reverse chronological order. For example:

EMPLOYMENT
Winnebago County Accounting Office   9-92 to present
   Rockford, Illinois
Swanson Bookkeeping Service   1-91 to 9-92
   Elmhurst, Illinois
Mutual Insurance Company   12-89 to 12-90
   Chicago, Illinois
Agricultural Research Institute   Summer 1989
   Dekalb, Illinois

Explain any significant gaps: "August to December 1989: Travel in Mexico."

5. Education   Since some of the skills and abilities you list may well come from your education, this section is usually treated briefly (and in reverse chronological order) with a simple listing of schools attended, location, dates of attendance, and degrees.

EDUCATION
Vassar College   Poughkeepsie, NY   BS Biology 1992
City College of New York   New York, NY
     45 hours of biology and computer science
Willamette University   Salem, OR   BA Psychology 1986

6. References   As with a chronological résumé, obtain promises of references from at least three or four sources, and make the last line of your résumé read "References available on request." Take several copies of your typed list of references with you to the interview.

Figure 17-4 shows an example of an analytical résumé for a student about to graduate. Notice how specific he is in listing his computer-related experience.

Advantages and Disadvantages of an Analytical Résumé

The advantages of an analytical résumé are that it

- tells an employer what you think your greatest strengths and abilities are

William Magnuson

Permanent Address	Campus Address
244 Jackson St.	385 S. 10th St. #309
Houston, TX 77004	Houston, TX 77004
(713) 213-6687	(713) 277-1092

Both permanent and temporary addresses make contacting the writer easier

OBJECTIVE	Computer Programmer/Analyst
EXPERIENCE	Languages: Pascal, FORTRAN, BASIC, 8080 assembler, C. Operating Systems: MSDOS, UNIX, VMS, Novell Word Processors: WordPerfect, MS Word, Wang Editors: XEDIT, EDT Platforms: Sun/UNIX, HP3000, Windows 3.0 Work: Developed payroll, quarterly, and profit/loss reports. Debugged and modified programs. Courses: Data Structures, Combinatorics

Skills are grouped into major categories

Instructional	Conducted training on HP3000, Wang, and Plexus systems. Tutored math, electronics, and English.
Technical	Maintained HP3000 computers. Courses: Microcomputer Repair, Solid State Circuit Design.
Operational	Performed data entry in an accounting office, an insurance department, and an agricultural laboratory. Backed up data bases, performed essential reports, and managed operations.
Communication	Wrote documentation for a speadsheet program and a list management report. Wrote formal technical report of 30 pp. on compact disk players. Courses: Technical Writing, Speech

Employment and education are summarized

Employment	Montgomery High School District, Westfield, TX 9-91 to present
	Bookkeeping Service, Bryan, TX 1-89 to 8-91
	Highlands Insurance, Highlands, TX 10-87 to 12-88
	Institute for Science, Houston, TX 6-86 to 9-87
Education	BS: Mathematics and Computer Science Anticipated in December 1992. GPA 3.73 Texas Southern University, Houston, TX

References available upon request

Figure 17-4 **Analytical Résumé for a Graduating Senior**

389

- relates specific skills to the job objective
- allows you to combine educational, work, and unpaid experience in one section
- may attract the reader's attention because it is not as common as the chronological résumé

One disadvantage of an analytical résumé is that its nontraditional organization makes it more difficult to write than a chronological résumé. Thus, if you do not clearly state your job objective, and then specify your skills, this résumé will not be effective. On the other hand, if you clearly support the job objective, an analytical résumé can be a powerful persuasive tool.

Checklist for Writing a Résumé

1. Do I know my own strengths and my specific work requirements?
2. Have I researched job opportunities and this company?
3. Which form of résumé is appropriate for this job?
   __ chronological
   __ analytical
4. Do I have complete information for each of the following parts of a résumé?
   __ personal data
   __ job objective
   __ education
   __ work experience
   __ skills and abilities
   __ references
5. Do I need a section that summarizes my experience?
6. Have I listed education and work experience in reverse chronological order?
7. Have I used active verbs and specific nouns?
8. Is my résumé easy to read? Does it lead the reader's eye to important information? Will it tell key information about me in 15–20 seconds?

17.4 Job-application Letter Form and Organization

The letter accompanying a résumé is a standard business letter that follows the general organization of a letter of request (see Chapter 16). It should be only one page long, typed single-spaced on good quality bond paper, be well placed on the page, and have absolutely *no* errors, erasures, or whiteouts. Before you send your application letter, have it edited by at least two people whose judgment you trust.

If you are responding to an ad or job posting, address your letter to the person and department indicated in the announcement. Otherwise, try to send your letter to the specific manager for whom you would be working rather than to the personnel department. You may have to call a company to learn the specific name and job title. If you must send your letter to the personnel department, obtain a specific name in that department. Do not simply address your application letter to the company. Think about the reaction you have to a piece of mail addressed to "Occupant" as opposed to a letter addressed to you by name.

Like a letter of request, an application letter has three parts.

The Opening Paragraph

In the first paragraph, you tell specifically what job you are seeking, including the title of the position and where you heard about it. You want this paragraph to attract the reader's attention and single your letter out from others, so you need to tailor it to the specific company and job. You can't write a standard opening paragraph that will work for 20 different companies. Experts in job searching recommend openers like these:

If the position was advertised

> Your advertisement for the position of _____ in _____ is of special interest to me, and I would like to apply.

If someone known to the reader told you about it

> Tom Jasper, Vice-President of Sales and Marketing at Starr Manufacturing, suggested that I contact you about the engineering aide position you have open in your quality assurance department.

If you have already talked to this person

> As you requested in our telephone conversation on Friday, October 5, I am enclosing my résumé for your consideration. I feel my background in bio-technology could fill XXX Company's need for a laboratory assistant.

If you are building on past skills and experience

> Having doubled the sales in my region during the past three years, I am looking for a challenge like that offered by your organization. Your advertised opening for a sales manager with a record of increasing sales appears to be suited to my capabilities and interests.

If you are prospecting for an unadvertised job

> Because of my three years experience as _____ , I believe I am qualified for the position of _____ at _____ .

> Can _____ use someone with five years experience in designing _____ ?

> I would like to apply for a position in _____ department with the goal of becoming _____ .

Middle Paragraphs

Paragraphs in the middle of the letter should highlight the abilities and skills that will be of most use to your employer. This information is already summarized in your résumé, but your letter should emphasize qualifications that fit this particular job at this particular company. For a new graduate, such qualifications might include courses of study, work experience, grades, or outstanding accomplishments. Try to see yourself as a potential employer would, and talk specifically about what you have to offer. Mention the company name in this section, instead of just talking about "your company," and tell why you want to work there. One executive says: "Make me feel like *I'm* your first choice, not one among fifty. The cover letter should be directed toward my company and its needs. Don't make it a fill-in-the-blank exercise" (Rogalia 1980). In other words, make sure that your cover letter *builds* on the résumé, tying the general résumé to the specific job.

The letter in Figure 17-5 was written by a person seeking to change careers. Notice how she relates the experiences in her background to the particular job she is seeking.

The Closing Paragraph

In the closing paragraph, you come to the point of the letter: your desire for a job interview. Make it easy for the reader to contact you by giving your phone number and times when you can be reached. (Your phone number also appears in your résumé, of course, but it pays to repeat it here.) Be specific but brief in this paragraph, and take care to project a tone of self-confidence without being pushy. Closing paragraphs might be written like these:

> I would very much like an opportunity to discuss my qualifications with you, and I will be available for an interview after June 7, 19__. Would you please contact me any weekday after 3:00 p.m. at (216) 282-4587? I am looking forward to hearing from you.

> I look forward to discussing with you my qualifications and the research department's career opportunities. I hope that you will find a match between your recruiting needs and my professional interests. To schedule an interview, please contact me at (515) 324-2919. You can reach me at this number on Tuesday and Thursday afternoons. Thank you for your consideration.

> If the experience listed above would be valuable to your department, please contact me for an interview. My telephone number is (713) 461-4130.

> I would like to join your test lab because I think I can make a real contribution. May I meet with you to discuss my qualifications in more detail? I will call you in two weeks to schedule an interview, or you can reach me at (404) 332-1180.

One of the biggest challenges in writing the job-application letter is telling about yourself without constantly beginning sentences with *I am* or *I have*. While the examples above do not avoid this problem entirely, notice the techniques you can use:

3580 Birchwood Terrace
Houston, TX 77082

March 13, 19___

Ms. Jan Babcock, Employment Manager
Network Systems Company
2560 North First Street
Houston, TX 77049

Dear Ms. Babcock:

Opening paragraph identifies the specific job and source of information

I would like to be considered for the position of **Senior Technical Specialist** advertised in the Houston Post on February 26, 19___. I believe the skills acquired from my six years of experience in programming provide me with the qualifications you are looking for.

Bold type calls attention to key words

As a **programmer/analyst**, I supported financial and other corporate systems; evaluated client needs; wrote, tested, and installed system and program architectures; and prepared systems and operations documentation. I participated in all phases of system development projects. I functioned as the primary support for the Data Design Associates accounts payable system.

Highlights of my achievements include: (1) establishment of the first off-site disaster recovery plan for the cash disbursements system and (2) development of a PC-to-mainframe upload of expense billing data—a first for the company.

I am a full time student at Rice University and will graduate with a B.S. degree in Business Administration (Dec. 19___). My concentration in **Information Management Systems** and my minor in **Technical Writing** further enhance the qualifications you seek.

Closing paragraph requests an interview

Ms. Babcock, may I meet with you personally to discuss the possibility of working for Network Systems Company? Please contact me at my home at (713) 749-8780 after 2:00 P.M. to suggest a mutually convenient time. I look forward to meeting with you.

Sincerely,

Artemisa Diaz Valle

Artemisa Diaz Valle

Enc.: Résumé

Figure 17-5 Job-application Letter of a Person Changing Careers

Begin with a clause or phrase.

> If my experience fits your needs . . .
> To contact me, you can call . . .

Shift from yourself as subject to your experience or education.

> My classes in package design . . .
> The experience in agricultural inspection . . .

Concentrate on the company's needs.

> Your job announcement asks for someone with experience in . . .
> Paul Smith told me that the mapping department needs . . .

Figure 17-6 shows a job-application letter from a student about to graduate from college. Notice how specifically he relates his experience to the needs of the employer.

17.5   Completing the Job Search; Interviewing

Completing and Reviewing Résumés and Letters

Sections 17.3 and 17.4 gave several examples of résumés and job-application letters; obviously, those documents follow standard forms, but they must also be distinctly yours. Do not simply choose an organization or phrase from the samples and plug in your personal information. As you read the samples, you may be uncomfortable with the tone of some of them, and if you react this way, the tone is wrong for you. The way you write and organize résumés and application letters should reflect you at your professional best. That's why it is also crucial to avoid errors in punctuation, spelling, or content. Errors of any kind in these documents reflect badly on your competence. Job-application letters should be individually typed. Buy a new typewriter or printer ribbon and use the best and clearest typeface you can find. Take care even with mundane matters like the envelope—make it accurate and neat. Remember that out of every 100 applicants, 99 don't get the job. The more careful you are with even little matters, the better your changes are of being the *one* that is hired.

Besides reviewing for accuracy, you should review your résumé and job-application letter for accessibility, asking "How easy is it to find the important information?" Remember that what you submit will have many competitors, and you must make your point in 15–20 *seconds* of reading time. Plan for quick and easy reading by using white space, list markers, bold type, and underlining as highlighters. Ask friends to review these documents for you—reading quickly at first to see if they find the important information, then reading more carefully to check for details.

Interviewing

If you target an employer's needs in your letter and show in your résumé that you have the required experience and abilities, chances are good that you'll be asked to

17990 Laurelwood Lane
Durham, NC 27708

October 1, 19__

Ms. Joanne McCully
Dept. PM81
Cooper Fabrication, Inc.
3420 Central Expressway
Raleigh, NC 27603

Dear Ms. McCully:

I'm very interested in being considered for the Software Engineer position posted at the Duke Career Center on September 29, 19——.

Letter emphasizes key items in the résumé

The enclosed résumé highlights the many skills I'm ready to lend to Cooper Fabrication. Among them you'll find prerequisites specified in the job description, including a B.S. in Computer Science from Duke University, two years of microcomputer experience (with IBM's Research Division), and strong communication skills enhanced by a technical writing course at Duke.

Details of work experience provide credibility

In addition to the qualifications listed above, I have experience configuring PCs for laboratory data collection, machine control, inventory tracking, and project management. By implementing these tasks on micro-based systems, I can dramatically improve the efficiency of your laboratories, thus giving your scientists more time to actively pursue their goals.

Based on my education and experience, I'm ready to provide your scientists with the professional expertise they need to get the most from their laboratory microcomputers. Since I've had experience providing senior level scientists with support (at both IBM and SRI International), I'm confident that I can provide positive and lasting contributions to the staff at Cooper Fabrication.

Names of major employers are highlighted

Closing paragraph requests action

I'd be glad to stop by any day this week to answer questions you might have, or you're welcome to call me at 919-779-7169. I'm looking forward to hearing from you soon.

Sincerely,

John C. Day

John C. Day

Encl: Résumé

Figure 17-6 Job-application Letter for a Graduating Senior

interview. By now you've probably participated in a number of job interviews, either for summer full-time work or for part-time employment, but you may be surprised by the breadth and intensity of interviews when you seek employment as a professional.

For example, if you interview through your campus career center, you may begin with a short on-campus interview with a recruiter. The second interview may also be on campus, or more often will be at the company's location. You may meet with only a human resources specialist, or you may have a series of interviews with hiring managers. On-site interviews often include a tour of facilities and lunch or dinner with an employee.

You will need to prepare for these interviews by finding out as much as you can about the company before you go. Also review your own qualifications so you can talk with confidence about your experience and abilities. If appropriate, bring along samples of your work, including writing samples. Dress professionally even if you know the company is informal; once you are hired, you can be more casual.

Then, when the interview is over, increase your chances of being hired by writing excellent follow-up letters to your contacts.

17.6 Follow-up Letters

Follow-up letters are as important to the job hunt as are your résumé and initial job-application letter. Many job applicants ignore these letters, so if *you* write them, you will increase your chances of being favorably noticed. Tom Peters, author of *In Search of Excellence,* says, "I think there's a strong correlation between the little thank-you notes I get and the business, fortune, and fame of those who send them. . . . We wildly underestimate the power of the tiniest personal touch" (1991). Peters is talking about thank-you letters specifically, but in the job search, you have several opportunities to show employers that you know how to communicate effectively.

If you submit a job-application letter and résumé and have had no response within three or four weeks, write a follow-up letter in which you mention your previous letter and the date you sent it, repeat your request for an interview, and enclose another copy of your résumé. One manager told me that he *expects* follow-up letters. He will not consider anyone who is not interested enough in the job to follow up on the original application. In order to do this you will, of course, keep copies of all your letters and maintain a file of responses.

If you are granted an interview, respond immediately and enthusiastically. Then, within two days after the interview, write a brief thank-you letter to the person who interviewed you. If you had more than one major interview, write to each key person. Since only one of every 100 job seekers takes the time to write thank-you letters, your letter will be noticed favorably. Use this letter to reinforce your value to the company and to follow-up on any questions or comments that arose during the interview. Mention people's names and the company name. Repeat your interest in the job.

The easiest letter to write is the one accepting a job; don't neglect it. Again, be personal, enthusiastic, and convey your thanks, but don't gush. Confirm details such as your starting date and moving arrangements.

A harder letter to write is one in which you must reject a job offer or ask to be withdrawn from consideration. In your happiness at accepting one offer, don't risk antagonizing another company that has seriously considered you. Who knows—you may want to work for it in a few years. Thank the company and the individual for their time and consideration, and offer your regrets. Do not say anything negative about the company you are rejecting. You need not tell them what offer you have accepted; you can simply say, "I have decided to accept another offer." Close with an expression of goodwill. One type of rejection letter is shown in Figure 17-7, in which the writer repeats his thanks for the interview, even though he had already sent a thank-you letter.

Letter opening is complimentary; tone is positive

Rejection

Expression of good will in closing paragraph

> 3139 North 47th Street
> Milwaukee, WI 53218
> June 13, 19___
>
> Mr. Donald Cook, Administrator
> GRA Corporation
> 2486 Anderson Drive
> Raleigh, NC 27608
>
> Dear Mr. Cook:
>
> I was pleased to receive your offer of employment as an electrical engineer with GRA. I greatly appreciate the opportunity I had to visit your facilities, and I enjoyed talking with members of your engineering departments. I have reviewed your literature on work types, benefits, and salary, and found them quite satisfactory; however, at this time I am unable to accept your offer.
>
> Thank you for the interest you have show in my career. You may be sure that I will keep GRA in mind in the future.
>
> Sincerely yours,
>
> *Robert P. Johnston*
> Robert P. Johnston

Figure 17-7 Job-rejection Letter

Checklist for Writing Job-application Letters

1. Have I determined the specific goals and skills to highlight in my application letter?
2. Have I studied the company so I know its background and needs?
3. Does my letter follow a standard letter format? Is it correct in every detail of spelling and punctuation?
4. Does the opening paragraph tell what job I am seeking? Is it tailored to the company?
5. Do the middle paragraphs highlight my skills and abilities that fit this job?
6. Does the closing paragraph motivate the reader to grant me an interview?
7. Have I made it easy for the reader to contact me?
8. Do I need to write any of these other job-search letters?
   ___ follow-up
   ___ response to an interview
   ___ thank-you
   ___ acceptance
   ___ rejection
   ___ withdrawal

EXERCISE

Study job-application letters A, B, and C for content, tone, and correctness of form. Rewrite each letter or edit to correct the errors. Be prepared to discuss any problems or strong points.

A.

> Paul Hillyer
> PO Box 1467
> St. Louis MO
>
> DEXI Corporation
> Professional Staffing
> 5100 Henry Drive
> St. Louis, MO
>
> Dear Sir:
>
> I am currently employed as a development engineer at the software deivision of Entry Company. My primary interest is in the development of software systems which involves a knowledge system programming, microprocessor, graphics, and/or computer network systems. I think that my experience is relevant to your needs, and that my record in adapting and learning would be valuable to you.

I have heard of the positive working envirnment at your company and would appreciate being considered for a challanging position allowing further career growth.

If you have any questions, please contact me at (314) 442-9012 during normal business hours, or leave a message at (314) 342-3254.

Thank you very much and I look forward to hearing from you.

Yours sincerely,

Paul Hillyer

Paul Hillyer

B.

Jackie Ash
113 Wanton Way
Pittsburg, PA

May 15, 19___

To Whom It May Concern:

I would really like to work there. The people I met on the tour were very nice, plus I'd like to know more about your company. I would like a job researching or technical writing. I can write and I can do research if you tell me what it is that you're researching. I can get along with people.

I can work 8 to 5, and I sort of like to have weekends and holidays off. Actually I will take any job you have open, because the bottom line is, I really need the money.

I will send you my résumé upon request.

Thanking you in advance,

Jackie Ash

Jackie Ash

C.

4356 64 Street
Milwaukee, WI 53102
January 5, 19__

A. P. Washowicz
Manager, Production
Clean-Line Research, Inc.
875 North Avenue
Milwaukee, WI 53108

Dear Mr. Washowicz:

I am responding to your ad in the MILWAUKEE JOURNAL on January 3, 19__, for an assistant engineer on a research vessel. I will soon graduate with a degree in mechanical engineering, and I have had experience working with both Caterpillar diesel engines and hydraulics.

Although I don't have the three years experience you ask for, I am a hard worker and learn quickly. I took college classes in hydraulics and fluid mechanics. My résumé gives details.

Please contact me at the above address for an interview.

Sincerely,

J. Dolan
J. Dolan

WRITING ASSIGNMENTS

1. Choose a real job for which you are qualified or will be qualified when you graduate, and write either a chronological or analytical résumé that will support your application for that job. You can find job descriptions in the classified section of your newspaper, at your college career center, or by calling local companies.

Make this a real résumé of your education, experience, and skills. Follow the suggestions in this chapter, but remember not to reproduce the forms shown exactly; your résumé must be distinctly yours. Keep the résumé to one page. Attach your planning sheet (see section 1.6) and the job description to your résumé when you submit it for review. If possible, meet with three to four other students for peer review of your résumé while it is in draft. Then follow their suggestions to revise before submitting it to your instructor.

2. Following the form of a standard business letter, write a job-application letter for a real job, either the job for which you prepared your résumé or another job advertised or listed at your career center. Do not copy the sample letters in this chapter. Your letter must sound like you and reflect your skills and experience. Submit the letter for review in draft form to a small group of your classmates. Listen carefully to their comments and revise your letter as needed, concentrating on specific content, effective tone, and correct form.

REFERENCES

Peters, Tom. 1991. Thank yous may bring welcome changes. *San Jose Mercury News*. 2C, 2 September.

Rogala, Judith and Laura Liswood. 1980. The briefcase, *San Jose Mercury News*. 29 July.

Chapter 18

Instructions and Manuals

How did you learn what to do when filing your income-tax form? When changing the oil on your car? When installing a modem on your computer? Some people learn these tasks by trying one option after another until they succeed, but most of us learn either by listening to someone who already knows how to do the task or by following written instructions. Good instructions can save time, prevent costly errors, and make users' tasks easier. Bad instructions can confuse and frustrate users, even keeping people from buying a product. Because instructions are used by everyone from generalists to specialists, and because they provide the primary way of teaching readers how to do something, instructions are one of the most important kinds of technical writing.

Many professional writers spend most of their time writing instructions and manuals; they are called "information developers" or "learning products specialists" as well as "technical writers." In many companies, these writers begin working with engineers, programmers, and other experts at the early stages of product development. Craig Johnson, a design engineer of electronic test and measurement equipment, says, "Some people have proposed writing a user's manual as a prerequisite to writing the design documents that determine a new product's specifications. Considering that user satisfaction is the most important ingredient in any product, that might be a very good idea." As you can see, the emphasis on writing instructions is on meeting the user's needs, effectively teaching that reader what to do.

As a writer, you can teach a reader either how to *do* something (instructions) or how something happens or *is done* (procedures). Both ways of "telling how" are sequentially organized: one step follows another in the order of performance. But notice the difference in purpose: instructions *teach* performance of specific activities, and procedures *give information* about ongoing or repetitive activities. For example, you might write instructions on how to properly dispose of organic industrial wastes like oil, toluene, and benzene; in this case, the reader will be carrying out the disposal. You could also write a procedural description of how such wastes are generally handled; the reader simply wants information and is not going to do the job. Chapter 19 discusses procedures: describing how a process occurs. This chapter explains instructions: telling how to perform some action. It also explains how instructions and other techniques and forms are combined in manuals, and it briefly discusses special techniques such as quick reference cards and online instructions.

18.1   Definitions of Terms

Instructions are a form of writing telling a reader how to do something—either how to find a place or person (also called *directions*) or how to perform some physical operation (also called *orders*). Because of their distinctive form, simple instructions are often found in a separate brochure or booklet. Figure 18-1 is an example of a simple set of instructions included in a newsletter to clients of an insurance company (Customer News 1991). Since most Americans drive a car, instructions telling how to use jumper cables to start a car meet the needs of many kinds of readers.

In contrast to these simple instructions, think of the sets of detailed manuals needed to operate and maintain an industrial complex like an oil refinery. In a

The safest way to use jumper cables

A jump start can be dangerous. An incorrect hookup could damage a car's electronic systems. Worse, a spark might ignite battery fumes, causing explosion or injury.

It's best to avoid doing a jump start if possible. But if you must, here's a reminder to tuck into your glovebox on how to safely use jumper cables.

Before starting, check your car's owner's manual for special precautions, and wear safety goggles if you can to protect your eyes.

1. Attach one end of the red (positive) jumper cable to the positive (+) terminal of the dead battery.
2. Connect the other end of the red cable to the positive (+) terminal of the good battery.
3. Connect one end of the black (negative) cable to the negative (-) terminal of the good battery.
4. Attach the other end of the cable to the engine block or other unpainted metal surface of the stalled car. Clamp it away from the battery, carburetor or fuel line to keep potential sparks from igniting any fumes. Be sure the cables are clear of all fans, belts or other moving parts.

5. Be sure everyone has stepped back from both cars. Start the car with the good battery, then start the car with the dead battery.

6. Remove the cables in reverse order, starting with the cable attached to the engine block or metal ground.

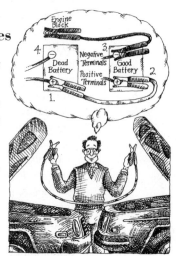

Steps are numbered

Each step begins with a verb

Figure 18-1 Simple Instructions for a Broad Range of Readers
Reprinted by permission of SAFECO Customer News © 1991.

complicated operation like that, each task must be clearly identified and then subdivided into component parts, and instructions must be designed for each part. The needed instructions will fill many manuals, with different manuals going to different types of readers. Complex instructions form the major part of manuals for installing, testing, operating, maintaining, and repairing equipment.

Readers come to instructions seeking direction and advice; they want to know how to do a specific task. Usually they are impatient to begin working and won't take time to read the instructions from beginning to end before they start. What's more, the saying "when all else fails, read the instructions" is too often true; as a writer you must plan for spot reading, hasty reading, and reading after the fact (when the reader has already done part of the task and may be in trouble). With that in mind, this chapter suggests ways to help you organize and write good instructions. Because they are often read *while* the reader is performing the action, instructions must be carefully organized, easy to follow, and very clear. Those requirements dictate the form used in successful written instructions.

18.2 The Form and Organization of Instructions

Good instructions generally consist of two parts: (1) an overview and (2) a set of individual steps that follow a sequential method of organization. The amount and variety of detail in the overview will vary depending on the reader's expertise and the complexity of the operation. The detail given in each step will also vary depending on whether the reader is doing the task for the first time or has performed similar tasks in the past.

The Title and Overview

The title defines the operation and its purpose, so it should include the words "How to" or "Instructions for" and the name of the operation. The title should be specific enough that readers know at once they have the right set of instructions. Titles might read

> **How to Assemble the Easy-Gro Spreader**
>
> **Cleaning Instructions for Automatic Coffee Makers**

The overview serves as an introduction to the instructions. Readers *should* study both the overview and the detailed steps before they begin, but many of them ignore the overview and begin immediately with step 1. Sometimes they get in trouble doing this, but they seldom blame themselves; instead, they blame the "lousy instructions." Your job as a writer, then, is to make overview information instantly accessible to impatient readers.

Required Overview Information

All overviews should include these four items:

1. An opening statement that, like the title, identifies the purpose and content of the instructions. For example (IBM 1981):

> **The Hardware Maintenance and Service Manual is the publication you use to isolate and repair any failure of a Field Replaceable Unit (FRU).**

2. A list of the materials needed. Materials are the items that will be used up while the job is being done, such as paint, oil, nails, and plastic tubing. For example:

> **For cleaning wood furniture, you will need:**
> - **gum turpentine**
> - **boiled linseed oil**

3. A list of tools or equipment required. Tools and equipment are items that are reusable and transferable from one job to another. For example:

> **Tools needed to hang wallpaper:**
> 1. **smoothing brush**   5. **blades**
> 2. **seam roller**   6. **plumb line**
> 3. **pasting brush**   7. **trim guide**
> 4. **razor knife**

4. A summary of the steps involved or a statement of the number of steps. For example:

> X-ray film developing includes these five major steps, all performed in darkroom conditions:
> 1. developing
> 2. rinsing
> 3. fixing
> 4. rerinsing
> 5. drying

You can help the reader keep on track by numbering the steps in the overview and then repeating those numbers as you explain the individual steps. You can also help the reader by using the same grammatical form for naming each step—the technique called parallelism that is described in section 6.4.

Optional Overview Information

Some instructions will need other information in the overview, such as:

- warnings and precautions that apply to the whole operation
- definitions of terms, especially for the first-time user
- discussion of the skills or expertise needed and, for beginners, perhaps assurance that the job can be accomplished with minimal skill
- directions on how to organize materials and set up equipment
- special conditions required, such as temperature or humidity range or a dust-free environment
- the time required to complete the whole task or certain segments of it
- troubleshooting information
- for very formal instructions, the theory of operations

The Individual Steps

Sequential Organization

The individual steps are the core of a set of instructions. They should be in sequential (chronological) order—that is, in the order in which they must be performed. If there are many steps, you should group related items under fewer than 10 umbrella terms so that the reader's short-term memory is not overwhelmed. (See section 4.2 for more on short-term memory.) Mark each major step by numbers, letters, or bullets, by indentation and white space, or by use of boldface type. Numbers are best because they help the reader follow the sequence. Start each step with the verb that tells the reader what to do. For example (Thames n.d.):

> To remove old finish:
> 1. *Place* the piece of furniture in a horizontal position so remover won't drip or run down.
> 2. *Apply* remover to one small section at a time—a leg, rung, or table leaf. *Protect* other parts with newspaper.

3. *Apply* a heavy coat of remover, brushing in only one direction if you use a wax-containing remover.
4. CAUTION: Immediately *wipe off* any remover that drips on parts, or stubborn spots will form.

Dangers, Warnings, Cautions, and Notes

Notice that the last instruction begins with the word *Caution.* Precautions warn of possible damage, and they should be inserted wherever the reader needs to be alerted. The terms used to indicate precautions are

- *Danger* Warns the reader of imminent hazard to people if proper procedures are not followed.
- *Warning* Warns of possible injury to the operator or others if proper procedures are not followed.
- *Caution* Calls attention to a situation or condition to prevent damage to equipment or materials.

Highlight warnings and cautions by boxing them or by using bold or red type to attract attention. If you box the precautions, center the word *WARNING* or *CAUTION* above the entry. For example (Varian 1986):

WARNING

The internal cryoarrays remain cold for some time after power is turned off. They could cause severe frostbite if handled when cold. Never disassemble the pump and handle unless you have determined that they are at room temperature.

CAUTION

Be sure not to bend the capillary tube at the braze joints when cleaning the hydrogen vapor bulb assembly because the tube will break easily.

In addition, you can use the heading *Note* to explain or add information. This heading is usually put at the left margin. For example:

Note: Both rods are inserted from the same side of the bracket.

MicroTrak ™

Chlamydia trachomatis
Culture Confirmation Test

**Test Procedure—
Coverslip Technique**

Simple graphics accompany each step

1 **Reconstitute reagent**
- Remove seal and rubber stopper
- Add 3.0 ml deionized* water
- Replace stopper
- Swirl gently to dissolve
- Equilibrate at room temperature for 15 minutes

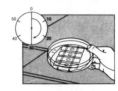

6 **Incubate**
- Incubate at 37° C for 30 minutes in **moist** chamber

2 **Fix monolayer**
- Remove medium from vial
- Immediately add ethanol
- Leave for 1 to 10 minutes
- Aspirate ethanol

7 **Aspirate excess reagent**
- Remove from incubator
- Aspirate excess reagent from monolayer

Hands in action poses help clarify steps

3 **Remove coverslip from vial**
- Remove coverslip from vial using forceps
- Touch edge of coverslip to blotting paper

8 **Rinse monolayer**
- Remove coverslip from slide using forceps
- Rinse in deionized* water for 10 seconds
- Touch edge of coverslip to blotting paper

4 **Place coverslip on slide**
- Place coverslip cell-side up on labeled slide

9 **Mount specimen**
- Place drop of mounting medium on clean, labeled slide
- Place coverslip, cell-side down, on top of drop

Instructions are grouped in meaningful units

5 **Stain monolayer**
- Moisten cells with deionized* water or PBS (if dry)
- Blot excess
- Add 30 µl of reagent to monolayer
- Ensure entire area is covered

10 **Examine**
- Examine monolayers using fluorescence microscope

*or distilled

Figure 18-2 Instructions with Line Drawings

Reproduced by permission of Syva Company, San Jose, CA.

Graphics

Help the reader by illustrating your words with plenty of graphics: photographs, exploded drawings, line drawings, and the other graphics explained in Chapter 11. Place the graphics next to or right after the explanation, and choose close-ups that show someone's hands performing the task if possible. Study the instructions in Figure 18-2 that show lab technicians how to perform a culture confirmation test. Notice how much the simple line drawings help explain the steps.

Lay out the steps on the page for easy readability. Don't try to save space by jamming all the instructions together; use white space as a frame around each step, making the steps easy to find when the reader is in the middle of the task. Use a type size that is large enough to read easily.

Writing Techniques

For clear instructions, write in the present tense, using the second person (*you*), active voice, and imperative (command) mood. That way you will be talking directly to the individual reader in a straightforward way. Imperative mood means that as writer, you are giving orders, not information.

Command	Remove the four screws on the battery case.
Information	The four screws on the battery case are removed.
	The installer removes the four screws on the battery case.

Note too that you can help the reader by carefully specifying "*the* four screws," indicating that there are *only* four screws. If you said "remove four screws," the reader might wonder if there were others not to be removed.

18.3 The Process of Writing Instructions

Planning

Before you can write instructions, you must understand clearly what readers will be able to do after reading what you write. Clarify this for yourself by writing either (1) a purpose statement or (2) a list of performance goals. You can also use these statements in the overview section of your instructions.

Purpose	to explain how to properly unpack and install XXX brand microwave oven
Performance Goals	After reading these instructions, the reader will be able to

   1. unpack the oven from the shipping carton
   2. safely install the oven in a desired location
   3. unlock the oven scale
   4. install the glass tray
   5. hook up the oven to grounded electrical service

Notice that by thinking through performance goals, you also begin to break the task into its component parts.

Next, learn as much as you can about your potential readers, because reader background and type will determine the vocabulary you can use and the skill level you can expect. Ask: What do they already know? What skills do they have? What terms do they understand, and what terms must be defined? Do they want to be led through every small detail, or do they simply want an outline of the general steps? Once you've characterized your readers, back up one step from the knowledge you think they have, and assume slightly *less* knowledge and skill. Those readers who are proficient will skip your added explanation, and those who need it will be grateful to have the details.

For instructions, the reader's purpose is simple: Readers want to be able to perform the action. Many instructions are written for generalists and first-time users, and you must explain any unfamiliar terms. In Figure 18-3, for example, you can see part of a glossary included in a brochure written for generalists on how to hang wallpaper. With these definitions, readers will more easily understand the individual steps.

Instructions must also specify all the tools, equipment, and materials needed to perform the task, so beginners will have everything they need before they start. If you leave out a piece of equipment, the reader can justifiably criticize your instructions.

Experienced readers—those who have performed similar operations—already know the basic equipment needed and are familiar with certain steps of the operation. They do not want definitions and excessive detail. The set of instructions in Figure 18-4 is written for technicians, and the writer assumes their familiarity with the parts of the pilot generator and the tools needed to perform the test. Those instructions would not be useful to a person who did not understand pilot generators. Thus, if you're unsure of the expertise of your readers, write for generalists and give them plenty of detail.

Remember to limit your topic in writing instructions. You need not describe the object, give theory, or even explain why the operation is carried out in a certain way. This type of added information clutters the instructions. The one exception is safety; it pays to explain *why* readers must do (or not do) something. For example:

CAUTION

DO NOT OPERATE THE MICROWAVE WHEN IT IS EMPTY. Microwave energy can damage the magnetron tube if the oven is operated empty for an extended period of time.

Even if you are writing very simple instructions for first-time users, your tone should be courteous and objective. Don't talk down to your readers or try to impress them with big words. To help you plan good instructions, use a planning sheet like that in section 1.6.

Note: The term "wallcoverings" in today's usage refers to any materials used to cover walls. "Wallpaper" is a traditional term used only for paper wallcoverings.

Adhesive — A paste for applying wallcovering to a wall.

Backing — A fabric, paper or synthetic material to which wallcoverings are laminated for strength and support.

Booking — The technique of folding the top and bottom of a wallpaper strip to the center, paste side to paste side. This allows the material to "relax" for several minutes so that it can assume its final dimension from the water in the paste. For booking time, refer to manufacturer's instructions.

Broad Knife — Wide putty knife used as a trim guide and also for scraping old wallpaper from walls. It may be used to spread spackle in order to fill holes or cracks in walls.

Butted Seam — To "butt" a seam, edges of wallcovering strips are fitted edge to edge without overlapping.

Deglossing — The breaking down of a glossy painted surface so that wallcovering will adhere to the surface properly.

Drop Cloth — A plastic sheet used to cover furniture or flooring for protection during pasting and papering.

Flock — A wallcovering with a velvet-like surface design.

Grasscloth — A handmade wallcovering, made by gluing woven grasses onto a paper backing.

Gre-Sof — A product used primarily for cleaning surfaces before refinishing or applying wallcoverings.

Figure 18-3 Glossary of Terms Needed to Understand Instructions
Reprinted with permission of Wallpapers to Go®.

Gathering Information

You need to know and understand the operation yourself before you can write good instructions. This means you must find and read specifications, descriptions, and procedures. As you read source material, determine the purpose of the source; for example, differentiate between general instructions (changing the oil in an automobile) and very specific instructions (changing the oil in a 1993 Ford Escort). If you are writing general instructions based on those specific to a certain product or piece of equipment, you will have to adapt and change details. If you are using instructions written for a product or operation similar to your model, don't simply copy that form and organization and plug in your new information—tempting as that might be. Think

FIELD TEST FOR SAFE IGNITION
(Turn Down Test)

WARNING

WITH PILOT GAS REDUCED TO LOWEST POINT WHICH WILL GENERATE THE MIILLI-VOLTAGE REQUIRED TO OPEN VALVE, THE MAIN BURNER MUST LIGHT OFF SMOOTHLY. IF IGNITION IS DELAYED DURING TEST, IMMEDIATELY SHUT OFF GAS. WAIT 5 MIN. BEFORE CONTINUING TEST.

1. Disconnect one pilot generator wire from valve terminal. Jumper thermostat terminals of valve.
2. Open pilot gas valve. Light pilot and adjust for MAXIMUM pilot flame. Check appearance to be sure flame is in proper position to ignite burner.
3. Open main burner gas cock. Touch and hold the loose pilot generator wire to valve terminal. Burner should light off smoothly within a few seconds.
4. Reduce pilot flame by adjustment valve until flame around cartridge is about $1/2$ of maximum. Wait two minutes for generation to stabilize. Touch and hold the loose pilot generator wire to valve terminal. If valve opens, burner should light off with no delay. Again reduce pilot gas flow slightly, wait two minutes, and test. Repeat until pilot flame is too low to produce sufficient millivoltage to open valve. Reposition pilot generator if necessary.
5. Remove jumpers from valve and reconnect all wires in their original positions. Make all wiring connections clean and tight. Readjust pilot gas flow for blue, non-blowing flame surrounding the generator cartridge. Be sure the main burner flame does not hit generator cartridge or snorkel.

Figure 18-4 Instructions for Technicians

Furnished courtesy of ITT General Controls. Copyright © 1977 ITT General Controls.

through the purpose, define the reader, and make sure that the approach is good before you imitate it, and even then, be careful not to plagiarize. Don't use the same wording or major ideas and pass them off as your own.

Besides using written material, a good way to get the information you need is by interviewing experts—the people who have designed, built, or tested the equipment. The most important and best way, though, is to observe, examine, or actually perform the task yourself.

Organizing

Instructions are relatively easy to organize because the individual steps follow a sequential method of organization: that is, what should be done first, what next, and so on. The trick is to *observe* the actions and anticipate what the reader must know before starting each one. Be careful to include all steps, and put them in the right order. You should not send the reader forward and then backtrack.

Not   Apply two coats of thin sealer 24 hours apart. Smooth with 3/0 steel wool, after the last coat has dried 24 hours.

But   Apply one coat of thin sealer. Wait 24 hours, and then apply a second coat. Wait 24 hours; then smooth the surface with 3/0 steel wool.

Writing, Reviewing, and Editing

Make your instructions easy to follow by keeping items of equal importance parallel as do the instructions in Figure 18-5 (Varian 1986). Section 6.2 explains parallel construction in detail.

Use numbers, letters, or subheadings to identify each major section, and begin with the verb indicating the action. As you write each step, think about the best way to illustrate what you are saying, and plan for graphics as you go. Notice how the exploded diagram in Figure 18-5 makes the parts and their relationships clear.

At the sentence level, follow these techniques to communicate most directly and clearly to the reader:

1. Write in the present tense and the command (imperative) mood, usually beginning each step with the verb that tells what to do. Write directly to the reader.
2. Use the active voice, so that the subject of the sentence is also the doer of the action.

   Not   Wing nuts are tightened securely.

   But   Tighten the three wing nuts securely.

3. Alert the reader to contingencies or conditions by putting that information first in the sentence and highlighting any warnings.

   Note: To distinguish the top of the panel from the bottom, locate the two holes drilled for casters in the bottom edge.

WARNING

If the burner does not ignite within 5 seconds, immediately shut off the gas.

6.5   PRESSURE RELIEF VALVE CLEANING/REPLACEMENT

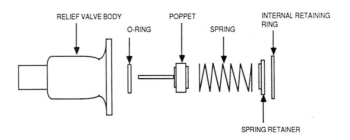

Exploded drawing identifies parts and shows the order of assembly

1. Remove the internal retaining ring located inside the relief valve body. Be careful not to lose parts as the assembly is spring-loaded.

2. Remove the spring retainer, spring, poppet valve and O-ring from the valve body.

3. Carefully remove the O-ring from its groove in the poppet valve.

4. Inspect the O-ring and discard if damaged.

5. Thoroughly clean the inside of the relief valve body making sure all surfaces are free of foreign matter and particulates.

6. Remove and clean the screen on the inside of the pump body. It is held in place by friction and should be replaced if damaged.

Notes are inserted where the reader needs the information

NOTE: The poppet pin guide and its retaining ring on the inside of the pump relief valve should not normally require removal. Visually inspect and replace if damaged or foreign matter is lodged around this area.

7. Reinstall the inside screen.

8. Apply a light film of vacuum grease to the poppet O-ring.

9. Install the O-ring on the poppet valve making sure it is in its groove and is not twisted or otherwise misapplied.

10. Attach the spring to the back side of the poppet valve so that the spring is held on the poppet recess groove.

11. Place the spring retainer (lip side toward spring) on the spring and insert the assembly in the relief valve.

12. Install the internal retaining ring into its groove in the relief valve body.

END

Figure 18-5 Parallel Construction in Instructions

Courtesy of Varian Associates. Copyright © 1986 by Varian Associates, Vacuum Products Division, Santa Clara, CA.

This construction puts the contingency at the beginning as a warning and the command itself near the end where it receives emphasis.

4. Give your instructions a clear title that identifies them as instructions and tells what the reader will learn. (See section 6.1 for more on writing titles.)
5. Do not generally omit the articles *a, an, the* in order to save space. Articles are important to help the reader understand what you are saying. (However, in some sciences, articles are conventionally omitted in instructions. Follow your company style guide in this matter.)

Table 18-1 summarizes these points.

Table 18-1 Techniques for Writing Instructions

Writer's purpose	Reader's purpose	Method of organization	Tense	Person	Voice	Mood
to tell how to do some action	to be able to perform some action	sequential	present: "Open the valve."	second: "(You) attach the heat shield to the vent."	active: "Insert the filter."	imperative (command): "Cover the surface."

Usability Testing

The best way to review your instructions is to have a typical member of your target audience perform the steps as you have written them, a process called "usability testing." Watch the operation in progress, noting when the tester backtracks, hesitates, or raises questions. At this time, do not give oral explanations to supplement your instructions; instead, note that what you have written is unclear. Later ask your tester where the problems occurred. Revise the instructions to clarify murky points. Then find *another* tester to try out the new version. Follow the same procedure to clarify and rewrite.

Checklist for Writing Instructions

1. Have I given my instructions a clear title that identifies them as instructions and tells what they are for?
2. Have I divided the instructions into an overview and individual steps?
3. Does my overview include the following sections?
   ___ a list of materials needed
   ___ a list of tools or equipment
   ___ a summary of steps involved or statement of the number of steps

4. In the overview, do I need to include any of the following?
  ___ dangers, warnings, and cautions applying to the whole operation
  ___ definitions of terms
  ___ a discussion of skills needed
  ___ directions for organizing materials and setting up equipment
  ___ a list of special conditions required
  ___ the time needed for completion
  ___ troubleshooting information
  ___ a theory of operations
5. In writing the individual steps, have I done the following?
  ___ arranged them chronologically
  ___ marked each step by numbers, letters, subheads, or white space
  ___ written steps in parallel structure
6. Is the document designed to do the following?
  ___ give dangers, warnings, or cautions where needed
  ___ use graphics to clarify
  ___ use white space to highlight information

18.4 Combined Forms in Manuals

As a person involved in a technical or scientific field, you will spend much time reading manuals—and perhaps writing them—because manuals provide the factual and theoretical backup for machines and procedures. Manuals answer questions like

- How does it work?
- How do I use it?
- What are its parts?
- How do I install it?
- How do I maintain it?
- How can it be fixed?

You may be familiar with the owner's manuals for your automobile, your stereo, or your other household appliances, and you may realize that these manuals are written for generalists. The owner's manual for your car, for example, tells you how to operate the heater and air conditioner, and gives you information about tire inflation and oil changes. But much about your car is not in that manual; in fact, a dozen or more other manuals probably exist to answer all the questions that designers, installers, and mechanics might pose.

Definition of a Manual

A *manual* is a document that gives both information and instructions. Many of the writing techniques discussed separately in this book come together in manuals: they usually combine definitions, descriptions, analyses, and explanations (all of which give information) with instructions and procedures (which tell how to do something or how something works). They are written for many purposes, including training, installation, operation, maintenance, troubleshooting, and repair. Figure 18-6 shows two pages of a manual that teaches writers how best to write manuals for users of Hewlett-Packard products. Notice how the manual combines information and instructions.

Types of Manuals

Manuals are written for first-time users and generalists as well as for experienced operators, technicians, and specialists. Thus, a single manual sometimes must provide (1) instruction for novices or first-time users, called "tutorials," (2) instruction by demonstration for more experienced users—sometimes called the "cookbook" approach, and (3) reference information for both. Alternatively, three separate manuals can provide this information.

Tutorials

Tutorials are task oriented, concentrating on what the reader must do, defining terms, clarifying concepts, describing and illustrating every move, and explaining the result of each action. Tutorials also provide positive feedback to assure novices that they can succeed and that the operations, if taken a step at a time, are not difficult. In other words, tutorials begin "at the bottom" with elementary concepts and definitions, and such simple operations as how to turn on a machine. Usually a tutorial will take a user through a sample or typical operation from beginning to end to illustrate the procedure, but will not go into all the possibilities of that procedure.

Demonstrations

Experienced operators or installers will be frustrated by a tutorial's detailed approach: they prefer a *demonstration* of a whole process "from the top down." This group of readers already knows in general what to do and wants instructions in what is sometimes called the "cookbook" format. If you think about how most recipes are written, you can understand the difference between a tutorial and a demonstration. A recipe written for an experienced cook assumes a great deal of kitchen knowledge. For example, consider these instructions: "Add two egg whites, beaten stiff but not dry." Such an instruction assumes that the reader knows how to separate the yolk from the white, how to beat eggs (with a spoon? fork? wire whisk? electric beater?), and what "stiff" means as opposed to "dry." By eliminating the details that such explanations would call for, demonstrations can cover much more material and make that information more available to the experienced reader.

These manuals are also task oriented, but they concentrate on involving users at once in meaningful tasks. In the computer industry, a new technique called *minimalism* encourages writers to eliminate lengthy introductions and step-by-step lessons. Readers are given enough instruction to get started and are encouraged to reason, improvise, and learn from their mistakes (Carrol 1990).

*The manual
provides
information,
definitions, and
examples*

The Learning Products
Installation Guide

A visual grid for hardware installation

Hardware installation is easier to understand when users can see clearly what is going on. Pictures are more efficient than words in this case. So the Editorial Design System provides a structure for pictures—the hardware task module.

A hardware task is organized in *panels*. These panels form a grid similar to a comic-strip grid. You can use the panels in this grid to provide three kinds of information:

● An *action panel* is a numbered instruction that tells the user to do something.

● A *go-to panel* is an unnumbered instruction that tells the user where to start and where to go next when the installation process is not linear. It is particularly useful for branching instructions.

● An *explanation panel* explains things the user needs to know to complete an instruction in an action panel. It may also explain branching choices.

> **If you have more than 256K of memory, follow these instructions to reset your memory.**
> **If you have less than 256K of memory, go to page 4.**

A go-to panel directs the user to the right set of instructions.

The combination of these three types of panels provides a flexible format for both simple, linear instructions and more complex, branching instructions. Also, the visual grid of this module is a page-saving presentation format. And because it allows more steps to appear on a single page, the user can get an all-at-once over-view of the steps—and thus feel more control over the process.

For software—tutorial or task modules

For most software installation, text works better than pictures. And a little text works better than a lot.

You can use the same tutorial or task modules that you use in a Task Reference to present the

50

Figure 18-6 Sample Manual Pages

The Learning Products
Installation Guide

Examples of panels show readers how to apply the information

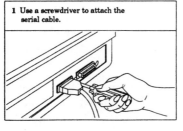

1 Use a screwdriver to attach the serial cable.

If the ROM version is earlier than A.01.06, you must install a ROM upgrade kit. See your owner's guide.

An action panel is one step in the installation process.

An explanation panel explains options or gives additional information for completing a step.

Pages are designed with white space to set off sections

installation process in manageable, easy-to-complete steps. Tutorial modules give you more room for explanations and step-by-step explanations. Task modules are best when the steps are simple and the results are obvious. You choose which is best for your product.

Chapters, sections, and go-to instructions aid complex installations

Of course, some installations are not simple at all. They may require several stages with lots of alternatives. So an Installation Guide can use chapters and sections to divide the process. Introductions to these chapters and sections provide essential back-

ground for the installation. The go-to panels in the hardware task module can also help direct the user through a complex maze of choices.

If the installation is not complex, however, it's better to avoid chapters and sections. They demand more reading and interrupt the flow of steps. If the installation is simple, the Installation Guide should be a simple map—in booklet style—through the installation process.

Figure 18-6 Continued

Reference Information

A good manual should also provide *reference information* about isolated topics—information that the reader may want out of sequence from the steps in a procedure. You can make reference information easy to find by including:

- a detailed table of contents
- a detailed index
- tabs or dividers for quick access to sections
- tables and figures like line graphs and exploded drawings
- lists

In writing reference information, you need to put yourself in the reader's place and ask where you might look for specific information and under what kinds of headings. Then you need to make sure that information is in the right location and cross-referenced if necessary.

Graphics in Manuals

Good manuals use graphics extensively because graphics can condense information and show relationships clearly. Manuals combine several kinds of writing, so different sections will also have different purposes. Thus, you need to understand the special purposes of particular graphics and choose accordingly. See Chapter 11.

Collaboration

The challenge in writing manuals is to meet the needs of all readers: to provide tutorials for the novice, demonstrations for the experienced user, and easily accessible reference material for both. To write a successful manual, you will need to examine existing manuals, study the reader and the reader's needs, familiarize yourself with the product, and work closely with product designers and developers. Once the manual is written, you will need to test its usability either by finding typical users and watching them perform or by working with a usability group at your company. In either situation, you can see that manual writing requires collaboration among many people.

18.5

Special Instructional Techniques: Quick Reference Cards and Online Instruction

Two special instructional techniques are becoming increasingly important in technical fields: the *quick reference card* (sometimes called a *quick start*) and *online instruction*. As someone preparing for a technical or scientific career, you need to know basic information about each technique.

Quick Reference Cards

A quick reference card is a brief summary version of a manual that provides the key information that users often need. It is usually on a single page and printed on durable card stock. It may be folded into a three- or four-page brochure, but is usually designed to be separate from the manual it summarizes. For example, the quick reference card for WordPerfect's IBM PC version 5.0 is a fourfold single sheet that lists features and their corresponding keystrokes. It also explains template colors; tells how to start WordPerfect; and how to save documents, clear the screen, and exit WordPerfect.

Online Instruction

Online documentation, according to Professor Henrietta Nickels Shirk, "is *not* paper-based documentation placed on the computer for access through a computer terminal. Rather, it is documentation written *specifically* for access *only* by means of a computer terminal" (1988). One type of online documentation is online instruction. Designed for display on the terminal screen, online instruction includes help facilities, computer-based training (tutorials), and reference information, including hypertext (nonlinear text).

Writing online instruction requires some different techniques than those presented in this chapter for paper-based instruction. Terminal screens cannot include as much information as can printed pages, and screens must be self-contained. Type size and style choices are different. Research is continuing, but experts currently recommend these techniques:

- *Chunking.* Online instructions need to be broken into small, manageable pieces. These pieces should also be limited in number to help short-term memory.
- *Queuing.* Instructions should be organized into a clear hierarchical structure. The hierarchy can be reinforced with clear headings and white space that separates sections.
- *Filtering.* Layering information for easy scanning is called filtering. Headings, lists, and screen layout help readers filter information.
- *Using visual cues.* Flowcharts, icons, and other graphics help readers find their way through online screens.
- *Creating consistent layout.* If screen designs are consistent, readers know what to expect from screen to screen.

As you continue in your career, chances are that whatever your major field, online instruction will be increasingly important. You can learn more about this technique by reading articles by experts like William Horton, David Jonassen, R. John Brockmann, and Ann Hill Duin.

EXERCISES

1. Sometimes you can best see the problems in instruction writing by analyzing instructions written for simple operations that most people know how to do and take for granted. The following examples come from an assignment to tell "How to Make a Peanut Butter and Jelly Sandwich." Parts of five student responses are shown. Analyze each example to determine its good and bad points, and be prepared to discuss your conclusions in class.

Example *A*.

3. Spreading the filling (filling = peanut butter and jelly)

Take a knife and dip into the peanut butter jar. Extract a wad of peanut butter about 1/2 as big as your fist. Place the wad in the middle of one of the slices. Take the knife and spread the peanut butter over the center toward the edges. Apply only enough pressure to spread the peanut butter without tearing the bread. Spread it so that the thickness of the peanut butter is about the same at the edges as at the center of the bread. . . .

Example *B*.

Stick the knife in the peanut butter and bring out some peanut butter on the knife. Spread the peanut butter on one side of a slice of bread. Put enough on the bread until the desired amount. Clean the knife. Put the knife in the jelly and bring out some jelly. Put this jelly on one side of the OTHER slice of bread. Spread jelly until desired amount. Take both slices of bread and put them together so that the peanut butter side and the jelly side of the bread touch. The tops of the bread should be together, but if they aren't, the sandwich should still be good.

Example *C*.

VI. Hold the knife at the dull end and dip the sharp end into the peanut butter jar and into the peanut butter.
VII. Remove about 4 handfuls of peanut butter and spread on bread.
VIII. Repeat steps VI through VII with the jelly.
IX. Now that the peanut butter and jelly are on one or both slices of bread, put two pieces of bread together with peanut butter and jelly in between.

Example *D*.

A. Open Bread and Remove Two Slices
1. This step is to prepare the bread used to contain the peanut butter and jelly. Opening the bread refers to opening the wrap that keeps the bread fresh.

2. To open the bread wrap, one must have the ability to remove the fastener that holds the bread wrap closed. Materials include bread wrapper and a plate. The plate is used to hold the bread and to contain any crumbs that may result from handling of the bread.

3. Removal of Bread
a. Holding the bread loaf in the left hand, grasp the fastener at one end of the wrapper and remove.
b. Be careful that the bread does not spill out of the wrapper.

    c. Reach into the open wrapper and grasp two (2) slices of bread. Note: If you skip the end pieces of the loaf, your bread will remain fresh in storage.

    d. Pull the bread through the open end of the wrapper and place on the plate in a side-by-side orientation. . . .

Example *E.*

I. Overview

    A. Definition. A peanut butter and jelly sandwich is a combination of peanut butter, jelly, and bread. By definition sandwich means to enclose, to be wedged between, to encase. In this case, the peanut butter and jelly are encased by two slices of bread.

    B. Intended Reader. These instructions are for the person who wants to make a peanut butter and jelly sandwich.

    C. Knowledge and Skills Needed

        1. How to use a knife to spread peanut butter and jelly.

        2. How to unscrew a cap-jar lid.

    D. Brief Overview of the Steps

        1. Peanut butter is removed from its container and spread on bread with a knife.

        2. Jelly is removed from its container and spread on bread with a knife.

        3. Sandwich is assembled.

II. Steps in Making the Sandwich
(Steps are detailed in this section.)

2. Write a set of good instructions for one of the following tasks. Be sure to follow the form and organization recommended in this chapter. Specify a purpose and reader.

how to cut up a chicken	how to load a dishwasher
how to trim a Christmas tree	how to change linens on a bed
how to fillet a fish	how to carve a jack o' lantern
how to build a campfire	how to shovel snow
how to split wood	how to prune a fruit tree
how to balance a checkbook	how to sew a button on a shirt
how to clean a chimney	how to wash a car
how to wean a calf	how to iron a shirt
how to get from where you live to your technical writing classroom	how to clean a vacuum cleaner
	how to plant peas and carrots in a garden

3. Analyze the following instructions written for customers of a large wholesale warehouse, and write a paragraph or two telling specifically what is good and what needs to be rewritten. Then rewrite as needed.

PROCEDURES FOR HANDLING SHORTAGES OR DAMAGES

SHORTAGE AT TIME OF DELIVERY

A. The driver will issue a credit request for merchandise that is verified by the driver as being short.

B. The driver may issue a credit request for average case cost for merchandise shortage which is not specifically identified. Credit will be issued for the cost of an average case of merchandise. After this credit is issued there will be no further adjustment made.

Note: The driver must report any shortage in excess of five (5) cases to our office immediately.

DAMAGE AT TIME OF DELIVERY

A. The driver will issue a credit request for identified damage only at time of delivery.

B. The driver will issue a credit request for damaged units within a case, rather than picking up the entire case.

DAMAGE AFTER DELIVERY

A. Merchandise found damaged after the time of delivery is returnable only if the case label is intact and the merchandise is in its original case. There will be no credit given for units that have been price marked or damaged at the store. The code number, invoice number, and the date on the case label must be on the request for pick up of damaged merchandise. The damaged product will then be picked up upon the next delivery.

B. Items over 30 days old will not be picked up.

---

4. Write a set of instructions telling a classmate how to make a paper airplane. In this case, do not use graphics but rely on the written word to convey your ideas to the reader. Limit your materials to basic school supplies such as regular bond paper, a ruler, a paper clip. Make two copies of the instructions. Turn in one copy to your instructor and a second copy to a classmate. On a launch day designated by your instructor, you will attempt to fly your classmate's airplane. Before that, follow the instructions to create the plane, and keep track of any problems you have. Write a memo to the writer, evaluating the instructions and pointing out good points and any problems. Turn in a copy of the memo to your instructor.

5. The instructions that follow were written for a generalist reader in response to exercise 2. Test the instructions by following them to wash at least one window. Make notes as you work about unclear steps or omitted information. Then write a memo to the writer, Travis Johnson, explaining what is done well and what needs revision.

METHOD FOR CLEANING THE GLASS OF WINDOWS

These instructions are for the proper way to remove any dirt or film that obstructs the visibility through glass windows on the inside and outside of the house.

PART 1: INTRODUCTION

Cleaning glass is a multistep process. Use of the steps listed below (in the order that they appear) will result in the desired effect: clean, clear glass windows with no streaks or smudge marks.

PART 2: MATERIALS AND TOOLS REQUIRED

Materials:

1. An ammonia based cleaning solution such as Windex®; at least one gallon of cleaner will be necessary
2. A roll of paper towels (use as necessary), four lint-free cloths, or a supply of newsprint paper (just use all of yesterday's paper, if you received one). The newsprint is used for the drying of windows and works just as well as the paper towels or lint-free cloths.
3. Water (use outside faucet)

Tools:

1. A rubber squeegee (a rubber strip attached to a rigid holder, used to remove liquid from the glass surface)
2. Plastic spray-bottle (about 16 oz. capacity) with the ability to spray fluids in a fine mist
3. 1 metal or plastic bucket (1 gallon capacity)
4. A ladder (if needed to reach high windows)
5. A large sponge (the kind used for washing cars)

PART 3: STEP-BY-STEP INSTRUCTIONS

A. Fill spray bottle with cleaning solution.
B. Pour about ½ gallon of cleaner in the bucket.

C. Since the outside windows are the dirtiest part, rinse them with water, using the hose and nozzle attached to an outside faucet.

D. Let excess water run off the window (outside windows only).

E. Let some of the cleaner (from the bucket) soak into the sponge, then squeeze out the excess so the sponge is thoroughly damp but not dripping with cleaning solution.

F. Use the sponge to clean the window surface. Move the sponge in a circular motion until the entire surface has been covered.

G. For small sections of glass, use paper towels, cloths, or newsprint to dry the window. Use a side-to-side motion of drying because when you use an up-and-down motion of drying on the other side of the glass, you will be able to see where the streaks are located.

H. For large sections of glass (i.e., big picture windows) use the rubber squeegee. After each drying stroke, wipe off the excess fluid with a paper towel or cloth.

I. On the inside portions of the windows, use the spray bottle to spray the cleaner in a fine mist over a manageable section of window (3 × 3 ft. sq.).

J. Dry windows with a polishing motion using paper towels or newsprint or lint-free cloths. Do this in an up-and-down direction for reasons noted in step G.

K. Check window for streaks. Use a little cleaner from the spray bottle and a clean cloth, towel, or piece of newsprint.

L. Put unused cleaner back in its original container and store away with other cleaning tools.

M. Throw away used paper towels or newsprint. If lint-free cloths were used, put them in the laundry to be washed.

6. Examine a manual from your major field or from the general consumer field. Analyze how well it

- defines terms
- describes objects and procedures
- gives instructions
- helps the novice user
- helps the experienced user
- provides reference information

Write a short evaluation report and be prepared to discuss strengths and weaknesses of the manual in class.

 WRITING ASSIGNMENT

Choose an operation in your major field of study that you know how to perform well. Write a set of instructions to tell a generalist how to perform this operation. Use source material, but do not simply copy someone else's information. Exchange instructions with a classmate. Then write a memo to your instructor, explaining the good and bad points of the instructions you read. Turn in a copy of the instructions along with the memo.

REFERENCES

Carroll, John. 1990. *The Nurnberg funnel: Designing minimalist instructions for practical computer skills.* Cambridge, MA: MIT Press.

IBM. 1981. *Hardware maintenance and service manual* 602S075. New York: IBM Corporation.

Shirk, Henrietta Nickels. 1988. Technical writers as computer scientists: The challenges of online documentation. In E. Barrett. *Text, context, and hypertext.* Cambridge, MA: MIT Press.

Thames, Gina. n.d. *Furniture restoration.* New York: Cornell University Cooperative Extension Service.

Varian Associates. 1986. *Cryopump operator's manual for CS 8 FA.* No. 87-400437. Santa Clara, CA: Varian Vacuum Products Division.

Chapter 19

Procedures

When oil fires rage in the desert, the news media explain how the experts extinguish such fires. When a famous professional quarterback has surgery to repair a ruptured disk, we learn how such surgery is carried out—complete with diagrams of the spinal column. These are *procedures*, providing information about how something works or the way in which something happens.

These procedures are simplified explanations for generalities, but they can also be complex when written for operators, technicians, and specialists. Jim Jahnke, for example, is an expert in air pollution control; he explains to plant managers how air pollution can be monitored and reduced. Ann Petersen, a medical writer, explains drug testing procedures to physicians and technicians. In the semiconductor field, Antonia Van Becker writes test procedures as a part of training manuals, and for a university agricultural extension program, Karen Zotz explains procedures in leadership and program development for low-income rural citizens.

Procedures are closely related to instructions because both kinds of writing "tell how." But procedures simply present information, whereas instructions, as explained in Chapter 18, specifically tell the reader how to do something.

19.1 Definitions of Terms

Procedures are a form of writing that explains how something happens or how it is done. The reader of the procedure is not likely to perform the operation but seeks to understand how it works. In a procedure, you describe the steps of a process much as you would describe the parts of a physical object, so procedures are often called *process descriptions*. For example, Part 1 of this book is called "The Process of Technical Writing," but it could also be called "A Procedure to Follow in Technical Writing." Likewise, section 19.3 is titled "The Process of Writing Procedures" because it describes a method that experienced writers use in explaining how something happens or is done.

The difference between physical descriptions and procedures is the method of organization. Descriptions are usually organized spatially, while procedures are organized chronologically. A procedure can be complete in itself in a letter, memo, brochure, or short report, or it can be part of a longer document like a proposal or feasibility report. When procedures are combined with definition and physical description, a reader can learn from one document what something is, how it looks, and how it functions. Combinations like these often appear in various kinds of manuals.

Procedures fall into three general groups: (1) those accomplished by machines or systems of machines, (2) those that occur in nature, and (3) those done by people. Table 19-1 lists typical procedures in each category. Notice how similar a procedure done by humans is to a set of instructions. Both involve people, but whereas instructions tell *how to do* something, procedures simply give information, telling *how something is done*. Readers of procedures also have a different purpose; they want information, not instruction.

Table 19-1 Types of Procedures

Machine Procedures	Natural Procedures	Human Procedures
how airbrakes stop a truck	how organic matter decays	how beer is made
how a chain saw cuts wood	how fog forms	how a plot of land is surveyed
how a sewage system operates	how cancer cells replicate	how an oil painting is cleaned
how a CPU powers on and self-tests	how shorelines change shape	how a calf is dehorned

19.2 The Form and Organization of Procedures

Procedures have three parts: (1) an introduction that provides an overview, (2) a step-by-step description of the action, and (3) a short conclusion that summarizes, discusses advantages and disadvantages, or relates the procedure to a larger ongoing project. Before you write a procedure, carefully assess your potential reader, because the reader's background and purpose will determine how much specialized vocabulary you can use, whether you can assume any familiarity with a similar procedure, and how much detail to include. Readers can range from generalists to specialists. Use a planning sheet like that in section 1.6 to help clarify your purpose and reader.

The Title and Introduction

The title should both explain the purpose and identify the procedure. You can identify the purpose with an *-ing* form of the verb and the procedure itself with key words:

> **Plotting a Cross-country Flight**
> **Fluoridating Municipal Water Supplies**

or you can use the word *How* and key words:

> **How a Laser Works**
> **How Scientists Predict Earthquakes**

or the word *process* or *procedure* and key words:

> **Startup Procedure for XXX Minicomputer**
> **The Process of Electroplating**

Notice that the title is different from one you would give a set of instructions. Instruction titles should begin with *How to, To,* or *Instructions for.*

The introduction should include four components: the purpose of the procedure, definitions of key terms, a list of tools and supplies, and an overview of the main steps. In addition, there may be optional components. Examples of each required component are given below.

Purpose

Begin with an opening statement that identifies the procedure and its purpose. For example:

> Coldheading is a widely used method of making screws; in coldheading a head of predetermined size and shape is formed or "upset" on one end of a cut blank of rod or wire (Clarke 1980).

> Crossdating is a method of dating wood (including trees, posts, and structural beams) by comparing the ring patterns in the older wood with the patterns in more recent wood (Phipps 1981).

Definitions

Be sure to define the process and the key terms essential to the process. Remember that your readers want information, so the procedure must begin with clear word definitions. For example:

> The electrocardiograph (EKG) is a very important test done in hospitals, doctors' offices, and other medical facilities; it helps in diagnosing and monitoring the effects of drugs, therapy, and illness on the heart. It is also used as an auxiliary test in the diagnosis and treatment of many other conditions, such as pregnancy and lung disease. The EKG is a graphic representation of the electrical activity of the heart. It is represented in the form of a time versus amplitude graph (Corbalis 1985).

For help in writing good definitions, see Chapter 9.

Equipment

Tell the reader what tools, supplies, or special conditions are required. You can do this in a list, as you would in instructions, or you can write a paragraph explaining what is needed and what each item is used for. You might do it this way (Bardellini 1980):

The only tools needed to perform a seismic refraction survey are the following:

- a seismograph—an instrument capable of measuring the time it takes a signal moving through a given medium to get from point A to point B.
- geophones   —small sensitive receivers of the signal (sound or percussion wave) as it passes by.
- seismic cable —the means by which the geophones pass the information on to the seismograph.
- shot point   —the point at which the signal emanates. The signal is produced by various means, for example, a hammer, explosives, or a mechanical device.

Overview

To provide an overview, indicate the main steps of the procedure. The best way is to name each major step, presenting the steps in a list if you have more than three. At least tell how many major steps there will be, so the reader can follow them as they unfold. For example:

> **The basic process of laser action involves four steps:**
> 1. excitation
> 2. spontaneous emission
> 3. stimulated emission
> 4. amplification

Optional Components

In addition to these four components in the introduction, you should add any of the following information that would apply:

- where and when the procedure takes place
- any special training or advanced preparation required
- how long the procedure takes
- the theory on which the procedure is based

Figure 19-1 shows how the theory behind a procedure might be explained (Leonard 1985). Note also how the theory grows out of the introduction and leads into the step-by-step description. Words in all capital letters are defined in the report's glossary.

Readers seeking information will probably read all of a well-written introduction; they are less likely to skip to step 1 than are readers of instructions. Even so, your introduction should be entirely functional; in other words, do not give details just because they are interesting to you.

The Step-by-Step Description

After the introduction, break the process into its component steps, keeping the number of major steps under nine. Within each step, you can subdivide; you will have to do that if the process has more than one thing happening at a time. Arrange the steps in a chronological method of organization and help the reader keep on track by devoting a separate paragraph or section to each major step. To show the sequence, you can number the steps, identify each step with a subheading, or if the steps are simple, use transitional words and phrases such as *next, then,* or *finally.* Unlike the steps of instructions, those of procedures are usually written as fully developed paragraphs.

The Conclusion

A good conclusion will help the reader understand either by summarizing the steps or by relating the procedure to a larger operation. With a long and complicated procedure, the reader needs a summary and reinforcement of the major steps. If the steps are relatively few and easy to understand, a stronger conclusion might

- discuss the advantages and disadvantages of the procedure
- show how this procedure is related to other work
- compare it to similar procedures
- list the results of the procedure
- indicate its importance or significance
- recommend this procedure or one procedure over another

In short procedures, the conclusion may be eliminated.

A well-written procedure will be a seamless document: careful chronological or sequential organization and judicious use of transitional words like *next, then,* and *finally* will help the reader follow the event as it unfolds. Like any good technical document, though, a procedure can be improved with access features like headings and numbered or bulleted lists, and design features like white space and type size. Because the content of a procedure can be very technical, you should use all the techniques at your disposal to make the writing and presentation clear.

Purpose of the procedure

Surgical Lasers Explained (Process of Laser Action)

A surgeon wants a tool that will cut and ablate (remove) tissue structures, coagulate vessels, and simultaneously sterilize the process. The laser, because of three special qualities, meets the surgeon's requirements. The three qualities that differentiate laser radiation from conventional radiation sources are listed below and illustrated in the figure.

Theory

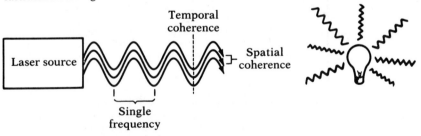

Laser radiation   Conventional radiation

Explanations of key terms

1. Single FREQUENCY. All laser radiation has the same frequency. This makes visible laser light MONOCHROMATIC. Some lasers produce invisible radiation; nevertheless, the beam is pure.

2. Temporal Coherence. All the wave fronts in a laser beam are in phase.

3. Spatial Coherence. The distance between waves in a laser beam remains virtually constant. This produces directed, or COLLIMATED, light as opposed to dispersed light.

Figure 19-1 A Procedure That Includes Theory

Overview

Pure frequency allows the surgeon to act on specific tissues. Temporal coherence allows the surgeon to deliver constant energy to the target tissue. Spatial coherence allows the surgeon to accurately focus the beam.

The process of laser action gives the laser these qualities. Although the laser beam is generated automatically within the laser device when it is turned on, a basic understanding of the process should give the surgeon added skill and confidence.

The basic process of laser action involves four steps:

1. Excitation
2. Spontaneous Emission
3. Stimulated Emission
4. Amplification

These steps all occur within the laser mechanism.

Step 1: Excitation

The first step is initiated when power is supplied. It is based on fundamental properties of matter.

In a normal state atoms exist as minute particles of matter each with a nucleus and a specific number of electrons. These electrons assume definite energy levels. In the absence of external energy, the electrons assume the lowest possible energy level. However, when energy is supplied, electrons can take higher energy levels. The energy absorbed must exactly equal the difference in energy between levels; there are no partial steps.

Electrons in elevated energy levels are called excited electrons. Thus, pumping energy into atoms excites electrons and initiates the excitation step.

Step 2: Spontaneous Emission

Steps are set off for easy understanding

Usually within one-hundredth of a MICROSECOND an excited electron will regain a lower energy level. Since the atom has less energy after the transition than before, energy must have been released. The released energy is exactly equal to the energy gained during excitation. The energy is given off as a PHOTON, a minute particle of light. This process is called spontaneous emission.

All the photons spontaneously emitted by similar atoms have the same frequency. They travel in all directions, however. Therefore, spontaneous emission alone will not create a collimated beam.

Step 3: Stimulated Emission

The step most crucial to laser beam generation is stimulated emission. In this step a photon strikes an electron that is already excited. When this happens, the energy of the photon is not absorbed by the atom as in excitation. Instead, the atom simultaneously releases the energy of the photon and the energy it would have released by spontaneous emission. The result is two photons with equal frequency. Not only do they have equal frequencies, but they also travel in the same direction.

Figure 19-1 Continued

Step 4: Amplification

As photons produced by stimulated emission travel, they can strike other excited electrons. If the process continues, the number of photons traveling in the same direction with identical frequency amplifies until a beam is produced.

Unless certain conditions exist, however, amplification cannot occur. Because excited electrons are very short-lived, a photon normally has little chance of striking one. The laser mechanism provides the necessary conditions.

Three basic components of the laser mechanism involved in the process of laser action are illustrated in the figure.

1. Energy source
2. Lasing medium
3. Optical resonator

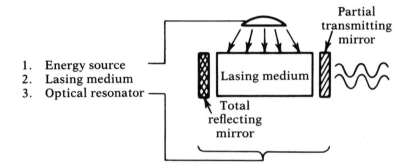

The energy source pumps energy into the LASING MEDIUM and excites electrons. Enough energy must be supplied to keep more electrons excited than unexcited. This condition is called a POPULATION INVERSION.

With abundant excited electrons, amplification can proceed. Photons traveling along the axis of the OPTICAL RESONATOR are reflected by the mirrors at each end. Then they travel back through the inverted population amplifying the beam in their direction.

After sufficient amplification, part of the collimated beam passes through the partially reflecting mirror at one end of the optical resonator. The reflected beam travels back through the medium stimulating enough photons to make up for the beam leaving the resonator. Thus, a sustained laser beam is produced.

Summary of Laser Action

Bringing all the steps together, the process of laser action occurs as follows. The energy source pumps energy into the lasing medium. Electrons within the medium become excited and they form a population inversion. At this stage conditions for stimulated emission exist, and a beam amplifies along the axis of the optical resonator. Part of this beam passes through the partially reflecting mirror and is available to the surgeon. The process is described by the word *laser*; it is the acronym of Light Amplification by Stimulated Emission of Radiation.

Figure 19-1 Continued

In Figure 19-2, a student writer explains the procedure of multitrack sound recording as part of a longer comparative report designed for generalists. Notice that the introduction defines the procedure, discusses needed equipment, and lists the three principal steps. Each step is then discussed in detail.

THE MULTITRACK SOUND RECORDING PROCESS

Definition of the procedure

Multitrack recording is the process of recording two or more audio signals onto separate tracks on a magnetic recording tape.

Purpose

The main purpose of multitrack recording is to simultaneously or consecutively record multiple sound sources so that they are synchronized on the recording tape, and so it is possible to individually adjust the level, equalization, and special effects on each sound source as it is being recorded or mixed down.

Multitrack recordings are generally made in a soundproof environment, such as a recording studio, so that no extraneous noises are recorded on the tape.

In making a multitrack recording, a recording engineer usually operates the recording equipment, while at least one performer is needed to produce the sounds. The sounds can be produced with musical instruments, voices, other recordings, or sound-effect devices.

Equipment

The equipment needed to make a multitrack recording is a multitrack tape recorder with a mixing section, and at least one microphone for each sound source. During recording, each sound source is picked up by a microphone, which converts the sound into an electronic audio signal. Each signal is fed into the mixing section of the recorder where the input level and tone are adjusted. The processed signals are then recorded onto their own separate channels or tracks on the recording tape. The resulting multitrack recording can then be mixed down onto a standard stereo tape for distribution.

Overview

The principal steps in making a multitrack recording are

1. setting up and adjusting the recording equipment
2. making a master recording
3. mixing down the master recording

1. SETTING UP AND ADJUSTING THE RECORDING EQUIPMENT

The recording engineer and the performers work together to set up the appropriate microphones for each sound source. The engineer decides the type and placement of these microphones. While the performers produce sample sounds, the engineer adjusts the input level of each signal as it is fed into the mixing section of the recorder.

Figure 19-2 A Procedure Written for Generalists

> Distortion results if the input level is set too high, and background
> noise results if the level is set too low. The engineer also adjusts the
> equalization or tone of each signal.
>
> ### 2. MAKING A MASTER RECORDING
>
> Once the recording equipment is set up and properly adjusted,
> the master recording is made. The engineer turns on the recorder and
> signals the performers to begin. During the performance, the engi-
> neer monitors the record levels and tone adjustments to make sure
> they stay within acceptable bounds. When the performance is over,
> the engineer stops the tape and rewinds it. If, after playing back the
> tape, the engineer and performers are satisfied with its quality, the
> tape is ready to be mixed down. Otherwise, the recording process
> may need to be repeated, or individual performers may do their parts
> over (overdubbing).
>
> ### 3. MIXING DOWN THE MASTER RECORDING
>
> Since most tape players can only play one or two tracks, master
> tapes having more than two tracks need to be mixed down onto more
> common two-track tape for distribution. The engineer uses a multi-
> track recorder and a second, mixdown recorder to accomplish this
> task. The master recording is played on the multitrack recorder, and
> the audio signals that are produced are routed through the mixing
> section. The signals are combined into a composite two-track signal,
> which is then recorded by the mixdown recorder. During this process,
> the engineer can again adjust the volume, equalization, and relative
> balance of all the tracks being mixed down from the master tape.
>
> ### CONCLUSION
>
> Multitrack sound recording can produce high-quality demonstra-
> tion tapes of specific musical techniques and sounds. The advantage
> of multitrack recording is in the degree of control the engineer has
> over the sound quality (Neher 1984).

*Numbered major
steps*

*Parallelism in
subheadings*

*Conclusion stressing
the advantage of this
procedure*

Figure 19-2 Continued

19.3   The Process of Writing Procedures

Planning

In planning a procedure, you need to keep in mind your reader's purpose in turning to
this piece of writing: the reader wants information. That means your purpose as writer

is to *explain,* not to give instructions. You will write clearer procedures if you remember that the reader will not be using what you say to actually perform the operation. It's confusing for readers if you mix the forms of instruction and procedures, so you need to have the differences clearly in mind.

You must also know who your readers will be and how much background information they already possess. One way to determine potential readers is by asking *where* this procedure will appear. If it is part of an article that will go to a professional journal or a general-purpose magazine, you can profile readers by looking at back issues and evaluating the vocabulary, level of detail, and breadth or depth of coverage in the published articles. If your procedures will appear as part of a report, you should find out where the report is going and to how many people. Fill out a planning sheet like that in section 1.6.

Many procedures are written to acquaint generalists with technological advances or to explain the results of scientific research. If you are writing for generalists, follow the suggestions in section 1.2 to keep the procedure interesting and easy to follow:

- avoid specialized language
- define terms
- use comparisons and anecdotes
- avoid equations and formulas
- keep sentences relatively short
- use graphics

Procedures may also be written for specialists and technicians—those who have extensive background in the general field but who may be unfamiliar with the procedure you are explaining. For these readers, you can rely on a shared vocabulary and on basic understanding of components. In Figure 19-3, the procedure of nondispersive infrared spectroscopy is explained as part of a handbook on air pollution measurement systems (EPA 1986, 3.0.9). Notice that the writer assumes understanding of basic terms and processes.

The tone you use in procedures will affect the reader's response. Readers who seek information and explanations should never be made to feel stupid because they don't already know. Your job as a technical expert and a communicator is to share your information objectively and confidently. The style you choose will influence your tone, so be aware that you *choose* a style—don't let it simply happen. Most technical procedures will be semiformal or formal in style, unless you are writing for children or popularizing science and technology; in those cases you can use an informal style.

Gathering Information

Accurate information is essential for procedures. Your readers come to you for explanations, expecting what they read to be correct and verifiable. That means you must go to original sources for your own information. You must read specifications, talk to experts, and observe the process in action whenever possible—taking notes as the various steps occur. You may have to spend much time gathering information:

Explanation of the key term

The procedure

Use of the procedure

> *Nondispersive infrared spectroscopy* utilizes infrared light in a limited range of the electromagnetic spectrum. The light is not s·anned or "dispersed" as with scanning laboratory spectrometers. In general, the light is filtered to select light wavelengths that will be absorbed by the molecules that are to be measured. The light passes through a gas cell that contains the flue gas extracted from the stack. A portion of the light from the lamp passes through a cell containing a reference gas that does not absorb the filtered light. A detector senses the amount of light absorption in the sample cell relative to the signal from the reference cell. Through proper calibration, the detector responses are electronically converted to pollutant concentration readings. A variant of this technique, called gas filter correlation spectroscopy, uses a reference cell that absorbs 100% of the light in the molecular absorption region of the pollutant.
>
> Infrared analyzers have been developed to measure gases such as SO_2, NO, NO_2, HCl, CO_2, and CO. The commercially available monitors differ primarily in the design of the detector and the level of rejection of interfering gases.

Figure 19-3   A Procedure Written for Specialists

taking measurements, timing operations, taking things apart, and studying the results of experiments. True specialists know much more about the subject than they can practically include in a written procedure. The additional information allows them to select and shape the material to meet the reader's needs more effectively.

Note that when you write procedures, you are explaining not a one-time phenomenon but a typical occurrence. Therefore, in gathering information you may have to examine many specific occurrences in order to generalize the typical procedure. For example, if you are writing "How Proposals Are Written to Secure Government Contracts," you must look not just at one sequence of events that resulted in a contract but at many such sequences.

Organizing and Reviewing

The three-part organization for a procedure combines hierarchical and chronological organization. The overall structure of a procedure is hierarchical; you divide the procedure into manageable parts with an overview, body, and conclusion. Within the body, however, the method of organization is chronological; that is, each individual step is discussed where it appears in the time sequence. Occasionally, the procedure is also causal—A causes B, which causes C, and so on. Even here, though, the organization is sequential and is best understood as a series of steps. Thus, a procedure is relatively easy to organize if the actions occur in a straight-line order.

If several actions occur simultaneously, the explanation in words becomes more complex, and you might need to supplement your explanation with a flow chart. In

Figure 19-4, the flow chart for installing a new computer system helps the reader understand the simultaneous actions that would occur in the 30 days between the pre-installation meeting and the point at which the system is functional.

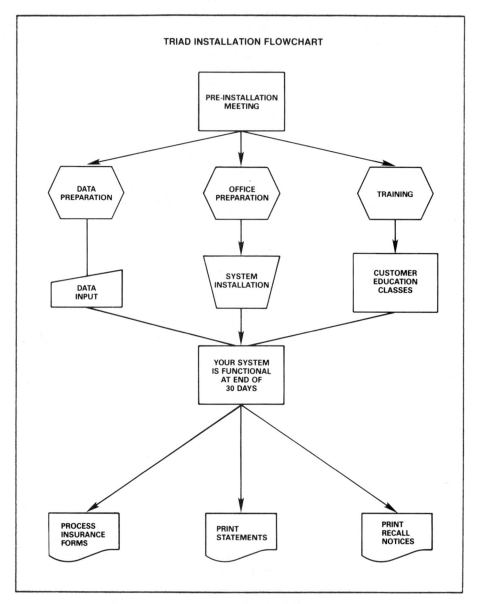

Figure 19-4 A Flow Chart to Help Explain a Procedure

Courtesy of Triad Systems Corporation. Copyright © 1984. Flowchart by Susan Wertz Constable.

With a visual representation of the procedure, you can take up one component at a time. You might discuss the simultaneous events in a left-to-right order:

- data preparation and data input
- office preparation and system installation
- training and customer education classes

Alternatively, you could break the procedure into six parts:

- data preparation
- office preparation
- training

and then

- data input
- system installation
- customer education classes

Once you have determined the organization you wish to use, review the details to ensure that you have not forgotten any steps. To check that your order of presentation is clear, ask someone unfamiliar with the procedure to look at your outline. Only after this kind of careful review are you ready to begin writing.

Writing

When you write, remember that you are giving information, not instructions. That means you should:

1. Write in the present tense and in indicative (information giving) mood. The present tense is most appropriate for the continuing actions described in procedures.

   > In planning a cross-country flight, the pilot locates the course line and draws it on the chart.

   > Convection heat transfer occurs when a fan blows heated air around food in an oven.

2. Keep the procedures objective and impersonal, using nouns and the third-person pronouns *he, she, it,* and *they*.

   > When the rotor is spun, the gyroscope acquires a high degree of rigidity, and its axle keeps pointing in the same direction, no matter how much the base is turned.

   > To use the locking clip, the operator first threads the vehicle's seat belt through the correct frame and buckles it. Then he or she pulls the shoulder belt portion until the lap portion is as tight as possible.

3. Use the active voice whenever you can. The active voice is effective when the procedure is performed by a person:

> **The materials testing technician (MTT) takes direct readings on the hammer gauge against vertical surfaces such as walls.**

> **Before the washing process begins, the operator selects the water level, cycle, water temperature, and speed of the motor.**

The passive voice may be used when the procedure is performed by a mechanism or by natural forces.

> **After the light has been gathered by the objective lenses and processed by the prisms, the image is focused and magnified by the ocular lenses.**

> **As the water enters the tube, air is forced into the air pressure dome.**

Even in these cases, though, you can usually use the active voice.

> **The objective lenses gather the light reflected from an object, and the prisms process that light. Then the ocular lenses focus and magnify the image.**

> **As the water enters the tube, it forces air into the air pressure dome.**

4. Write a title that clearly identifies what you have written as a procedure. Use function words and key content words like "How a Passive Solar System Works" or "The Process of Passive Solar Heating." (See section 6.1 for more on writing titles.)

Table 19-2 summarizes the techniques for procedure writing.

In addition, think of ways to enhance the words you are writing. Many procedures make good use of graphics, either to supplement the words or to shape the

Table 19-2 Techniques for Writing Procedures

Writer's purpose	Reader's purpose	Method of organization	Tense	Person	Voice	Mood
to tell how something happens or is done (information)	to learn how it works or how it happens	sequential chronological hierarchical	present: "The operator opens the valve."	third: "A technician attaches the heat shield" or "The heat shield is attached."	active or passive: "A mechanic inserts the filter" or "The filter is inserted."	indicative (giving information): "The surgeon covers the surface."

text. When you write for generalists, you will probably want to include graphics. In Figure 19-5, the flow chart (American Waterworks 1984) is cartoon-like because the writer wants to make the presentation enjoyable as well as informational. Even so, the information is serious and important, and the tone is objective, not condescending.

If the procedure is very complex, graphics can also enhance explanations for more technically oriented readers. In Figure 19-6, the writer summarizes the steps involved in digital radar landmass simulation (DRMLS). Notice how the simple graphics help clarify the explanations.

Reviewing, Editing, and Rewriting

To test the effectiveness of your procedure, have both experts and novices read it. Experts can tell you if what you say is accurate and if your chronological sequence is appropriate. Novices can tell you if your explanation makes sense to someone unfamiliar with the procedure. Apply what your reviewers say and rewrite any inaccurate or unclear passages. Make sure that you define any terms your readers don't know, and edit for correct punctuation, spelling, and sentence construction.

Checklist for Writing Procedures

1. Does my title clearly explain the purpose and identify the procedure?
2. Does the introduction include these four segments?
   ___ the purpose of the procedure
   ___ definitions of the procedure and any key terms
   ___ a list of needed equipment and supplies
   ___ a list of the major steps or an indication of the number of steps
3. Should I also include any of these optional segments?
   ___ the time and place of the procedure
   ___ special training or advanced preparation required
   ___ the time needed for the procedure
   ___ the theory on which it is based
4. In writing the individual steps, have I done the following?
   ___ arranged them chronologically
   ___ marked each step by number, letter, subhead, or paragraphing
   ___ subdivided if necessary
   ___ usually written in paragraphs
5. What type of conclusion is best for this document?
   ___ summary
   ___ advantages and disadvantages
   ___ relationship to other work
   ___ comparison with a similar procedure
   ___ results of the procedure
   ___ its importance or significance
   ___ a recommendation

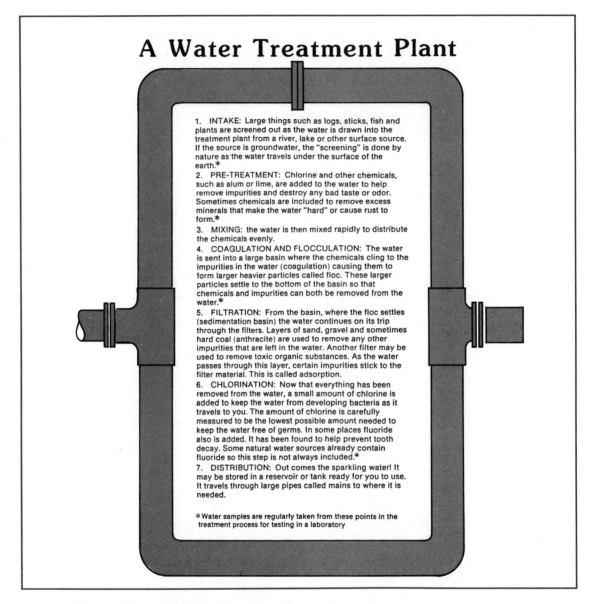

A Water Treatment Plant

1. INTAKE: Large things such as logs, sticks, fish and plants are screened out as the water is drawn into the treatment plant from a river, lake or other surface source. If the source is groundwater, the "screening" is done by nature as the water travels under the surface of the earth.*

2. PRE-TREATMENT: Chlorine and other chemicals, such as alum or lime, are added to the water to help remove impurities and destroy any bad taste or odor. Sometimes chemicals are included to remove excess minerals that make the water "hard" or cause rust to form.*

3. MIXING: the water is then mixed rapidly to distribute the chemicals evenly.

4. COAGULATION AND FLOCCULATION: The water is sent into a large basin where the chemicals cling to the impurities in the water (coagulation) causing them to form larger heavier particles called floc. These larger particles settle to the bottom of the basin so that chemicals and impurities can both be removed from the water.*

5. FILTRATION: From the basin, where the floc settles (sedimentation basin) the water continues on its trip through the filters. Layers of sand, gravel and sometimes hard coal (anthracite) are used to remove any other impurities that are left in the water. Another filter may be used to remove toxic organic substances. As the water passes through this layer, certain impurities stick to the filter material. This is called adsorption.

6. CHLORINATION: Now that everything has been removed from the water, a small amount of chlorine is added to keep the water from developing bacteria as it travels to you. The amount of chlorine is carefully measured to be the lowest possible amount needed to keep the water free of germs. In some places fluoride also is added. It has been found to help prevent tooth decay. Some natural water sources already contain fluoride so this step is not always included.*

7. DISTRIBUTION: Out comes the sparkling water! It may be stored in a reservoir or tank ready for you to use. It travels through large pipes called mains to where it is needed.

*Water samples are regularly taken from these points in the treatment process for testing in a laboratory

Figure 19-5 Graphics to Help Explain a Procedure to Generalists

Reprinted from *Story of Drinking Water*, by permission. Copyright © 1984, American Water Works Association.

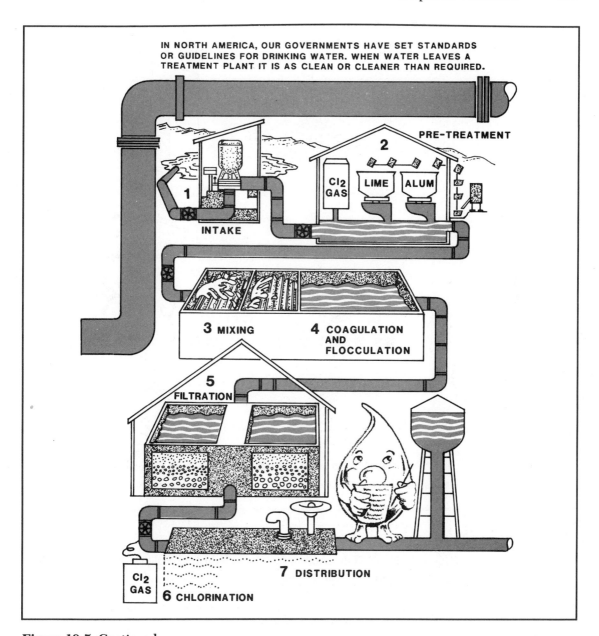

Figure 19-5 Continued

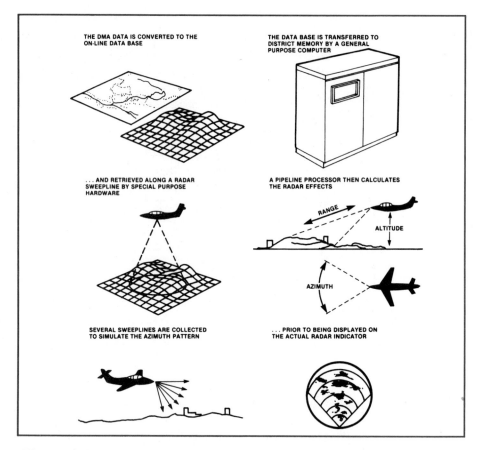

Figure 19-6 Graphics to Help Explain a Procedure to Specialists
Reprinted by permission. © The Singer Company, Link Flight Simulation Division.

 EXERCISES

1. The following procedure was written by an industrial design student (Liang 1986). The intended reader is an office manager who wants to use a desktop copier but does not understand the process involved in producing the copies. Determine whether this procedure has all its needed parts and whether it would be clear to the reader. Be prepared to discuss your evaluation. Rewrite it if so advised.

> **COPYING PROCESS**
>
> Electrophotographic copying system is based on electrostatic charging and photoconductivity; it is a complex copying process which office copiers today are based upon and can be classified into six distinct steps:

CHARGE: A corona discharge caused by air breakdown uniformly charges the surface of the photoconductor.

EXPOSE: Light reflected from the image discharges the normally insulating photoconductor and produces a latent image, a charge pattern on the photoconductor that mirrors the information to be transformed into the real image.

DEVELOP: Electrostatically charged and pigmented polymer particles, called toner, are brought into the vicinity of the oppositely charged latent image; they adhere to the latent image, transforming it into a real image.

TRANSFER: The developed toner on the photoconductor is transferred to paper by giving the back of the paper a charge opposite to that on the toner particles.

CLEAN: The photoconductor is cleaned of any excess toner using, for example, coronas, lamps, brushes or scraper blades (Burland 1986).

FUSE: The image is permanently fixed to the paper as it passes between two heated rollers or under fusing heat.

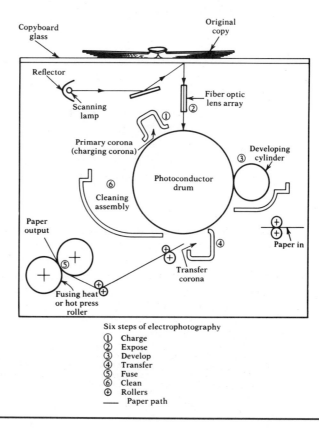

Six steps of electrophotography
① Charge
② Expose
③ Develop
④ Transfer
⑤ Fuse
⑥ Clean
⊕ Rollers
—— Paper path

2. Following the suggestions in this chapter, write a good procedure for one of the following or another approved by your instructor.

Procedures done by people:

how a fish is filleted   how yarn is spun
how copper is mined   how CPR is given
how roses are properly pruned   how vinegar is made commercially

Procedures done by machines:

how an electric razor works   how shock absorbers work
how a squirt gun works   how a snorkel works
how a fire extinguisher works   how a snow blower works

Procedures in nature:

how a tooth grows   how funnel clouds form
how sound is transmitted   how a skunk emits its charge
how glaciers form   how wood becomes petrified
how frost forms   how leaves turn color in autumn

3. Choose a procedure that is carried out in your major field and write a brief explanation of it for junior high school students. Use simple definitions and tie it to their experience with some kind of analogy. For help with definitions, see Chapter 9.

 WRITING ASSIGNMENT

As part of your formal research or feasibility report, write a procedural description. Check your outline to see if you have placed it effectively. If not, revise your organization. Seek review of the procedure. If you are working as a team, assign one person to write any needed procedures. Have that person work with the graphics specialist to supplement the procedure with a flow chart or other illustration.

REFERENCES

Bardellini, David. 1980. *The process of seismic refraction.* Unpublished report. San Jose State University.

Clarke, Donald, ed. 1978. Screws. *Encyclopedia of how its made.* New York: A & W Publishing.

Corbalis, Mary. 1985. *How an electrocardiograph works.* Unpublished report. San Jose State University.

EPA. 1986. *Quality assurance handbook for air pollution measurement systems*, vol. 3. Research Triangle Park, NC: Environmental and Support Laboratory.

Leonard, David. 1985. *An evaluation of Nd:YAG surgical lasers for Allied Hospital.* Unpublished report. San Jose State University.

Liang, Alice. 1986. *Desktop copiers*. Unpublished report. San Jose State University.

Neher, Jonathan. 1984. *The multitrack sound recording process*. Unpublished report. San Jose State University.

Phipps, R. L. and J. McGowan. 1981. *Tree rings*. No. 361-618/114. Alexandria, VA: U.S. Geological Survey.

Chapter 20

Abstracts and Summaries

Jamie Oliff works for a pharmaceutical company, Pam Patterson in aerospace technology—occupations different in content and scope. Yet both these professionals write abstracts and summaries as an important part of their job assignments. Other professionals—in fields ranging from graphics software to nuclear power, from semiconductors to test equipment, from probation departments to air pollution control—also list abstract and summary writing as a major assignment. In addition, studies show that when given a long proposal or report to read, *everybody* reads the abstract or summary, while only 15 percent read the body (Westinghouse 1970). Obviously then, abstract and summary writing is an important skill to acquire.

Abstracts and summaries condense material from a longer, major document, allowing busy readers either (1) to grasp the main ideas without reading the whole thing, or (2) to review, for clarification, the main points they have just read. Both abstracts and the several kinds of summaries depend on the main document for their content. Because they must be comprehensive as well as succinct, they are not easy to write, but readers of technical documents depend on them for help in understanding complex subjects.

20.1 Definitions of Terms

Abstract and *summary* are general terms, each having several variations. In both, the material in a longer document is condensed or abridged, but the essential points are preserved.

Abstracts

An *abstract* is a condensation of a longer document that is written primarily for a technical reader like a specialist or technician. It is independent of the main document, appearing either in a collection of abstracts (for reference purposes) or on a separate page at the front of a proposal or formal report. Abstracts are of two types: descriptive and informative.

Descriptive Abtracts

A *descriptive abstract* describes what the report is about: its topic, methodology, major results, and the type of conclusions. It does not include the main points of information. It is very short—usually three or four sentences—and is written for technical readers. Since it does not include the key information of the report, a descriptive abstract has somewhat limited value. Therefore, if the abstract is the only condensation provided, experts suggest adding to the description of the report the key information in it, thus creating an informative abstract (ANSI 1979). However, if the document includes both an abstract and an executive summary, the abstract may be descriptive. See section 20.2 for an example of a descriptive abstract and an explanation of how to write one.

Informative Abstracts

An *informative abstract* condenses the key information in the report, including the major facts, the conclusions, and the recommendations. It follows the general organization and style of the report itself but omits examples, illustrations, and references. Abstracts do not quote from the main document but condense and paraphrase; therefore, they do not use quotation marks. Informative abstracts are also short, usually no more than 200 to 300 words. They should, however, be fully intelligible pieces of writing, so when you write one, do not leave out the articles *a*, *an*, and *the* in order to save words. Section 20.2 includes an informative abstract.

Key Words

Since many abstracts are now filed in computer databases, you may need to supply at the end of your abstract from three to six *key words* that would help a reader locate that abstract in an index or database. In choosing key words, you need to think like a potential reader, deciding what words your readers would use to search for the information in your abstract, and (by extension) in your report or proposal. Think of the computer database as an index in the back of a book in which your abstract is included. What cross-reference words would you supply? Develop the habit of listing the key words below your abstract under the heading "Key Words."

Summaries

Technical writers use three types of summaries to condense information for readers: executive summaries, report and document summaries, and overviews. Each type meets a slightly different need, and some documents include all three. Report summaries can also be independent documents.

Executive Summaries

Executive summaries, like informative abstracts, condense the main information in a report or proposal—including facts, conclusions, and recommendations. They differ from informative abstracts in that their intended readers are executives or managers: people who are not oriented to technical matters but to planning, allocation of resources and personnel, costs and profits, and the broader concerns of the organization. Therefore, the executive summary helps the decision-making process by providing the manager with an accessible and concise condensation of the report. It may also include the following information about the project or the document:

- the purpose
- the scope
- the methods
- definitions of key terms and concepts
- the relationship to larger management concerns

An executive summary often covers more material than does an abstract, always looking at the subject from the manager's point of view. Necessary technical terms are defined; those that the manager doesn't need for understanding are omitted.

Executive summaries should be written after the body of the report is finished. They are usually placed immediately after the table of contents. Because of the necessary additions, they may be longer than abstracts, but they usually are confined to one or two pages. Like abstracts, executive summaries should be concise, but without leaving out connective words. See section 20.2 for an example.

Report and Document Summaries

When you write on technical subjects, your job is to help your reader understand complex subject matter in every way you can. One good way to help is by repeating and summarizing the main points at the end of a long document or a section of a document.

Because these *report*, or *document, summaries* appear in the concluding section, they are sometimes called "Conclusions," but a true summary includes no material not in the document itself. A true conclusion section, by contrast, draws logical inferences from the material and presents these to the reader as findings—sometimes continuing to recommend specific action. Typically, conclusions sections first summarize, then conclude, and then recommend. Another way to summarize is with a checklist like those at the end of each chapter in this book.

Summaries that are not part of a longer document include product summaries, clinical summaries, and literature summaries. Like abstracts, these summaries condense or abstract the main points of a document for busy readers who need key information without the detail of the original. Executives and decision-makers (for example, physicians, company presidents, and public figures like senators) rely on these summaries to save time and still be informed. See section 20.3 for an example of an article summary.

Overviews

Overviews are like abstracts in that they come *before* the main document. Generally, overviews simply present the main points, which helps the reader see both the organization and the key words in those main points. Sections and subsections can also have overviews—each one becoming more detailed. Overviews are explained in section 6.3. You can also find examples at the beginning of each chapter in this book.

Table 20-1 summarizes these differences.

20.2   The Form of Abstracts and Executive Summaries

Document summaries and overviews vary in form and organization, but abstracts and executive summaries have a traditional form and format:

- They are written in paragraph form, using complete sentences.
- The title of the original document and its author's name appear at the top.
- Sometimes the page is labeled "Abstract" or "Executive Summary." If the abstract is attached to a report, no further identification is needed.
- If the abstract appears in a listing of abstracts, the location of the article or report is also given: the journal or publication, volume, publisher, and date.
- To assist in indexing, many abstracts are followed by key content words.

Table 20-1 Types and Characteristics of Summaries

Type	Purpose	Primary reader	Location	Readers' use	Length
Descriptive abstract	to describe the contents of the main document	specialist or technician	in indexes of abstracts or attached to front of report	to see if they want to read report	short—3 to 4 sentences
Informative abstract	to condense the key information in the main document	specialist or technician	in indexes of abstracts or attached to front of report	to see if they want to read report or to see major points	fairly short—200 to 250 words, or less than 10 percent of the main report
Executive summary	to aid in decision-making by summarizing information and providing background for recommendations	manager or generalist	attached to front of report	to understand major points in order to make decisions	1 to 2 pages
Report and document summaries	to review the main points	all types	at end of report or as a separate document	to understand through repetition of major points or to summarize for a busy reader	fairly short, but varies
Overview	to preview the main points	all types	at beginning of report or document	to see key information and organization that will follow	fairly short, but varies

The informative abstract in Figure 20-1 condenses a 25-page report.

Formal proposals also require abstracts. The descriptive abstract in Figure 20-2 is part of a student collaborative proposal for aluminum recycling. This proposal also includes an executive summary, so the abstract describes the proposal contents; it does not include key information.

The two examples in Figure 20-3 are an abstract and an executive summary written for the same document, a "Report of Findings" by a geological consulting firm after the removal of an underground storage tank. By comparing the two, you can see how each one meets the needs of different reader types.

Cheri Porter. Computer Workstations Designed for Productivity and Health.

Elliot Corporation; 9 December, 19___.

Computers proliferate in business offices across America today. Increasingly employers are realizing that providing properly designed workstations in well-ventilated, pollution-free environments makes good business sense. This report shows how several aspects of the computer workstation environment affect employee productivity and health. Recommendations are given for designing adjustable workstations. Distances from potentially harmful video display emissions, protection from monitor glare, and proper monitor and keyboard heights and angles are discussed and illustrated. The report provides guidelines for lighting and selection of wall and carpet colors that will promote employee productivity. Properly functioning ventilation systems supplemented by green plants are recommended for maintaining air quality.

Key words: carpal tunnel syndrome; ergonomics; lumbar support; sick-building syndrome; VDT emissions; workstation productivity.

Figure 20-1 Informative Abstract for a Recommendation Report

L. Flora, J. Schonholtz, G. Snoddy, M. Veal. Proposal for an Aluminum Recycling Program at Central State University. May 19___.

This report proposes an aluminum recycling program to be located at the dormitories of Central State University. Both environmental and economic impacts of aluminum recycling are examined. Basic recycling techniques and current programs at other universities are reviewed. Implementation and maintenance procedures are defined. Conclusions include estimates of fiscal impact to the university and recommendations for future areas of growth for the program.

Key words: recycling, aluminum, waste management, pollution, scrap metal, foundry, environment, consumption, conservation.

Figure 20-2 Descriptive Abstract for a Proposal

Identification of the project

Description of the work done

Description of the report contents

Identification of the project

Details of the soil samples

Explanation of the analysis

Conclusions of the report

Title: Report of Findings: Underground Storage Tank Removal
Prepared for: Mr. John Fletcher, Fletcher Realty
Prepared By: Environmental Consultants, Inc. (ECI)
Date: March 24, 19___

Abstract

On February 27, 19___, Environmental Consultants, Inc. (ECI) removed one 550 gallon steel, single-walled, underground storage tank (UST) from the subject property located at 1223 Central Ave., Midland, Michigan. The scope of our work included completing and submitting the tank closure permits, as required, to the Midland County Health Dept. (MCHD); removing the tank and associated product line; collecting appropriate soil samples from the tank pit excavation and product line areas; and properly disposing of the tank and the associated product line. This Report of Findings summarizes the history of the tank, the results of the inspection of the tank, subsurface sampling methods, and analytical results of the samples collected after removal of the tank.

Executive Summary

On February 27, 19___, Environmental Consultants, Inc., (ECI) removed an unleaded gasoline underground storage tank (UST) from the subject property located at 1223 Central Ave., Midland, Michigan. Three soil samples were collected and sent to a state certified analytical laboratory: one sample, T-1, was collected at 6 feet below the surface grade (under the western end of the base of the tank); one sample, T-2, was collected as an extra sample at 6 feet below the surface grade (under the eastern end of the base of the tank); and one composite sample, CGS-1, was collected from the excavated soils pile. Mr. Wayne Yip of the Midland County Health Dept. (MCHD) supervised the collection of the soil samples. He requested later that the extra sample, T-2, be analyzed. All samples were analyzed, following MCHD and state sampling guidelines, for TPH (as gasoline), [TPH(g)]/ Benzene, Toluene, Ethylbenzene, and Total Xylenes [BTEX], EPA methods 8015/8020. Analytical results for samples T-1, T-2, and CGS-1 indicated no concentrations of TPH(g) or BTEX above test method detection limits of 1.0 ppm and 5.0 ppm, respectively. Upon inspection, the UST showed no evidence of holes, cracks, rust, or pits. No staining and/or odor in or around the tank excavation area was present.

Figure 20-3 Abstract and Executive Summary for the Same Document

Another example of a descriptive abstract appears in Figure 23-1; informational abstracts appear in Figure 22-2 and in the report in section 22.3.

20.3 **The Process of Abstracting and Summarizing**

Abstracts and Executive Summaries

Often you will be asked to write an abstract and an executive summary of your own documents, and occasionally you must condense someone else's document in this way. Fortunately, you can follow the same seven steps in either case.

1. Determine the purpose of the main document and identify the primary readers and their level of technical expertise. Choose the type of abstract or summary you will write based on this information: if the readers are technical specialists, you will probably write an abstract, but if the readers are managers, you will write an executive summary. If a single document goes to different reader types, write both an abstract and an executive summary to accompany it.

2. Get an overview by reading the whole document carefully but rapidly. If you are abstracting your own writing, you must first finish writing the main document.

3. Read the document a second time, marking or listing subheadings, key words, and topic sentences. In a well-written document, subheads and topic sentences will help you understand the organization.

4. Determine the key information from the subheadings, key words, and topic sentences. Write a rough outline of those major points.

5. Using the outline, write the abstract or executive summary in complete sentences. Do not omit articles or adjectives in your attempt to keep it short. Try to keep an abstract between 200 and 300 words (or less than 10 percent of the original), but do not leave out important information. Try to keep an executive summary to one or two pages, again in proportion to the document. Keep your reader's needs clearly in mind. Unless you are writing both an abstract and an executive summary for the same document, write an informative abstract. As a condensation that includes key information, the informative abstract is more useful to the reader than the descriptive abstract's simple explanation of the contents.

6. Read what you have written and check it against the document to make sure it covers all key information. Edit and correct any errors.

7. Extract from three to six key words and list them below your abstract.

Summaries and Overviews

You will usually write a summary of your own work for the concluding section of documents like reports and procedures. It's important to finish the document first so you can then go back and pick up the main points to include in the summary. Use your outline and the subheadings of your document to guide you in choosing the main

The Susceptibility of Young Adult Americans to Vaccine-Preventable Infections: A National Serosurvey of U.S. Army Recruits, by P. W. Kelley, B. P. Petruccelli, Paul Stehr-Green, et al: <u>JAMA</u> 1991; 266: 2724–2729.

U.S. Surveillance reports showed a greater than 98% decline in the incidence of measles, mumps, and rubella (German measles) between the late 1960s, when vaccines against these diseases were first licensed, and the middle 1980s. However, a significant resurgence of these three diseases is now becoming evident, particularly in adolescents and young adults.

To estimate the percentage of young adults with little or no immunity to measles, mumps, rubella, varicella (chicken pox), and poliovirus, the authors evaluated the antibody status of 1547 male and female army recruits. These recruits, inducted during September and October 1989, came from 49 states, the District of Columbia, Puerto Rico, the Virgin Islands, and Guam. They range in age from 17 to 35 years, with over 77% younger than 21 years. Recruits with prior military service were analyzed separately for measles, rubella and poliovirus antibody status because recruits have received measles and rubella vaccines since 1980, and they have received a poliovirus vaccine since 1955.

Of the 1504 recruits without prior service, 17.2% were seronegative (lacked antibodies) for measles, 16.2% were seronegative for rubella, and 2.5% were seronegative for poliovirus type 1, 0.7% for poliovirus type 2, and 12.7% for poliovirus type 3. The 43 recruits with prior service showed substantially better antibody status, with only 7.0% susceptible to measles, 2.3% to rubella, and none to poliovirus. Seronegativity rates were 13.6% for mumps and 8.8% for varicella.

The authors attribute these seronegativity rates to the following: reduced exposure to natural infection as children, lack of immunization due to incomplete coverage by school immunization laws, primary vaccine failure due to factors such as immunization too early in life, and the waning of vaccine-induced immunity.

Figure 20-4 Summary of a Journal Article for Busy Professionals
Reprinted from *Syva Monitor*© 1992, Syva Company, by permission.

points. Write the summary in paragraph or list form. Double-check your work by comparing your summary with your outline and topic sentences.

You'll place an overview in an introduction so the reader sees it *before* reading the main document, but you may find it easier to write *after* you have written your report, procedure, instructions, or proposal. That way you'll know exactly how much to include and which main points to stress. For more on this, see Chapter 6.

If you are asked to summarize a long document for a decision-maker or a busy professional, follow steps two through six above, using the document's subheadings and organization to guide your writing. Remember that your reader depends on you for a clear, accurate rendition of the contents. In this case, you might want to include direct quotations if they are controversial or central to the document. Figure 20-4 shows the summary of a five-page article that was reduced to about 275 words by medical writer Ann Petersen (1992). Summaries like these are written for specialists and technicians, in this case readers of a newsletter section called "In The Scientific Literature" that summarizes long articles in a variety of professional journals.

Checklist for Writing Abstracts and Summaries

1. Who is my primary reader?
   ___ technician or specialist
   ___ manager or decision-maker
   ___ operator or generalist
2. What type of abstract and summary will be most appropriate for my purpose and that reader?
   ___ descriptive abstract
   ___ informative abstract
   ___ executive summary
   ___ document summary
   ___ overview
3. Do I need to include key words? If so, which ones?
4. In writing my abstract or summary, did I carry out the following steps?
   ___ read the original document to get a sense of the whole
   ___ reread, noting the title and subheadings to get the main ideas
   ___ marked or highlighted key words
   ___ marked or highlighted the topic sentence in each paragraph
   ___ followed the general organization of the document and wrote the summary, omitting examples and illustrations
   ___ listed below the summary three to six key words
   ___ checked the summary against the original document for accuracy

---

EXERCISES

1. Read the following short report and write (1) a descriptive abstract and (2) an informative abstract or executive summary of the same material. The intended readers for this report are county and city planners and elected officials. To help you understand the report, the major terms are defined in the following glossary.

> *basin*: in geology, an area in which underground strata dip from the margins toward a common center
> *groundwater*: subsurface water available for use
> *overdraft*: using more water than is replaced
> *seawater intrusion*: contamination of groundwater by salt from seawater
> *acre-foot of water*: the amount of water that would cover an acre of land, one-foot deep

aquifer: a geological formation that contains water; especially one that supplies wells or springs

recharge: the amount of water needed to replace that which is used

---

GROUNDWATER MANAGEMENT STUDY OF PAJARO BASIN

The Pajaro Basin Groundwater Management Study, sponsored by the Association of Monterey Bay Area Governments (AMBAG), examines groundwater conditions, uses, demands, recharge, and management. The report was prepared by H. Esmaili & Associates with guidance from the Technical Advisory Committee, which represented municipal, local, and regional interests.

BACKGROUND

Water flows into the Pajaro Basin from Monterey and Santa Cruz Counties. The east side of the basin rests against the San Andreas Fault and the west against the Pacific Ocean. To the north is the Soquel-Aptos Groundwater Basin, and to the south, the Salinas Valley Groundwater Basin.

In most coastal valleys, groundwater is recharged within stream channels by water of excellent quality. In the Pajaro Basin, rainfall percolating in the surrounding hills is the principal source of recharge, and the Pajaro River is the main source of poorer quality groundwater.

The City of Watsonville is the only major user of surface water in the region. Groundwater, drawn for the most part from depths of less than 700 feet, supplies over 90 percent of the total water demand in the basin.

The following groundwater-related problems have been identified:

- an overdraft of between 6,000 and 18,000 acre-feet per year;
- seawater intrusion, extending up to three miles inland at some locations;
- water quality problems related to pollution, and;
- potential loss of natural recharge due to development.

The emphasis of this study was on methods for protecting and enhancing natural recharge, and on new sources of water to halt the seawater intrusion.

STUDY FINDINGS

Groundwater Table Levels

The amount of water pumped from the groundwater table annually is expected to increase slightly during the next 15 years. Groundwater level declines in the inland portions of the basin do not justify any major water projects at this time.

Seawater Intrusion

The seawater intrusion problem can be addressed by reducing or halting pumping from wells in the affected areas. If pumping were stopped in the seawater intruded areas, 6,500 acre-feet of replacement water would be needed to control the intrusion. The only water supply projects which would meet estimated supplemental water needs in the basin are Pescadero Creek Dam, the San Felipe Project, Arroyo Seco Dam, and deep aquifer wells.

If no action is taken, seawater intrusion will lead to the need to drill deeper wells, relocation of wells inland, and finally to the removal of the land from agricultural production.

Loss of Natural Recharge

The sand hills surrounding the Pajaro Basin, in their undeveloped state, have unusually low runoff rates. This corresponds to a high recharge rate of the aquifer. Conversion of these watersheds to residential and agricultural use greatly increases runoff rates.

The increased runoff has resulted in erosion, sedimentation, and damage to drains and sensitive lowland habitats. Runoff due to conversion to residential use amounts to the equivalent of one quarter to one fifth of the annual rainfall.

The study indicates that an integrated water management approach for the sand hills could improve runoff control and recharge protection.

RECOMMENDATIONS

The study recommends a project to supply 6,500 acre-feet of water per year to halt the seawater intrusion. It recommends the following drinking water supply alternatives in the following order:

1. Deep aquifer wells in the Pajaro Valley
2. The Arroyo Seco Project
3. The San Felipe Project
4. The Pescadero Creek Dam

The study recommends selected management of specified groundwater problems in the Pajaro Valley by a proposed groundwater management agency.

This agency would guide a project to supply 6,500 acre-feet of water per year to the seawater intruded areas.

Four additional small studies should be made in the sand hill areas to assist in the development of specific programs to deal with runoff and lost recharge. Recharge source protection measures should be taken in these areas.

> Additional groundwater monitoring and a groundwater pollution control program are advised.
>
> **COSTS**
>
> Damage to roads and drainage facilities currently costs the individual counties from $200 to $300 per homesite in the sand hills areas. Inaction in dealing with seawater intrusion could cost from $255,000 to $725,000 each year. The costs for the management approach to problems in the Pajaro Basin are estimated to be $256,000 annually for administration. There would also be an initial cost of from $105,000 to $210,000 for research and the construction of a monitoring well.

2. Look in a textbook or journal for your major field of study and find an abstract or overview. Determine what kind of abstract or overview it is. Read the article or chapter condensed by the abstract and see if any major information is missing. Rewrite the abstract or overview if necessary.

3. Interview a professor in your field to find out what kind of abstracts and summaries are required and what these abstracts and summaries accomplish. Report back to your class what you learn.

 WRITING ASSIGNMENTS

1. When you have finished your term project, write an informative abstract to attach to the report. Even though it is your own writing, in your abstract describe the report objectively, as though someone else had written it. Submit either for peer or instructor review. If this is a team project, assign one person to write an abstract of the finished document. Review the abstract in your team meeting to ensure its accuracy.

2. If you will have two types of readers for your term project report (for example, a professor in your technical field and your writing instructor or classmates), write both an abstract and an executive summary—one for each type of reader. If you are doing a collaborative feasibility study, assign one or two team members to write the abstract and executive summary. Have other team members review for accuracy and reader appropriateness.

REFERENCES

ANSI. 1979. *American national standard for writing abstracts.* American National Standards Institute (ANSI), 239.14.

Association of Monterey Bay Area Governments. 1984. *Groundwater management study of Pajaro basin.*

Flora, L., J. Schonholtz, J. Snoddy, M. Veal. 1991. *Proposal for an aluminum recycling program.* Unpublished report. San Jose State University.

Porter, Cheri. 1991. *Computer workstations designed for productivity and health.* Unpublished report. San Jose State University.

Westinghouse. 1970. *What to report.* Internal document.

Chapter 21

Proposals

Gordon Anderson is a civil engineer who spends about 20 percent of his workday on writing tasks. His company provides "technical and management services to develop, manage, engineer, build and operate installations . . . to improve the standard of living and quality of life worldwide" (1991). The way the company secures much of its business is by preparing winning proposals, and part of Anderson's job is to write these proposals.

Complex proposals are by necessity collaborative projects; they require expertise from many people. At Anderson's company, a group consists of the project manager and a group leader from each discipline involved, such as civil engineering, structural engineering, and electrical engineering. Each person is responsible for writing a particular section of the proposal; then the entire group reviews drafts and makes suggested changes.

Thousands of people in technical fields are like Gordon Anderson; part of their job is to seek out, plan, and write proposals. These proposals can either be to provide services or to supply equipment and materials; they can be in any field, including research, development, sales, training, and evaluation. Proposals range from simple suggestions for a change within a department to complex bids for the design, construction, and launch of an orbiting space station. Proposals are such an important part of the technical and business world that sometime in your career you will probably need to write one. This chapter will help you.

21.1 Definitions of Proposal Terms

A *proposal* is a written offer to solve a problem or provide a service by following a specified procedure, using identified people, and adhering to an announced timetable and budget. If you study each part of this definition, you'll have a good idea what's involved in writing a proposal.

1. A proposal is an *offer or a bid*—in other words, a sales or persuasive document. Proposals are much like résumés in that both sell your skills and abilities; however, proposals focus on your ability to do a single task better than anyone else. Because a proposal is a sales document representing you and your organization, you must take special care to make it well designed, organized, and error-free.

2. The proposal offers to *solve a problem or provide a service*. This means you must clearly state the problem, after making sure that your understanding of it is the same as your reader's. You often need to contact the reader in advance to clarify the problem, or to understand exactly what the reader's needs are. Proposals can be for commercial business, like the purchase of a product, service, or idea. They can also be for educational, research, artistic, or other nonprofit activities; these proposals request *grants* from governments or foundations.

3. The proposal spells out a specific *procedure* or plan of attack for solving this problem or providing this service. The point is to convince readers that it's the best procedure or plan. You may have intense competition for this job, with many other companies or individuals bidding against you, and you need to be innovative, clear, and complete in telling *how* you'll solve this problem or *what* the service will be.

4. In a proposal, you identify the *people* who are going to do the job and explain their expertise. To show that you (and your group) are better than the competition, you discuss past accomplishments, experience on similar projects, and successful completion of previous jobs within time and budget limits. You can also provide résumés of key personnel to assure readers that the leaders know what they're doing.

5. An effective proposal includes a carefully planned *timetable.* You break the task into its segments and project the time needed to complete each segment. To do this, you will need some background in a similar project or advice from an expert. You must consider contingencies and make sure that the projected timetable is both possible and competitive.

6. If the project is for a *fee,* the proposal must forecast both the hourly cost of workers and the cost of equipment and supplies. You must figure in a profit but maintain a competitive edge. It's not easy to do all this, but it can be done. The first few times you prepare proposals, seek help from experienced technical people in your field. When you win a contract or a grant, keep careful track of costs as they occur and use those figures to help write the next proposal. If you don't win, find out why and file that information so you'll be better prepared next time. Only a small percentage of proposals are accepted, but you can also learn from your rejections.

21.2 Types of Proposals

Proposals can be classified in two ways:

1. by origin—whether unsolicited or solicited
2. by the intended reader—whether internal or external

In both classifications, proposals can be presented either informally or formally; form and organization are discussed in sections 21.3 and 21.4.

Unsolicited and Solicited Proposals

If you see a problem that needs solving, a change that needs to be made, or a service that needs to be performed, you may write an *unsolicited* proposal. For example, suppose that your job requires you to test a complex procedure with a computer program that has a long run time. At present, you must either drive to your office at night to check on the program, or you must wait until morning to make corrections or proceed to the next stage. If you had a terminal at home, you could monitor the program easily and be more efficient. You mention the idea of a home terminal to your manager. "Put it in a proposal," you're told. Or suppose that your company specializes in the detection and prevention of methane gas migration. You learn that a local developer plans to construct a large park and outdoor amphitheater over a filled city dump. You recognize the potential for methane gas problems, and you write a proposal for monitoring and construction procedures to ensure safety in the park. When your proposal is unsolicited, you must take special care in defining the problem, since you must convince the reader that the problem does exist and is worth solving.

You write *solicited* proposals in response to a Request for Proposal (*RFP*). Companies and governments at all levels regularly issue RFPs for problems they need solved. Many RFP announcements appear in the *Commerce Business Daily,* a publication of the U.S. Department of Commerce. Others appear in newspapers or are mailed to prospective bidders. An announcement in the *Commerce Business Daily* may be for either an RFP or an Information for Bid (*IFB*). An IFB is usually for physical objects like electric hospital bed parts or shaft and gear assemblies, and the requirements are very specific. An RFP is usually for services or studies, and the proposer has more latitude in designing a solution. Figure 21-1 shows two announcements of RFPs from the *Commerce Business Daily* (1992). These announcements are summaries of the actual RFPs; notice that the reader is instructed to write for the RFP.

Figure 21-2 shows an RFP announcement that appeared as a newspaper ad. When you respond to an RFP, you must be sure to meet all the requirements that it lists, and often you must follow the method of organization that it specifies.

Announcements from the federal government inviting grant proposals appear in the *Federal Register,* a publication that outlines each type of grant, its intention, the population to be served, evaluation criteria, and so on. Information on private foundations that award grants is available from *The Foundation Directory,* which is national in scope. In addition, some states have listings of state and local grant providers.

Internal and External Proposals

Internal proposals go to management within your company, so you can write a short proposal in standard memo form (see Chapter 15). However, long internal proposals,

---

NOAA, WASC, Procurement Division WC3, 7600 Sand Point Way NE, BIN C15700, Seattle, WA 98115
B – ASSESSMENT OF ENVIRONMENTAL HAZARDS ASSOCIATED WITH OIL AND HAZARDOUS MATERIALS RELEASE SOL 52ABNC200043 DUE 031992 POC For a copy of the solicitation, call Karen Bruce at 206-526-6262. For further information, call Contract Specialist Jan Sullivan at 206-526-6032. Provide the necessary services to NOAA and other government agencies in coastal physical processes, environmental sensitivity, protection and cleanup strategies involved in the assessment of environmental hazards associated with oil and hazardous materials realeases. This includes (1) providing environmental assessments, on short notice, during releases of oil and hazardous materials in coastal areas of the United States, (2) conducting training in the area of environmental sensitivity, (3) identifying and developing information sources for advancing the state-of-the-art in response operations and (4) developing information to support spill preparedness and planning activities, (5) providing on-site scientific advice to Federal On-Scene Coordinators and response personnel during approximately 75 oil and hazardous waste site investigations and (6) preparing Environmental Sensitivity Index (ESI) mapping for coastal areas of the United States. The period of performance will be for one year with an option to extend the performance for four additional one-year periods. All responsible sources may submit a proposal which will be considered by the agency. (0024)

Federal Bureau of Prisons, Office of Procurement and Property, 320 First Street, N.W., Washington, D.C., Attention: Community Corrections Center Contracting
G – COMMUNITY-BASED SUBSTANCE ABUSE TREATMENT SOL RFP 200-115-W DUE 032692 POC Bonnie Sinsel, Contracting Officer 202/514-5439. Community-Based Substance Abuse Treatment. A program of transitional services for the Federal offender. This notice seeks response from concerns having the ability to provide services related to substance abuse treatment for Federal offenders who will reside at Community Corrections Centers in the Phoenix, Arizona area. Interested parties are requested to give written notification (including telephone number and a point of contact) to the contracting officer listed in this notice. Services required include intake assessment, urinalysis testing, individual counseling, group counseling, and family/couples counseling. The services described shall be in accordance with the Bureau of Prisons Statement of Work. Estimated total number of offenders requiring these services is 15 for the 24-month base period. Estimated total number of counseling hours per offender is 96. RFP issue date is o/a February 25, 1992. Estimated closing date is March 26, 1992. Contract period is 24 months. A requirements contract is anticipated. All responsible sources may submit a proposal which shall be considered by this Agency. No collect calls will be accepted. No telephone calls for solicitations will be accepted. (0031)

Figure 21-1 Announcements of RFPs in the *Commerce Business Daily*

NOTICE TO SANITARY SEWER ENGINEERING CONSULTANTS
CITY OF SAN JOSE
REQUEST FOR PROPOSALS

The City of San Jose is seeking the services of an engineering consultant firm to provide expertise for various stages of fiscal year 1992-1993 sanitary sewer rehabilitation projects. These include, but are not limited to, rehabilitation projects on: Almaden Road, Fifth Street, Market Street, and San Antonio Street. The tasks include: video inspection, recommend rehabilitation methods, provide project design, review of plans and specifications, provide inspection and testing guidelines, construction monitoring, and obtain permits.

If your firm wishes to be considered for the above work, please apply for the RFP package at address given below:

> City of San Jose
> Department of Public Works
> 801 N. First Street, Room 300
> San Jose, California, 95110
> Attn: Mr. Ken Salvail.

Proposals must be received by the City before 12:00 noon Friday January 31, 1992. Following review of the proposals, an oral interview will be conducted with the three leading candidates in mid-February, 1992.

If you have any questions about this RFP, please contact Mr. Ken Salvail at (408) 277-4638.

Figure 21-2 Announcement of an RFP in a Newspaper

or those that must go many levels up in the management hierarchy (perhaps to corporate headquarters) are usually written as formal proposals, using a specific form like that explained in section 21.4.

External proposals go outside your organization to prospective clients or granting agencies like the Ford Foundation. Thus, short external proposals can be written as standard business letters (see Chapter 16). Long external proposals, however, are written as formal proposals and accompanied by a letter of transmittal.

21.3 Informal Proposals: Organization and Examples

Whether presented internally as a memo or externally as a letter, an *informal proposal* must answer these questions:

1. What is the problem or need?
2. What is the proposed solution and by what procedure will it be accomplished?
3. Who will solve the problem?
4. What is the timetable?

5. What is the cost?

6. Is this solution acceptable (request for approval)?

To answer these questions effectively, most informal proposals are divided into separate sections, with each section providing the answer to one question. However, section divisions are not rigid, and you can always adapt the form and organization to the needs of a particular reader.

Despite their name, informal proposals are not written in informal style but in semiformal or formal style. "Informal" refers to the length (usually fewer than five pages), the flexibility of the organization, and the lack of front and back matter. Informal proposals generally include

1. an introduction
2. a proposed procedure or plan
3. information on staffing and cost
4. an action request or request for approval

Organization of Individual Sections

Introduction

The introduction identifies the problem or need specifically and states the proposed solution and the plan for achieving it. In addition, it may give background information, telling who has studied the problem and what they have found. It may list the goals or objectives of the project—both short term and long term. Finally, it may discuss the significance of the need and the importance of the proposed procedure. The most important function of the introduction, though, is to spell out clearly what the problem is. In an unsolicited proposal, you must convince the reader that a problem really exists; in a solicited proposal, you must convince the reader that you understand the problem and all its ramifications. (Often you do this by careful repetition of the language used in the RFP.)

In the following introduction (Lane 1991), for example, the writers discuss both the need for the study and the specific objectives.

> The purpose of this proposal is to develop a campus-wide aluminum recycling program for Central State University (CSU).
>
> Central City and Hayward County face a near-term shortage of landfill capacity. In 1985, there were 23 landfill sites in use, but at the present rate of waste disposal, there will be only 6 unfilled sites by the year 2000. At the present time, students discard aluminum cans in regular trash cans. Recycling aluminum will significantly reduce the volume of waste generated at CSU. In addition, recycling will save energy and natural resources and will be cost effective, since the aluminum can be sold.

The following introduction (Long 1991) defines the problem in order to indicate the need for the study.

> Subject: The comparison of Top-Rated 1040 Tax-Preparation Software Programs for Professionals

As we discussed in our telephone conversation of June 12, I propose to compare the top-rated 1040 tax-preparation software packages for professional tax preparers. This comparison will be ready for publication in the Enrolled Agents newsletter by September 1. It will provide enrolled agents with both program features and buying recommendations.

Problem Definition

Professional tax preparers, both certified public accountants (CPAs) and enrolled agents (EAs), turn in increasing numbers to the use of tax-preparation software programs. From 1987 to 1989, use of these programs doubled; from 1989 to 1991, use doubled again. Tax programs have several benefits:

- Calculations of tax, depreciations, and other factors are automatic, reducing errors and decreasing time on task.
- Programs are periodically updated, keeping preparers abreast of changes
- Worksheets, financial summaries, and IRS-approved forms are printed as part of the package.

These benefits make such software packages attractive investments, especially for beginning tax preparers with small practices or established EAs with growing practices.

However, there are approximately 30 professional tax-preparation programs on the market, far too many to research for an individual who wants to make a buying decision before the next tax season. This comparison deals with five of the top-rated 1040 tax programs used by EAs in the eastern chapter according to a membership survey in March 19__. It will assist tax preparers in software buying decisions because it evaluates programs based on the criteria they defined in a second survey on May 12, 19__.

Proposed Procedure

At the heart of any proposal is the *proposed procedure* or *plan*. In this section, you describe how you intend to solve the problem or provide the service—detailing the procedure or methodology. You may want to divide this section into three segments:

- a description of the approach (how the problem will be solved)
- a description of the evaluation (how you will judge success)
- a detailed schedule

Staffing

The staffing section in an informal proposal may be very short, especially if only one person is involved. In this section, you discuss the credentials or qualifications of the people involved to verify their ability to do the job. Remember always that a proposal is a sales document, and one way to succeed is to sell your expertise or previous track record. As part of an informal proposal in letter form, here is a brief statement of a project director's qualifications:

The project director will be Dr. A. B. Cristos, whose résumé is attached. For 10 years, Dr. Cristos has been involved in contamination control at Allied Laboratories, and for the last three years he has directed

the contamination control laboratory. In 19__, Dr. Cristos chaired a sympo-
sium on Contamination Control in Semiconductor Manufacturing in
Raleigh, NC.

See section 21.4 for more detail on staffing considerations.

Cost

The cost section presents the budget for the project, including costs for staff,
equipment, and material. This information is usually presented in a table for easy
reference.

40 hours research @ $40 per hour	$1,600
10 hours report preparation @ $15 per hour	150
.5 hours computer time @ $120 per hour	60
supplies	100
TOTAL	$1,910

Remember that you must abide by your cost estimates, so make them accurate. If your
proposal is accepted by a client, you are entering into a contract and have legal
obligations to fulfill it. Any cost overruns will come out of your profits.

Request for Approval

Many informal proposals close with a request for approval. This section is similar
to the action section that closes a letter or memo of request; it should remind the
reader of the advantages of accepting the proposal, and its tone should be courteous
and enthusiastic. For example:

> The plan detailed above will improve test procedures and schedules by
> 200 percent. If you have any questions, please contact me at extension 434.
> I look forward to your favorable response.

> May I have your approval to pursue this study? Please notify me of your de-
> cision at your earliest convenience.

Example of an Informal Proposal

The informal proposal that follows is written in memo form because it is an internal
document. The writer is responding to a request, issued by a subcommittee, for a study
of scanning electron microscopes that will recommend one for purchase by the
department. Notice how the proposal's introduction gives the background leading up
to the request, then sets up a proposed procedure to accomplish the study. Three
sections address the procedure itself: proposal (methodology and content), evaluation,
and schedule (including a Gantt chart showing how the procedure will be carried out).
The writer sells herself as a likely candidate to conduct the study by explaining her
experience with the department's needs, the subject of the study (scanning electron
microscopes), relevant course work, and interaction with other staff members.

To: Dr. Connie Martinez, Department Chair

From: Kim Rosetta *K. R.*

Date: September 10, 19___

Subject: Comparative report for the Smith University Department of
Biological Sciences on scanning electron microscopes (SEMs)
compatible with image enhancement (IE).

Summary

The summary is like an abstract

I propose to write a comparative report on SEMs compatible with IE
as part of my research assistantship. I will submit the report by December 15, 19___ to Dr. Bernard Colson, Chair, Subcommittee on SEM
Purchase, Equipment Committee, Department of Biological Sciences,
Smith University. My report will compare SEMs specified by the subcommittee, with emphasis on features the subcommittee selected for
further study. The report will form a basis for selection of a new SEM,
to be purchased with funds from Federal Grant #1067A.

Introduction

The need for the proposed study

The Department of Biological Sciences has determined the need for
an SEM with recently available features, including IE. On August 15,
19___, the Department was notified that the new microscope would be
funded through Federal Grant #1067A. In light of funding approval,
the Department Equipment Committee, chaired by Dr. Hilda Johnson,
met with SEM users on September 1, 19___, to discuss guidelines for
purchase. The following guidelines were established by the Committee.

GUIDELINES FOR SEM PURCHASE

Price Range, including IE	$120,000-$140,000
Electron Gun	tungsten; possibly also LaB_6
Saturation	auto
Focus	auto/manual option
Magnification	$\times 250,000$
Stigmator	auto/manual option
Brightness/Contrast	auto/manual option
Dynamic Focus	standard
Gamma	standard
Tilt Correction	standard

Further detailed study was delegated to a temporary subcommittee
on SEM Purchase. Through preliminary inquiries, the subcommittee
identified eight SEMs, manufactured by five companies, for further
study. The committee felt that further study should focus on the following features:

Criteria

FEATURES FOR COMPARISON

SEM

- Lens system
- Vacuum systems
- Resolution
- Specimen chamber size
- Specimen stage-movement in x, y, and z axes; rotation; and tilt
- CRTs-number, size, type, and number of slow-scan lines for direct record
- Installation requirements
- Maintenance requirements
- Service agreements

IE

- Number of pixels for digitized image record
- Grey scale-number of bits
- Pseudocolor-available?
- Image processing-standard features available

Subcommittee members noted that the SEM is an expensive long-term purchase; also, they recognized that the SEM must fulfill the varied research requirements of over 30 professors and graduate students. In order to make an informed recommendation of a model for purchase, the subcommittee felt that a detailed comparative report delineating differences between the eight models should be prepared. The subcommittee decided that a careful review of these models by someone familiar with both SEMs and departmental research was necessary. Because such in-depth study will require considerably more time than subcommittee members can allocate, members felt that the study should be incorporated into a research assistantship. The graduate student assigned the study will present the report to the subcommittee at their December 19, 19__, meeting.

Proposed procedure

Proposal

I propose to perform the comparative study needed by the SEM Subcommittee and provide a detailed report by December 15, 19__.

I will request information on the eight models identified by the subcommittee from the following companies:
- AMRAY
- Cambridge Instruments
- Carl Zeiss, Inc.
- Hitachi
- Jeol USA, Inc.

The requests will be by phone to initiate the study as quickly as possible, with a written follow-up to ensure receiving the correct information. I will clearly specify the model numbers and information needed, according to the SEM subcommittee's decisions.

I will review the information received, compile it, and arrange it in table format. To obtain information not provided in company brochures, I will conduct discussions with company representatives by phone. I will also invite representatives to visit our department; I will notify all interested members of the Department of Biological Sciences of these visits.

In addition, I will send a one-page survey to department SEM users to clarify and confirm departmental needs. I will compile a list of preferred features and indicate how many members designated each as helpful, and how many designated each as essential.

My report will include the sections listed below:

(1) A table of information pertaining to features for comparison for the eight SEMs identified by the subcommittee
(2) Notes on special aspects of the SEMs that may be of interest to the subcommittee
(3) A list of SEM features preferred by departmental members, showing, for each feature, the number of SEM users who judged the feature as essential or helpful to their research
(4) A one-page summary of the advantages and disadvantages of the three models that, in my opinion, will best serve departmental members, according to priorities identified in the survey
(5) An appendix containing literature received, and notes from company presentations, organized by company and model number (in primary copy only)
(6) An appendix containing the names, addresses, and phone numbers of company representatives contacted
(7) An appendix containing evaluation comments on the report by department members

Evaluation

Method of evaluation

I will provide a draft of the report to interested members of the department by December 1, 19—, for their review. Members will be asked to evaluate the report for completeness in terms of specifications given in this proposal. In addition, they will be asked to assess the value and usefulness of the report.

Schedule

Schedule

The following dates are presented as a tentative schedule:

Sept. 19-27, 19—	Telephone companies
Sept. 26-Oct. 2	Write follow-up letters
Oct. 3-9	Write one-page survey for departmental SEM users

Oct. 10-14	Approve survey with SEM Subcommittee Chair
Oct. 17-21	Send out survey to departmental SEM users, to be returned Nov. 4
Oct. 17-Oct. 31	Review information; compile preliminary table "Features for Comparison"
Oct. 24-Nov. 11	Obtain information still needed for above table through discussions with company representatives
Nov. 7-13	Compile data from departmental SEM users survey into list form
Nov. 7-13	Complete table of information "Features for Comparison"
Nov. 14-20	Write report—first-draft stage
Nov. 21-25	Enter report into departmental word processor
Nov. 26-29	Edit report—second-draft stage
Nov. 29-30	Print and make 10 copies of second draft
Dec. 1	Give second draft to interested department members for evaluation, to be returned by Dec. 5
Dec. 5-11	Prepare final report
Dec. 5-11	Prepare appendices
Dec. 12-14	Print and copy final report
Dec. 15	Give final report to Chair, subcommittee on SEM Purchase
Dec. 19	Present final report to subcommittee at their meeting

Gantt chart shows overlapping of tasks

Time Line

1. Telephone companies
2. Write letters
3. Write survey
4. Approve survey
5. Send out survey
6. Review information
7. Obtain information still needed
8. Compile data from user survey
9. Compile table
10. Write first draft
11. Enter into word processor
12. Edit report to second draft stage
13. Print and copy second draft
14. Draft for evaluation
15. Prepare final report
16. Prepare appendices
17. Print and copy final report
18. Give final report to chair
19. Present report to subcommittee

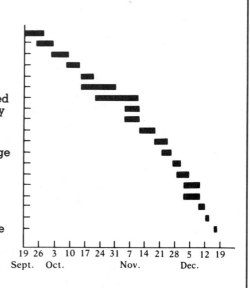

Staff/Qualifications

I plan to complete personally all work specified in this proposal.

I am a fourth-year Ph.D. student in microbiology, working under Dr. Pat Georges, Professor of Microbiology. My research work on the filamentous bacterium *Actinoplanes philippinensis* has demanded extensive SEM work. I have spent an average of fifteen hours/week in the SEM laboratory during the past two years. Thus, I have become familiar with Departmental SEM projects, and I have assisted many of the newer graduate students.

My formal training includes 15 quarter units of physics and 140 quarter units of biological sciences; courses in both SEM and transmission electron microscopy (TEM) were part of my graduate work. I also attended a one-week seminar in SEM-IE through Microscopy Seminars, Inc. I served as a teaching assistant for the Smith University SEM course during both my second and third years as a graduate student.

Cost

TIME

Obtain information from companies	10 hours
Write survey	2
Review and compile information from companies	10
Review and compile information from surveys	6
Write report	14
Type report	8
Edit report	4
Organize appendices	5
Printing and copying	2
Presentation	2
Total time	63 hours × $10.00/hour = $630.00 (out of research assistantship hours)

COPYING

10 copies—second draft	15.00
15 copies—final report	23.00

TELEPHONE

Average of two 10-minute long-distance calls for each   50.00
    of five companies

TOTAL COST $718.00

Request for Approval

Action request

I request approval to conduct the research and prepare the comparative report as described in this proposal.

21.4 Formal Proposals: Organization and Examples

Like the recommendation reports explained in Chapter 22, a long *formal proposal* is divided into sections for the convenience of different types of readers. The sections correspond to those discussed for informal proposals, but each section is presented more formally and may be further divided into subsections. A formal proposal also includes additional material at the beginning and end called *front matter* and *back matter*. Here are the typical parts of a formal proposal in the order they usually appear.

Front matter

> letter of transmittal
> abstract
> title page
> table of contents
> list of figures and list of tables
> executive summary

The proposal itself

> introduction
> procedure or methods
> evaluation
> schedule
> personnel
> budget

Back matter

> appendixes

Each proposal will modify this organization to some extent because it must meet the reader's particular needs. Also, even though the sections appear in this order, they are usually not written in this order. You will probably want to begin writing with the introduction or procedures section. The following paragraphs explain requirements in each section; an example of a formal proposal follows the explanations.

Organization of Individual Sections

Letter of Transmittal

A *letter of transmittal* is a formal business letter directed to the person who will approve or accept the proposal. Like a job-application letter, the transmittal letter should tell how you learned about the problem (for example, an RFP or a letter from the client), highlight the strengths of your solution or approach, and acknowledge any contact or help you have received from the reader or the reader's organization. It should close by offering to answer questions and by requesting action (a response to the proposal). The letter should be a separate document from the proposal but attached to it. For correct letter form, see Chapter 16.

Abstract

The *abstract* should be informative—that is, a brief condensation of the proposal's key information—and aimed at specialist readers. The abstract appears at the beginning of the proposal, but it should be written last—after you know exactly what the proposal includes. See Chapter 20 for details on writing abstracts.

Title Page

The *title page* includes the proposal title, the name of the requesting organization, the number of the RFP or other announcement, the date of submission, and perhaps the name and address of the author or submitting division. Most of this information is repeated on the cover of the proposal when it is bound.

Table of Contents

The *table of contents* shows the organization of the proposal and provides entry points to the information. The table of contents can be constructed from the outline you create before writing the proposal. Since proposals often do not have an index, a good table of contents is essential.

List of Illustrations

The *list of illustrations* (or separate lists of figures and tables) is also vital because figures and tables often condense key information in an easily accessible form. A proposal might be rejected if no good tables or figures were included to help readers find the details they need. Tables and figures are explained in Chapter 11.

Executive Summary

Most proposals contain both an abstract written for specialists and an *executive summary* written for managers. An executive summary should be free of specialized language and should summarize whatever background the reader needs to understand the proposal. Chapter 20 explains executive summaries. They can be placed either before the title page or before the introduction.

Introduction

The *introduction* can be subdivided into the following segments:

- *A statement of the proposal's purpose.* The purpose statement discusses objectives or goals of the project, both short term and long term. It may detail the project's significance and its benefits to the company or to society in general. Proposals can be either for research projects (to discover information) or for technical applications (to build, analyze, apply, or administer projects). The purpose statement should identify which type of proposal you are making.
- *A definition and description of the problem or need.* If the proposal responds to an RFP, that fact is noted here with an identification of the RFP. If the proposal is unsolicited, the problem must be clearly presented to convince the reader that a need truly exists. The problem often needs to be presented in considerable detail and should be supported by objective data such as monetary losses, numbers of defective parts, and low performance ratings.
- *An overview of the proposed project.* An overview allows the reader to see the whole picture and how the parts relate to the whole. It should tell what is included in the proposal and describe the order of presentation of the major points.

Procedure or Methods

The *procedure,* or *methods, section* presents the work plan or the method of solving the problem. Sometimes this section is called "Scope of Work." It describes, in chronological order, how you intend to achieve your purpose. It is often written like the procedures discussed in Chapter 19; the difference is that instead of using the present tense, it uses the future tense to refer to activities (*will design* instead of *designs*). The procedure section is usually written in first or third person and the active voice. In this section, you tell not only what will be done but how it will be accomplished; thus, this section often contains specifications, technical description, and functional description. When your proposal is evaluated, this section will be reviewed by specialists in the technical field, so before you submit the proposal, check the procedure section carefully for technical accuracy.

One portion of the procedure section may cover progress reports; some companies and funding agencies will not accept proposals unless regular intermediate evaluations are provided as well as a final report. Evaluations can be done by (1) measurement, test, or audit; (2) questionnaire and expert review; or (3) progress and final reports. If you propose to evaluate by testing or measuring at regular intervals, indicate in the procedure section the kinds of tests that you will administer, or specify the measurements that you will take. If you propose questionnaire or expert review, provide a sample of a questionnaire or identify the expert reviewers by name, job title, and professional affiliation. If you propose progress reports, specify the points at which these reports will be submitted—such as at the conclusion of the design phase, modeling phase, and test phase. Most of the time, your proposal will be written to conform to specific requests made in the "Statement of Work" section of the RFP. However, readers will generally respond most favorably to proposals that have detailed and specific plans, so even if the RFP does not ask for details, the proposal should provide them.

A schedule can form part of the procedure section, or it can appear in a section by itself. Decide whether to provide a separate section by the length and amount of detail of the proposed solution. One of the ways to sell a proposal is to show careful planning, and a detailed schedule provides such evidence. Schedules can be presented in a list of specific dates, in estimated hours or weeks to completion of each segment, or in a task-breakdown graph (Gantt chart) like that shown in the informal proposal example in section 21.3.

Personnel

In a formal proposal, the *personnel* section must include (1) the credentials of project leaders, (2) the size and qualifications of the support staff, and (3) organizational support and resources, like laboratories, computer facilities, and instrumentation. Often the key information on credentials is summarized in the personnel section and supported by résumés placed in an appendix. Support staff and organizational resources can be simply listed; you can judge by the RFP how detailed this section must be.

Budget

Since most formal proposals are bids to obtain contracts or grants, the *budget* is obviously a key item—justifying item by item the total proposed cost. You need to prepare the budget carefully because if other factors are equal, cost is usually the deciding factor. Budgets include personnel costs (salaries, benefits, and overhead), material costs, and expenses for such things as computer time, travel, and consultants. The budget section should also include the length of time the quote is valid and terms for payment.

Appendixes

Tailor the number and content of *appendixes* to the needs of the proposal. You might include résumés of project leaders as well as endorsement letters, examples (or a listing) of previous work, and technical material (including graphics) that supports the procedures section but may not interest all readers. A review of relevant literature can be included here, or it can appear as part of the introduction.

Example of a Formal Proposal

The formal proposal that follows proposes a major program change in a university department. It is formal—rather than informal—because it is long, complex, involves many people, and must be approved by many levels of administration. As you study the proposal, notice how different sections are directed specifically to different reader groups: the letter of transmittal to the dean, the abstract to expert readers, and the executive summary to higher administrators unfamiliar with the background leading to the proposal. Notice too the care taken to verify the need, document the details of the solution, and explain the advantages of making the change.

Letter of transmittal

**Department of Aviation
Northern State University
Waterville, Iowa 51304**

May 8, 19__
Dean K. C. Parkhurst
College of Applied Arts and Sciences
Northern State University
Waterville, Iowa 51304

Dear Dean Parkhurst:

As you requested in your letter of February 6, 19__, we have prepared a formal proposal for a Professional Pilot Training (PPT) concentration in the Aviation Department.

This proposal assesses industry need for the next 10 years, notes the large student base already in the Aviation Department, and documents the interest among present aviation students in a program that would better prepare them to enter the industry after graduation.

We list the components of a successful PPT concentration and then examine three possible configurations. We recommend a concentration that combines in-house administration and ground school preparation with contracted flight school instruction. This configuration ensures university control over the quality of the program. It also makes the program self-supporting from the department standpoint and keeps flight instruction costs as low as possible for the students.

Since the Aviation Department already has excellent on-site facilities and a faculty well qualified to design new courses and administer the program, we believe that the PPT concentration is both feasible and attractive.

Please let us know how we can help you in moving this proposal toward adoption.

Sincerely,
PPT Task Force

Allison Gates

Allison Gates, Chair

Abstract

Condensation of the proposal information

ABSTRACT

The need for professional pilots will increase dramatically over the next 10 years. Aviation students at Northern State University and industry professionals show great interest in a Professional Pilot Training (PPT) concentration within the aviation major. This proposal describes the necessary 11 components for a successful concentration, explains three possible configurations, and recommends a combined program: administration and classroom instruction in-house combined with flight training contracted to one or more flight schools in the area. Needed new courses, staffing, equipment, and costs to the department and students are detailed. The program is intended to be self-supporting.

Key words

Key Words: professional pilot, flight instruction, aviation careers, pilot training

Title page

A PROPOSAL
FOR A
PROFESSIONAL PILOT CONCENTRATION

in the Aviation Department
at Northern State University

prepared by
The PPT Task Force

Professors Jim Stokes and Allison Gates
Students Stan McDaniel, Mario Recino, Jon Riddle

May 8, 19__

TABLE OF CONTENTS

EXECUTIVE SUMMARY

Airline expansion, decreased military pilot training, and accelerated pilot retirements will create the need for 52,000 to 62,000 professional pilots over the next 10 years. No other university in the state offers a Professional Pilot Training (PPT) concentration, and Northern State University is uniquely qualified to do so. The university already has excellent facilities at the Municipal Airport, a large student base in the Aviation Department (600), and the demonstrated interest in such a concentration by 70 percent of the students.

The proposed PPT concentration will combine in-house administration and classroom teaching with flight training contracted to one or more flight schools at nearby airports. This configuration will meet each of the 11 components of a successful program. It will allow the Aviation Department to supervise and administer the program, while keeping the program self-supporting. Students will be ensured excellent flight training under FAR 141 regulations (guaranteeing high-quality instruction and fewer required hours of training), will have access to a wide variety of aircraft and qualified instructors, and will be assured the lowest possible cost for the full range of flight training. Costs to students are estimated at $17,500 to $18,000 for the full flight training program.

Existing faculty are well qualified to design the new courses and supervise the program. The new flight-simulator laboratory will provide excellent simulator instruction with no further expenditure. Some additional computer-based training (CBT) stations will be required, but they will be phased in over the next two years. Support from industry and professional organizations can be expected once the program is in place.

A PPT concentration will attract motivated students to the Aviation Department, will help meet industry needs, and will allow Northern State University to fulfill its mission of educating students for the workplace and society.

Need for the proposed program

Description of the proposed program

Advantages

Summary of reasons to adopt the proposal

PPT PROPOSAL   PAGE 1

INTRODUCTION

Topic of the proposal

Aviation students and industry recruiters agree that a Professional Pilot Training (PPT) concentration is needed to enhance the current undergraduate degree in aviation at Northern State University. Such a program will provide aviation students with practical as well as theoretical education, and it will improve students' chances to compete for positions as professional pilots upon graduation.

Request for the proposal

Background information

A memo on February 2, 19__, from Dr. J. E. Smallwood, Aviation Department chair, noted the interest among faculty, students, and industry recruiters in a PPT concentration and asked for a proposal for such a program. During the past three months, a task force consisting of Professors Jim Stokes and Allison Gates and aviation students Stan McDaniel, Mario Recino, and Jon Riddle has investigated the problem and now presents the following proposal.

Overview of the proposal parts

This proposal documents the need for professional pilots over the next 10 years, describes the best method for implementing a concentration in Professional Pilot Training, and describes the benefits for students, for the Aviation Department, and for the university.

The Problem

The Northern State University (NSU) Aviation Department currently has approximately 600 students pursuing a B.S. degree in Aviation. Of that number, only 30 percent currently possess any type of pilot's certificate, althqugh a student survey on March 1, 19__, indicated that 70 percent of the majors would pursue pilot certification if it were available. (See the Appendix for a copy of the survey and results.)

Documentation of the need for the program

For students, the present decade is an excellent time to pursue a career as a pilot. Predictions are that in the next 10 years 52,000 to 62,000 pilots will be needed to fill openings in commuter, national, and international airlines and other jet operations, an increase over the last 10 years of between 12,000 and 22,000 openings. (See the Appendix for additional statistics.) Increased job opportunities come from airline expansion and from increases in the number of pilots reaching the mandatory retirement age. In addition, fewer trained pilots are leaving the military, and the number of student pilot startups has remained constant, meaning that fewer experienced pilots are available for airline positions.

No other university in the state offers a Professional Pilot Training concentration; in fact, of the four surrounding states, only one offers any pilot training.

Finally, the obvious hiring preference for degreed pilots is clear; in 1988, 64.9 percent of all pilots hired in professional positions had a four-year degree. For the major airlines, the percentage was even higher: 74.9 percent of all pilots hired were college graduates. (See Appendix.)

Overview of the Proposal

Acknowledging the importance of a combined aviation degree and professional pilot training, this proposal will

Map of the major sections

1. Describe the 11 components needed for a successful concentration
2. Explain three possible configurations for a PPT concentration
3. Explain the best program and describe how it will work
4. Detail the needed personnel, time, and budget allocations

COMPONENTS OF A SUCCESSFUL PROFESSIONAL PILOT TRAINING CONCENTRATION

The following 11 components will be essential to a successful PPT concentration:

1. **Location.** Both ground instruction and flight instruction must be easily accessible to students and faculty.
2. **Hours of operation.** For greatest flexibility, flight training must be available 12 months a year and 7 days a week.
3. **Aircraft rental.** Planes must be available beyond actual training hours so that students can build time and maintain currency.
4. **Number and variety of aircraft.** An adequate number and variety of aircraft are needed to provide different training exercises (spin recovery, soft and short-field landings, mountain flying, multi-engine training, and so on).

Detailed criteria

5. **Job possibilities.** As students gain ground- and flight-instructor ratings, they should be eligible to secure jobs as ground, simulator, and flight instructors. This will supply income, further training, and flight experience for the students.
6. **Examiner on staff.** An FAA flight examiner is needed on staff to reduce costs, improve scheduling, and ensure communication between the FAA and the university.
7. **FAA written tests.** FAA written tests should be administered within the program so they are available when students are ready to take them.
8. **Quality control.** The Aviation Department should control the quality of ground and flight instruction to ensure consistency and completeness.

9. **FAR 141.** The flight school should operate under FAR 141, to attain high standards and to reduce the number of hours required of each student, thus reducing cost.
10. **Simulators.** State-of-the-art flight simulators should be available to increase training efficiency, reduce the demand for airplanes, provide for training during inclement weather, and reduce student costs.
11. **Startup costs.** Costs for procuring aircraft, simulators, and staffing must be as low as possible without compromising a quality program.

THREE OPTIONS FOR A PPT CONCENTRATION

Overview of the options

The three most common methods of providing flight training are to

1. Conduct all training on campus with Aviation Department equipment and staff (in-house option).
2. Contract out all training to a local flight school (contracted option).
3. Combine these two methods in a way that best meets the needs of both the students and the Aviation Department (combination option).

While both the in-house and contracted options have distinct advantages for students and the department, these options also have serious problems.

In-house Option

For students, problems in this option include the limited number of hours and aircraft available for instruction and practice. Students would also have restricted job opportunities; as property of the state, the aircraft could not be used for revenue-generating charter, banner towing, or glider towing. For the department, the biggest problem would be the extremely high startup costs: purchase of aircraft, insurance, maintenance, parking space, administration. A second problem would be obtaining FAA approval of an in-house program.

Contracted Option

For students, problems with this option include the lack of high-quality simulators (Frasca 141 and 142, for example) at local flight schools, and the fact that students would have to leave the campus for ground school as well as for flight training. For the department, the biggest problem would be accreditation and the lack of quality control: aviation professors would not be involved in ground school or flight training on a daily basis.

The recommended option is given last

Combined Option

By combining the two options, NSU can gain the advantages of both programs, avoid high startup costs, and ensure quality pilot training by having control over and interaction with the program. In addition, students can earn college credit for pilot training.

THE PROPOSED PROGRAM

Under a combined option, the Aviation Department itself will undertake the following responsibilities:

1. administration of the program, design of new courses, and determination of course credit
2. ground school instruction
3. simulator instruction
4. scheduling and administration of FAA written examinations
5. provision of an FAA flight examiner on staff

The Aviation Department will contract the flight instruction segment with one or more of the flight schools at the nine airports within a 30-mile radius of the NSU campus. The flight schools will provide:

1. certification as an FAR 141 school to ensure high quality training and to reduce the number of required student hours of instruction
2. certified flight instructors (CFI) for all rating levels and types of aircraft
3. available rentals of both single-engine trainers and multi-engine and high-performance aircraft
4. an agreement to consider hiring qualified students as CFIs and for charter work, banner towing, and so on

New Courses

Details of the proposed program

New courses, as shown in Table 1 and Table 2, will be developed to support the PPT concentration. All courses will be designed by

Table 1. New Courses Taught by Aviation Department

Code	Title	Units
AV 76	Commercial Pilot Certification	3
AV 183	Flight Instructor Certification	3
AV 184	Instrument Flight Instructor Rating	2
AV 185	Flight Instructor: Multi-Engine Rating	2

Table 2. New Courses Taught by Contracted Flight Schools

Code	Title	Units
AV 2L	Private Pilot Certification Laboratory	1
AV 76L	Commercial Pilot Certification Laboratory	4
AV 183L	Flight Instructor Certification Laboratory	4
AV 184L	Instrument Flight Instructor Rating Laboratory	1
AV 185	Flight Instructor: Multi-Engine Rating Laboratory	1
AV 192	Instrument Rating Laboratory	4
AV 194	Multi-Engine Rating Laboratory	1

department personnel; those in Table 1 will be taught within the university; those in Table 2 will be taught by the contracted flight school.

Personnel

Faculty members who would teach the proposed in-house courses and supervise the program include:

Dr. John Wingood
Prof. Richard Clairy
Dr. Eugene Smalley
Prof. Jim Stokes
Dr. Allison Gates

Résumés detailing their rank, academic background, and experience appear in the Appendix.

Flight instructors under contract will be required to have flight hours and skills substantially in excess of FAR 141 requirements. Faculty members will document programs and performance of students and will inspect flight instructors' qualifications and records.

Support staff will be minimal; one half-time employee will be needed for record keeping, verification, and for preparation of new courses.

Equipment Budget

NSU already has excellent aviation facilities, including office, classroom, and laboratory space on site at Municipal Airport. This concentration will require no additional space. Recent funding from a $362,000 FAA Airways Science Grant has provided for installation of a new flight-simulator laboratory.

New courses will require the following equipment for the first two years:

Subheadings guide readers to specific information

Year 1   2 Computer Based Training (CBT) stations
           Video study materials
           2 Work/Study video stations
                                      Total Cost $18,200.00

Year 2   2 CBT stations
           Video study materials
           2 Work/Study video stations
                                      Total Cost $17,900.00

Legal and Insurance Costs

Contracted flight schools will provide $1 to $3 million in liability insurance for each student in the flight instruction fees. Other insurance is already provided by the university.

Student Flight Costs

Anticipated student cost for the required flight instruction would be approximately $17,500 to $18,000. Details of cost on a course-by-course basis appear in the Appendix. This cost is in line with costs for similar programs at other universities, and the cost is substantially lower for students than in private programs. In addition, there is considerable interest in this concentration from major airlines and the Airline Pilots Association. Indications are that the program would receive both money and equipment from those sources. We anticipate that a scholarship program will be established to assist students with flight instruction costs. The pilot training portion of the program will be self-supporting from its inception.

CONCLUSIONS

Summary of the need

Major airlines and the airline industry show increasing need for professional pilots over the next 10 years. Since 70 percent of the 600 students currently majoring in Aviation are interested in pursuing professional pilot training, the university already has a potential student base for this concentration. In addition, we estimate that the concentration would attract at least 25 new students each semester to the Aviation major.

Advantages of the proposed program

The Aviation Department will benefit by its ties with industry and with the profession; Northern State University will benefit by meeting a need in the entire four-state region; and students will benefit by obtaining classroom and flight training of a high caliber at the lowest possible cost.

[The Appendix follows. To save space, the material in the appendix has not been included.]

PPT PROPOSAL   PAGE 7

21.5 The Process of Writing a Proposal

Planning and Gathering Information

In proposal writing, planning centers on defining the problem and determining the purpose of the proposal. Because you and your company need considerable background in the field in order to do these two tasks, gathering information is often part of the planning process. Remember that a successful proposal will demonstrate your understanding of the problem, describe your proposed solution and the scope of your work, and convince the client that you can carry out your proposed plan. You may want to divide information gathering into three parts:

1. technical information needed (what will be done)
2. management and personnel information needed (who will do the work and how they will do it)
3. schedule and cost information needed (how long it will take, what equipment is required, and what personnel costs will be)

As you plan, you must always keep in mind that *specific people* will approve your proposal. To be successful, then, you must understand your potential readers: how they perceive the problem and what they find important. One way to do this is to talk with them as you develop the proposal. Another way is to look at proposals they have approved. You always need to understand reader type, background, and purpose when you write; in proposal writing, however, such understanding is not just important—it's crucial. To see how a company works toward understanding its readers and develops a proposal from an RFP, read the following description of the process, written by a student intern after working in the proposal department of a large company that specializes in projects for the federal government (Mawk 1985).

> Proposal writing is a complex, specialized field of writing. The marketing department of a company is responsible for obtaining information about potential customers and competitors and serves as the interface between the company and the customer. All government contracts are obtained through a proposal/bidding process; therefore, a company that secures business this way must develop strong proposal-writing skills. The internship with this company provided an opportunity to observe the process of preparing proposals—from receipt of requests for information (usually before the customer begins preparation of the request for proposal) to shipment of the final proposal. I learned several things about the proposal process:
>
> - In order for a company to receive information relative to upcoming government contracts, it must demonstrate its technical ability through presentations and demonstrations to potential customers. Such activity ensures that a company will be on mailing lists to receive requests for information (RFIs), requests for proposal (RFPs), and other such information.

- The proposal process is an interactive one, with much information exchanged between the customer and bidders. Prior to preparing an RFP, the government will frequently send out an RFI to obtain information about current technological capabilities. Also, the government often releases a draft RFP and solicits comments and questions from potential bidders before releasing the final RFP.
- Proposal writing is both an art and a science. The government is usually looking for specific things in a proposal. A successful bidder has learned the customer's idiosyncracies and prepares the proposal with these in mind.
- The draft proposal goes through many reviews before it is released. There are technical reviews, reviews by those versed in government procedures, reviews by private consultants, and reviews by high-level management. Note that the reviewers don't necessarily always agree.

Learning to read and interpret an RFP is an awesome and sometimes overwhelming task. An RFP can consist of a few pages to several hundred pages; all instructions, specifications, and requirements need to be addressed. One of the tasks of the proposal editor is to prepare a cross-reference of proposal paragraphs to requirement paragraphs, a tedious, time-consuming job.

Once a proposal has been sent to the customer, the job has only begun. Then a series of customer inquiries follows, necessitating "change pages." This process continues until all questionable areas are clarified. After all of this, you can only hope that your company wins the contract.

Organizing and Reviewing

Many proposals follow a standard form and are, in fact, partly composed by a "cut-and-paste" procedure. In other words, certain standard information like the background of the company or the division, résumés, and related experience can be written once and then used in more than one proposal with only minor changes. Having such standard segments (which are called *boilerplate*) is an advantage, because most proposals are written under severe time pressure and with inflexible deadlines. However, the danger of the cut-and-paste method is that the proposal will lack coherence—segments that seem unrelated may weaken the impact of the document as a whole.

Students and workers writing their first proposals have a related problem. It's tempting to adopt an existing proposal model without evaluating whether its organization will work in a new situation and with new material. Before adopting any model, study its organization and the contents of each section. If they are not appropriate for your needs, change to a new organization or adapt what you have. Don't just copy the model. Use the various brainstorming techniques in section 3.1 to gain a fresh perspective on the material.

Many times you must, of course, follow your instructor's or your company's form and organization. Sometimes that method of organization is even specified in the RFP. But when you and your colleagues review the organization, make sure that the parts contribute to a unified whole. The proposal should present a logical, coherent, and clear argument for your plan.

Figure 21-3 shows the outline that one semiconductor company asks its employees to follow when they propose in-house projects to their managers. This is a "résumé" proposal; that is, it's deliberately designed to take no more than one page. Notice how the headings help to limit the amount of writing.

Another method for writing proposals, especially long and formal proposals written by a team, involves *storyboarding,* a technique also called Sequential Topical Organization of Proposals (STOP) (Tracy 1983). Storyboarding is a technique borrowed from the film and advertising worlds; in this approach, the proposal outline is broken into two-page storyboard segments. Individual writers are assigned specific sections. The left page of each storyboard contains a thesis sentence for that segment plus a list of the major points of each paragraph that will support this thesis. The right page contains rough sketches of graphics that will support the points on the left page, plus captions for those graphics.

Writers complete their assigned storyboards and then pin them in order to the walls of a large room. Comment sheets are placed below each storyboard. The team then reviews the proposal by literally "walking through it." Questions of organization, accuracy, and style can be easily addressed, and team members quickly begin to see both what the persuasive message is and how their parts fit into the whole. Storyboarding works well in collaborative situations and under time pressure—both significant elements of proposal writing.

---

1. Name

2. Description

3. Key Objectives
   Economic Benefit

4. Strategy
   Description of Tasks and Approach

5. Resources
   Time Required
       Engineering
       Technical
   Equipment
   Facilities
   Major Expense

6. Proposed Starting Date
   Proposed Completion Date

Figure 21-3 Outline for an Informal In-house Proposal

Writing, Reviewing, and Editing

Writing a good proposal calls for nearly all the techniques covered in Chapters 9 through 13. Proposals often include definition, description, and analysis. The procedure section is a systematic description of the plan of action, and the whole proposal employs interpretation and persuasion. Longer and more detailed proposals usually include graphics, so you must examine the specific needs within sections of the proposal to see how graphics can help clarify information. The whole proposal needs to follow document design principles so it is easy to read.

Once the proposal is written, it needs several kinds of review. Specialists in the technical field should review the introduction and the procedure section (including graphics) to verify technical accuracy. Management representatives should review personnel and budget sections to ensure accuracy of time and cost estimates. Copyeditors should review the writing itself for clarity and correctness of syntax, diction, and punctuation. As a student, you must sometimes take on these reviews yourself, or you may have access to peer review. Use outside reviewers whenever you can. You may be too involved to see the material objectively.

Some agencies or companies "grade" a proposal and comment on its strong and weak points as well as telling why they have rejected it (if they do). If your proposal is graded, value that information for helping you learn what it takes to succeed in this challenging form of writing.

Checklist for Writing Proposals

1. Have I fulfilled all the needs of a proposal? Does my proposal meet the following requirements?
   ___ solve a problem or provide a service
   ___ follow a specified procedure
   ___ use identified people
   ___ adhere to an announced timetable
   ___ project a specific fee
2. What type of proposal is this?
   ___ solicited
   ___ unsolicited
   ___ internal
   ___ external
   ___ informal
   ___ formal
3. If the proposal is informal, does it include the following sections?
   ___ an introduction giving the need or the problem and background
   ___ a proposed plan for solution (procedure) that describes the approach
   ___ a timetable
   ___ staffing
   ___ cost
   ___ a request for approval

4. If the proposal is formal, does it include these elements?
   ___ a letter of transmittal
   ___ an abstract
   ___ a title page
   ___ a table of contents
   ___ a list of figures and tables
   ___ an executive summary
   ___ an introduction
   ___ the procedure or methods (plan for solving the problem)
   ___ personnel (staffing)
   ___ budget (cost)
   ___ appendixes

EXERCISES

1. The following proposal was written by students to establish a campuswide recycling program for aluminum cans (Lane 1991). Analyze it to see if it includes all the parts of a formal proposal; persuasively makes its case; is organized effectively. Be prepared to discuss your findings. (Note: Only the body of the following proposal is included. The transmittal letter, front matter, and back matter have been omitted to save space.)

A PROPOSAL FOR A CAMPUS ALUMINUM RECYCLING PROGRAM

INTRODUCTION

The purpose of this proposal is to develop a campus-wide aluminum recycling program for Central State University (CSU).

Central City and Hayward County face a near-term shortage of landfill capacity. In 1985, there were 23 landfill sites in use, but at the present rate of waste disposal, there will be only 6 unfilled by the year 2000. At the present time, students discard empty aluminum cans in regular trash containers. Recycling aluminum cans will significantly reduce the volume of waste generated at CSU. In addition, recycling will save energy and natural resources and will be cost effective, since the aluminum can be sold.

OVERVIEW OF PROPOSED PROJECT

The first step in getting the money for the project is presenting the proposal to SAFER. Once SAFER accepts the project, it will apply for a grant of $1056 to purchase the bins, locks, pushcarts, and hand

---

crushers. After the money is granted to SAFER, the SAFER official who is responsible for the project will purchase the materials.

The materials will be purchased from the following companies. Approximately eight bins will be purchased from Common Ground Garden Supply at $99 per bin. Eight locks will be purchased from Home Depot at $10 per lock. Two pushcarts will be purchased from The Natural Order at $45.95 per pushcart. Four hand can crushers will be purchased from Area Action at $12.50 per crusher.

PROCEDURE AND METHOD

Drawing Up the Recycling Contract

A contract between SAFER, Recycle America, and the Urban Ministry will be drawn up. The contract will contain all information regarding the disposal containers, type of material to be picked up, and payment for the material to be picked up.

Putting Bins in Place

The bins will be installed by a member of CARe, a member of SAFER, and volunteers from the Urban Ministry. This way all parties involved will know where the bins are located.

The bins will be placed at the entrances to all major buildings on campus, next to the newspaper recycling bins. This will give students the ability to properly dispose of all their recyclable materials at the same time.

Picking Up Aluminum

Materials will be picked up weekly by volunteers from Urban Ministry. Saturday is the most desired day for pick up because there is less student traffic than during the week. The bins will be locked, so an official must be present to unlock the bins for material removal.

The cans must be crushed before the material is put into the disposal container. This is a requirement of Recycle America because crushed cans take up one third of the volume that uncrushed cans require. The crushing will be done by using the hand can crusher or by stepping on the cans.

According to the schedule, Recycle America will pick up the materials to be recycled. The Urban Ministry will then receive payment from Recycle America. The money will be used for a program for the homeless.

2. Following the form given in this chapter, write an informal proposal to the chairperson of your major department suggesting a change in a course or procedure. Justify the reasons for the suggested change. Submit it to classmates for review; then revise as needed.

WRITING ASSIGNMENTS

1. Follow your instructor's RFP and write a proposal to solve a problem in your community. Be sure to specify the type of reader. Proposals may be directed to a grant foundation, a government agency, a board of directors, upper management, a dean or department chair, or the like. Ideas for proposals might include any of the following or one of your choice:

- purchase of equipment for a laboratory, department, dormitory, or library
- change in a procedure or program
- addition of a requirement or course of study
- change in an environment (closing a street, demolishing a building, adding a park, adding walkways)
- inauguration of a new program

2. If you have not yet written a proposal for your term project in response to the writing assignment in Chapter 1, follow your instructor's RFP and write the proposal now. Include a planning sheet like that in Chapter 1. Submit the proposal for review as instructed. If this is a team project, discuss the proposal content in a group meeting, and then assign one or two people to write the proposal. Review it in the group before submitting it to your instructor.

REFERENCES

Anderson, Gordon. 1991. Letter to author. 8 October.

Lane, J., D. North, S. Patel. 1991. *A proposal for a campus aluminum recycling program.* Unpublished document. San Jose State University.

Long, Barbara. 1991. *Top-rated 1040 tax-preparation software programs for professionals.* Unpublished document. San Jose State University.

Mawk, Kenna. 1985. *Internship report.* San Jose State University.

Rosetta, Kim. 1989. *Informal proposal* written by A. Rosenthal. 10 September.

Tracy, J. 1983. The theory and lessons of STOP discourse. *IEEE Transactions on Professional Communication.* PC-26.

U.S. Dept. of Commerce. 1992. *Commerce business daily.* Issues PSA-0519 and PSA-0524. 28 January, 4 February.

Chapter 22

Comparative, Feasibility, and Recommendation Reports

Before she requests new graphics software, publications manager Anne Rosenthal conducts a comparative study of the competing brands. After she analyzes research information from several sources, medical writer Jamie Oliff writes a clinical study that draws conclusions. Using his expertise in pollution monitoring and control, consultant Jim Jahnke writes a recommendation report to a client company. In these reports, each writer presents—in an organized way—information that will help readers understand what has happened and make decisions about what to do next.

As a student, you too have been writing something called "a report" since the third or fourth grade, when you laboriously copied out paragraphs on whales from the encyclopedia. Since then you've written book reports, science reports, history reports, and committee reports. In college, you've struggled with research papers and laboratory reports. By this time, you may wonder if there is anything new to learn about the process or the form of report writing. In fact, there probably is. Few students have ever written the kind of formal report that is the subject of this chapter: a report that examines products, situations, or optional courses of action and recommends one purchase or specific action. Yet because these reports are widely used in technical and scientific fields, you need to know what they are, what form they take, and how they are written.

Reports that examine and recommend are often long and complex, requiring the expertise of many individuals. Thus, many reports are collaborative projects, both to draw on the knowledge of several experts and to get the job done faster. Usually each team member writes a relevant section, and the entire team collaborates on the review.

22.1 Definitions of Types of Reports

The word *report* comes from a Latin word meaning "carry back." In a report, then, the writer carries back to the reader the information learned through research or analysis. The writer has become an expert on some subject or problem and shares that expertise through a report. Reports can be classified in three general categories: research, completion, and recommendation.

Some reports present information from original research (in the laboratory or field) or from a review of existing information (gained through library research or interviews). These are called *research reports,* or scientific and investigative reports. They are often published in scientific and technical journals. The writer reports on "What it is," "What happens," "How it works," or "What happens if. . . ."

A second category of reports tells the reader what was done. These are *project* or *completion reports,* written at the end of a project to document the work that was accomplished, the problems that were encountered, and how those problems were solved. Project reports are similar to the progress and status reports discussed in Chapter 15.

In the third category are reports that analyze and interpret information, come to conclusions, and recommend action. These may be labeled *comparative, feasibility,* or *recommendation reports.* Those that recommend action are also sometimes called proposals. All three terms imply studying a subject carefully or matching one option against another to reach a conclusion. The writer asks "Will it work?" "Can we do it?" or "Which option is best?" and the report is structured to answer those questions. A *comparative* report examines two or more products or options and matches one against the other or against a set of criteria. A *feasibility* report looks at options, ideas, or products to see if they can be done or used and which one is preferable. A *recommendation* report analyzes a situation and suggests a decision or course of action. All three types of report in this category usually include recommendations and are, therefore, sometimes called *recommendation* reports. This chapter considers reports that examine alternatives in this way—reports that are written either for internal company use or for external clients or customers. For simplicity, I will call all three types recommendation reports, though sometimes conclusions are presented without recommendations.

Recommendation reports are usually divided into two large classes based on style and form—what we call informal and formal reports. Informal reports are often written in conversational style and may appear as letters and memos. They are discussed in detail in Chapter 15. Formal reports are written in a more objective, impersonal style, and also have an established form—a standard arrangement of parts to convey information efficiently for both specialists and nonspecialists.

22.2 The Form and Organization of a Formal Recommendation Report

Like proposals, formal recommendation reports are divided into discrete parts that are identified by subheadings. The various parts meet the needs of different types of readers and provide a number of ways to get at the information. Not everyone reads a report from beginning to end. For example, some managers read only the executive summary; others want to see the conclusions and recommendations. Specialists might be primarily interested in the abstract and in procedures or description. Even those who do read a whole report might not read it from beginning to end, but might move from abstract to conclusions to introduction to an appendix. Therefore, one of the writer's major tasks is to make the information in the report easily accessible.

Like most other forms of writing, formal recommendation reports are divided into three large sections: introduction, body, and conclusion. In addition, they have peripheral material at both the beginning (*front matter*) and the end (*back matter*). Listed below are typical parts.

Front matter

> letter or memo of transmittal
> abstract
> title page
> table of contents
> list of figures
> list of tables

The report

> executive summary
> introduction
> body
> conclusions (sometimes placed earlier)
> recommendations (sometimes placed earlier)

Back matter

> reference list or works cited
> glossary (optional)
> appendixes with optional additional information:
>> technical data such as specifications, calculations, tables, and graphs, or
>> supporting data such as letters, brochures, and sample questionnaires

Front matter provides a quick look at the contents and organization of the report and shows the readers how to access information. The abstract tells specialists what is in the report.

The report itself begins with an executive summary that appeals especially to managers, who want the broader view without all the details. For these readers, the writer may also want to move the conclusions and recommendations in front of or just after the introduction. The introduction, body, and conclusion interest those readers who are specialists or technicians and who want to see how the conclusions and recommendations were reached. If these are the primary readers, the conclusions and recommendations are more likely to appear at the end of this section. They will also be summarized in the abstract.

Back matter can appeal to two types of readers. Most often the reader interested in technical back matter is the specialist, who wants detail, calculations, and supporting data. If the specialist is the only reader, this material can be in the report body. Both specialists and nonspecialists consult back matter for the glossary and the reference list of sources.

Not all reports include every part, but those parts discussed in the next sections are important ones. After each explanation, you will find an example of that part (or a reference to an example) from a report on surgical lasers. Then, in section 22.3, you can study a complete team-written report to see how the various parts fit together. Be aware that a formal report is *presented* in the following order, but it is not *written* in this order. See section 22.4 and Chapter 5 for suggestions on what to write first.

Front Matter

Letter or Memo of Transmittal

The *transmittal document* follows standard memo or letter form, as explained in Chapters 15 and 16. It is called a "transmittal" document because the writer uses it to transmit, or transfer, the report to the intended reader. It is a personal communication: a letter for external communication and a memo for internal communication. The first paragraph should include the title and purpose of the report and the reason it was written. Middle paragraphs briefly mention the contents, highlighting key information or conclusions and discussing problems or limitations. As in a letter of request, the closing paragraph should suggest action, such as implementing the recommendations. In addition, it can convey thanks or note specific accomplishments. The tone should be confident, positive, and businesslike. The transmittal document is not numbered as a part of the report; it may be bound at the front or clipped to the cover. Figure 22-1 shows a letter of transmittal for a comparative study.

Abstract and Executive Summary

All reports should include an *abstract* that provides a condensation of the report. Abstracts are frequently circulated separately to potential readers; if an abstract looks promising, the reader can order the report. Abstracts are also available in some technical indexes either on the computer or in books in libraries. Abstracts can be either descriptive or informative, though informative abstracts are becoming more common. If the report is written for both specialists and managers, it often includes both an abstract and an executive summary—the abstract for technical readers and the summary for managerial readers. The abstract is not usually numbered as a part of the report; it may be bound at the front or left free so it can be detached. If bound, it usually precedes the title page, although in some companies it appears on the title page or after it.

An *executive summary* often precedes the introduction; it is written to managers or generalists, contains background information, and assumes less technical knowledge than the abstract. For details on writing abstracts and executive summaries, see Chapter 20. An executive summary appears in the sample report in section 22.3. The abstract of the report on surgical lasers is in Figure 22-2.

Title Page

In addition to the title itself, the *title page* displays the author's name and the date. It can also include the name and title of the person for whom the report was written and, if appropriate, a document or contract number. Some or all of this information is repeated on the report cover, most reports being bound in a sturdy cover of some kind. The title page of the laser report appears in Figure 22-3.

Table of Contents

The *table of contents* provides both a detailed outline of the report contents and a way of accessing information. It can be written from the draft outline and may include three or four levels of headings as well as the page on which each is found. The headings in the table of contents should match the headings in the report exactly. The table of contents may be in outline form with numbered points, or it may simply use indentation to show subpoints. Since reports often do not have an index, the table of contents is an important aid to the reader. Entries should use key words and phrases that give information and tell the purpose of the entry. For example, "Body" is not a

16594 - 47th Drive SE
Everett, WA 98204
May 6, 19__

Dr. Eleanor Swayze
Surgical Staff
Allied Hospital
Seattle, WA 98102

Dear Dr. Swayze:

Purpose and subject of report

As you requested on January 30, 19__, I have conducted an evaluation of surgical lasers for Allied Hospital.

Models evaluated

After submitting the original proposal, I met with Allied representatives Thomas Li and Roxanne Miller. They recommended that a study limited to Nd:YAG lasers would be most appropriate for Allied Hospital. Thus, the four models I have evaluated are

> Lasers for Medicine: FiberLase 100
> Cooper LaserSonics: Model 8000
> MBB-Angewandte Technologie: MediLas 2
> Laserscope: OMNIplus

Criteria

I evaluated these models on versatility, cost, safety, mobility, and ease of operation.

Recommendation

I hope that you and other members of the surgical staff will find this report helpful in making your final purchasing decision. The report recommends the FiberLase 100 for Allied Hospital as cost effective, safe, versatile, and easy to operate. However, you will note that the appendix contains a list of surgeons and hospitals who have experience with these surgical lasers and who are willing to discuss their strong and weak points. Members of your staff may want to contact these individuals.

Appreciation

Thank you for the opportunity to pursue this investigation and for your time and excellent suggestions.

Sincerely,

David Leonard

David Leonard

Figure 22-1 Letter of Transmittal

ABSTRACT

An Evaluation of Nd:YAG Surgical Lasers
for Allied Hospital

by David Leonard

Condensation of report information

This report evaluates four Nd:YAG surgical lasers for use at Allied Hospital:

- Lasers for Medicine: FiberLase 100
- Cooper LaserSonics: Model 8000
- MBB-Angerwandte Technologie; mediLas 2
- Laserscope: OMNIplus

Knowledge of surgical laser action, application, and market competition is needed to choose the right laser. The evaluation is based on five criteria: versatility, cost, safety, mobility, and ease of operation. Each model is analyzed by these criteria; comparison emphasizes key differences among the models.

The FiberLase 100 is recommended for Allied Hospital. It costs about $20,000 less than its nearest competitor, the LaserSonics Model 8000. Its audible fault indicators and self-calibration system combine to give it the best overall safety features of any model. Additionally, its broad delivery options and push-button input panel give it creditable versatility and easy operation. Following behind the FiberLase, and each with its own advantages, are the LaserSonics Model 8000, mediLas 2, and OMNIplus.

Figure 22-2 Abstract

AN EVALUATION OF Nd:YAG SURGICAL LASERS

FOR ALLIED HOSPITAL

Prepared for the Surgical Staff

by David Leonard

May 6, 19__

Figure 22-3 Title Page

useful entry in a table of contents, but "Analysis of the Four Calcining Processes" is. Front matter is not listed in the table of contents, which usually begins numbering the report with the executive summary or the first page of the introduction. Back matter, however, is listed; the pages may be numbered in sequence with the report, or each section of back matter may be numbered separately as appendixes: A-1, A-2, B-1, B-2, and so on. Notice in Figure 22-4, from the surgical lasers report, how much information about the contents and organization you can get from a good table of contents.

List of Figures and List of Tables

Graphics should be listed separately for the reader's convenience; that is, there should be both a *list of tables* and a *list of figures*. Both lists can be on one page titled "Illustrations." Some readers are more interested in the graphics than in the text, and they should not have to search through the table of contents looking for either figures or tables. Tables are numbered consecutively and given titles that identify the purpose and content of each. All other graphics, such as drawings, bar graphs, and diagrams, are considered figures. They are also numbered consecutively, but separately from tables, and given explanatory titles. In the lists of figures and tables, indicate the pages on which they appear. See Chapter 11 for detailed information on tables and figures. The lists of tables and figures for the surgical laser report appear in Figure 22-5.

The Report

Executive Summary

In some reports, an executive summary will be placed before the introduction. In others, the executive summary will be part of the front matter. See Chapter 20 for details on writing an executive summary, and note the example of an executive summary in section 22.3.

Introduction

Each report has a slightly different kind of *introduction*, dictated by its purpose, intended readers, and topic. The introduction should launch the reader into the report by giving background and explaining the problems to be solved; it can also provide an overview of what will follow. Introductions can include any of the following information, though most will not have all these parts. Also, while the order given here is standard, you can use other sequences.

- definition of the subject or problem to be solved
- background information
- purpose of the report and purpose of the study
- significance of the subject
- scope or focus: what is included, what is left out
- literature or historical review (sometimes included with the background information)
- procedure or method of analysis
- definitions of key terms (can also be placed in an appendix)
- overview of the report

Table of contents shows report organization

Titles contain key words and organizational indicators

CONTENTS

Figure 22-4 Table of Contents

Table and figure titles are understandable without the text

ILLUSTRATIONS

Figure 22-5 List of Illustrations

As you think about the whole report, keep the length of the introduction in proportion. It should not exceed one-fourth of the total report, so you may want to move parts you must treat at length into the body. However, all introductions should include:

- a statement of the problem
- the purpose
- the topic and scope
- a brief overview of the report

See section 5.4 for details on writing introductions and an example of the introduction to the report on surgical lasers.

Body

The central part of the report is the *body*. Divide it into sections that fit the report content, and give each section a meaningful title; do not call it "body." The body of a formal report can include:

- background
- theory
- description
- definition
- procedure or methods
- discussion of criteria
- evaluations of products or options
- comparisons
- results
- analysis and evaluation of results

The body of the report on surgical lasers, for example, contains the following sections. Numbers refer to the sections as presented in the table of contents (see Figure 22-4).

2 LASERS IN THE MODERN HOSPITAL

[This section is mostly background for the nonspecialist reader. It contains definitions, technical descriptions, a process description, an explanation of a complete hospital laser facility, and a brief discussion of the competition among hospitals and the role of the government in laser regulation. You can find the definitions from the laser report in Chapter 9 of this book, the description in Chapter 10, and the process description in Chapter 19.]

3 PURCHASING A SURGICAL LASER

[This section explains how the hospital will justify the need for a surgical laser and how the decision was made to study Nd:YAG lasers. Most of the section, however, is devoted to a detailed explanation of the five criteria chosen for evaluation, why each criterion was chosen, and how each criterion can be applied in evaluating the various brands.]

4 FOUR GOOD MODELS

[In this section, the writer first describes the group as a whole and then provides a technical description of each model. Details of the key differences of each model are summarized in the table on page 511, which appears in the report as Table 1.]

A recommendation report is a decision-making document, and the decisions must come from the analysis and evaluation that take place in the body of the report. Thus, as this writer did, you must set up and discuss criteria or standards for judgment. How can you choose such criteria or judgment guideposts? Sometimes the criteria will be given to you along with your assignment:

Determine whether we should switch to a solar hot water system for the design laboratory. Look at installation cost and maintenance cost vs. our current cost for the gas-fired heaters. Also find out how long the system would be inoperable if we made a change.

Table 1. Key Differences among Surgical Models

Category	Fiberlase 100	LaserSonics 8000	mediLas 2	OMNIplus
versatility power (W)	15-100	1-110	15-100	0-15
approval	GI, ENT	GI, neuro	GI	investig.
delivery systems	nonfocusing, focusing, and malleable pieces with range of focal lengths	curved and straight pieces, interchangeable lenses	extensive range of pieces, endo-scopes, guides with liquid or gas circ.	nine specific delivery systems
cost (list)	$79,000	$99,000	$105,000	$117,000
safety	audible fault self-calibrate short focal lengths	audible fault and delivery no kill switch	audible delivery	calibrate delivery devices no kill switch
mobility	660 lb.	800 lb.	400 lb., laser head projects	600 lb.
ease of operation	push-button controls	He-Ne and white aiming light	no high flow gas circ. needs ext. gas	remote operation, self-diagnostic increase/decrease touch sense controls
availability	U.S.	U.S., California based	Germany	U.S., California based

Table summarizes information for ease of comparison

This is a feasibility assignment: the reader wants to know if it's feasible to change to solar hot water. As the writer, you know that your primary criteria will be the two kinds of cost and the time lapse for installation.

Often, as the person who has (or will have) the most technical knowledge, you must choose the criteria yourself. One way to do that is by brainstorming at the planning stage for a list of all the criteria you can name. Then you can refine the list by considering the purpose of your study and the intended reader. As you gather information, you will be able to refine the list further.

For example, the writer evaluating surgical lasers brainstormed a list of the following criteria—only some of which he used in his final report.

adaptability	cost
ability to be upgraded	height
safety	warranty
weight	ease of use
accessibility of parts	repair record
training availability	technical support
projected life span	quality of written instructions

When you have criteria, you use them to compare alternative methods or products. You must objectively analyze the advantages and disadvantages of each option, for only then can you reliably conclude and recommend. The comparison section can be done in two ways: the point-by-point method or the block method.

In the *point-by-point* method, the criteria are the major organizing factors. Each criterion becomes a subsection in which the various options or products are compared. For example:

Cost Criterion
    Option 1
    Option 2
    Option 3
Reliability Criterion
    Option 1
    Option 2
    Option 3

In the *block* method, the options or products are the major organizing factor. Each option or product becomes a subsection in which the various criteria are applied. For example:

Option 1
    Cost
    Reliability
    Safety
Option 2
    Cost
    Reliability
    Safety

You also need to think about the order in which you present criteria. You may want to put the most important criterion first and complete the list in descending order. This is probably most common, but if you have a reason for doing it, you can follow a different order. See Chapter 4 for more on methods of organization.

Conclusions and Recommendations

The concluding section should briefly summarize what has been said and done and then discuss and interpret the results. Conclusions and recommendations are the reason for writing this kind of report, so they are often moved in front of the body, where they can be found and read easily. They can be combined in one section or treated separately. Both conclusions and recommendations should be stated confidently, for the whole report exists to back them up. *Conclusions* are the results of the analysis and interpretation—telling what the data means. For example:

> When the results are tabulated (see Table 1), the Hearthstone I emerges as the clear winner of the four wood-burning stoves evaluated. Its soapstone exterior is beautiful, durable . . .

A *recommendation* is like a call to action—telling what should be done about the results. For example:

> Metal Technology Corp. should purchase Anvil-4000 software and HP 9000 hardware for the new CAD/CAM system. These systems are clearly superior to the others evaluated.

If your presentation is thorough, your data complete, and your analysis sound, the reader will accept your conclusions and recommendations as valid.

Don't feel, though, that you must always support one action or one option. For example:

> I recommend the SL-1 office telephone system because it provides all the required features and has a better service record than the other systems. However, the SL-1 system is also the most expensive. If the cost is restrictive, I would recommend the NEAX 2400 system as a second choice.

Your conclusion might also be that the data is insufficient and more study is needed; it might be that no product or option meets the criteria and, therefore, other products or options need study. It might even be that the criteria need revision. Remember that through your study you have become an expert on this subject. You have an obligation to the reader to share whatever you have learned as clearly and objectively as possible.

In the report on surgical lasers, conclusions and recommendations are combined in section 5:

5 A RECOMMENDATION FOR ALLIED HOSPITAL

[This section summarizes what the writer has learned, recommends one brand of surgical laser, and presents the reasons for that choice. It also summarizes the strengths and weaknesses of the three other models. The writer uses Table 2 to summarize the relative performance of each laser.]

Table 2. Relative Performance of the Models Evaluated

	Versatility	Cost	Safety	Mobility	Ease of Operation	Total
FiberLase 100	3	4	4	2	3	16
LaserSonics 8000	3	2	3	2	4	14
mediLas 2	4	2	2	2	2	12
OMNIplus	2	1	2	2	2	9

Table summarizes the conclusions

Key: 4 = Best, 3 = Good, 2 = Fair, 1 = Poor

Back Matter

Back matter in a formal report consists of information that is important but peripheral to the central content. Even though the report is written in sections, it should form a unified whole; thus, you place in the back matter those details that would clutter the main report, sidetrack the reader, or interest only some types of readers. Back matter can all be put in one section (called an appendix) or in separate sections (appendixes), and the order of the sections can vary. If you choose to label each section as a separate appendix, identify each one with a letter of the alphabet and a title. Typically, back matter will include the following elements.

List of References or Works Cited

A *list of references* or *works cited* may also be called *references consulted.* It can include interviews, tapes, and speeches as well as company reports, journal and newspaper articles, and books. If you annotate the references, you will help the reader see which sources contributed what information. Usually this list is the first section in the back matter. See Chapter 13 for details on constructing a list of references.

Glossary

A *glossary,* or list of definitions of terms, may appear in the back matter or the front matter, but the longer the glossary, the more likely it is to appear at the end. In the glossary, you should alphabetize terms for easy reference and define them with phrases. You may also include symbols in the glossary, or you may have a separate section for symbols and abbreviations. A glossary is especially important if the report will have several types of readers, and if some readers are generalists or managers. Terms should also be defined the first time they are used in the text, or the words should be highlighted (by italics or boldface, for example) and the reader referred to the glossary. See Chapter 9 for more on definition. The first five entries in the laser report glossary look like this:

> *aiming light*   an accessory component of a surgical laser system that indicates the target spot with its own beam
> *bronchoscope*   a delivery component of a surgical laser system used for examining the air passage to a lung

collimation   the process by which divergent rays are converted into parallel rays

continuous wave   a mode of laser operation in which the laser beam is emitted continuously as opposed to pulsed mode

doubled Nd:YAG   a Nd:YAG laser beam whose frequency has been doubled and its wavelength halved by passing it through a special crystal

Technical Data

Specialists and technicians often want *technical data,* including specifications, calculations, tables, copies of lab reports, and physical or procedural descriptions. This material should go in the body if it is crucial to understanding the text, but in the appendix if it provides only support or added information. The laser report, for example, includes a list of hospitals using each of the four kinds of surgical lasers and the names of physicians who can be contacted by interested professionals.

Supporting Literature

Material in the *supporting literature* section might include copies of relevant letters, proposals, and brochures. Such material helps validate the report's information, but it does not directly relate to the content. The laser report includes a copy of the request letter written to laser manufacturers and a copy of the original proposal for the study.

The report in section 22.3 shows you how all of the major parts fit together.

22.3 An Example of a Formal Recommendation Report

The following report is a collaborative study of the feasibility of using wide-area assistive listening systems for the hearing impaired at a major university. The report studies the problem, examines the feasibility of several possible solutions, and recommends the best type of system to solve the problem. This is written as a formal report because it is long, complex, and intended for three different reader groups: Disabled Student Services, the Student Union board of directors, and the director of the Instructional Resources Center. As you read the report, notice how the various parts fit together. Notice too that there is some repetition of information, taking into account the fact that not all readers will read the entire report—or read it from beginning to end.

May 8, 19__

P.O. Box 3001
Los Altos, CA 94022

Dr. Martin B. Schulter, Director
Disabled Student Services
San Jose State University
One Washington Square
San Jose, CA 95192-0168

Dear Dr. Schulter:

We are sending you a copy of the feasibility report on wide-area Assistive Listening Systems (ALS) for use at San Jose State University. The report analyzes the need for an ALS, compares four types of Assistive Listening Systems, and evaluates each one according to seven criteria. A recommendation is given for the system which best meets all criteria.

Since the provisions of the 1990 Americans with Disabilities Act must now be met, the university should move to meet the needs of hearing-impaired individuals. This report will help to guide your office, the Student Union, and the Instructional Resources Center in meeting those needs. Additional copies are being sent to the Student Union board of directors and Ron McBeth, director of the Instructional Resource Center.

Thank you for your time and assistance in helping us understand and research this problem.

Sincerely,

Anne M. McLaughlin

Anne M. McLaughlin
Project Editor

enc.

ABSTRACT

This report describes the effects of hearing impairment on the individual, the need for Assistive Listening Systems (ALS), and four of the technologies that enable the hearing-impaired to participate in social and cultural activities despite hearing loss. The report evaluates the four assistive listening systems on seven criteria: portability, range of reception, flexibility, components, personal receivers, expandability, and cost. The FM radio system is recommended as the best choice for the needs of the hearing-impaired at San Jose State University. Shared funding and responsibilities by Disabled Student Services, the Student Union, and the Instructional Resources Center are also recommended.

Key Words: Assistive Listening System, ALS, hearing-impaired, audio induction, infrared, FM radio, AM radio, transmitter, receiver

Abstract condenses report information, including conclusions and recommendations

Key words are listed for indexing

Title page

**A FEASIBILITY STUDY
OF USING
WIDE-AREA ASSISTIVE LISTENING SYSTEMS
FOR
SAN JOSE STATE UNIVERSITY**

Prepared for

Disabled Student Services
The Student Union
Instructional Resource Center

by

Anne McLaughlin
Artemisa Diaz Valle
Wendi M. Velez

May 8, 19___

CONTENTS

Major divisions are fewer than nine to assist short-term memory

ILLUSTRATIONS

TABLES

FIGURES

Titles are understandable without text

1.0 EXECUTIVE SUMMARY

Hearing impairment affects more than 22 million Americans severely enough to deprive them of participation in the social, cultural, and educational activities that they could enjoy with normal hearing. Because hearing loss is a barrier to communication, it can cause more psychological problems than other physical disabilities. Frustrated by their inability to communicate, hearing-impaired persons often withdraw into isolation.

An Assistive Listening System (ALS) is any device designed to improve a hearing-impaired person's ability to assimilate verbal communication, but in this report, we use the term to refer to systems that transmit sound directly from the speaker to the listener while masking distracting background noise. These systems work through audio induction, infrared, and AM or FM radio. They can be as compact as a single transmitter and receiver carried in a small case, or large enough to encompass an auditorium (wide-area type).

These systems are not designed to aid the Deaf (who use sign language as their major means of communication and who require the services of a skilled interpreter). We are discussing ways to aid the person who loses partial hearing after language skills are developed, but who relies on spoken rather than sign language. For these people (about 37,000 in the San Francisco Bay area), the kinds of systems that we assess provide amplification and clarity to help them use their residual hearing.

Both federal and state laws mandate reasonable access for the hearing-impaired. In January 1992, the Americans with Disabilities Act went into effect. It requires auxiliary aids and services to ensure effective communication with individuals who have hearing impairments. In addition, Title 24 of the California State Building Standards Code requires that buildings which were remodeled after January 1, 1990, and which have a permanently installed public address system must provide ALS for the hearing-impaired.

This report describes four types of wide-area Assistive Listening Systems: audio induction, infrared, AM radio, and FM radio. The evaluation is based on seven criteria: portability, range of reception, flexibility, components, personal receivers, expandability, and cost.

The report concludes that using a wide-area ALS is feasible; that it will meet the requirements of the law and the needs of students and others served by university facilities; and that the best system would be an FM radio ALS.

Recommendations are (1) to request bids for an FM radio ALS, (2) to obtain funds through joint purchase by Disabled Student Services, the Student Union, and the Instructional Resources Center, and (3) that these three organizations share use and responsibility for the ALS.

2.0 INTRODUCTION

Purpose

The purpose of this study is to determine the feasibility of using a portable wide-area Assistive Listening System (ALS) at the auditoriums at San Jose State University (SJSU) to meet the needs of the hearing-impaired.

Intended readers

This report is intended for readers who are involved with providing communication accessibility to the hearing-impaired.

Definition of the target group

The hearing-impaired are those persons who lose partial hearing after language skills are developed. This group uses spoken language as the major means of communication rather than sign language, which requires the services of skilled interpreters. People with residual hearing would benefit the most from a wide-area ALS.

Scope

This report is concerned only with wide-area ALS. The university already has personal ALS available for students and employees. These ALS serve the individual in a one-to-one conference or classroom setting. However, the personal ALS does not work well in auditoriums; in this setting, a wide-area ALS that hooks up to a public address system and then transmits directly to the listener better meets the need. In this report, the term *ALS* refers to the wide-area system unless otherwise noted.

Procedure

We used the following methods of investigation: interviews, library research, and letters to companies for specification sheets.

- Interviews with experts in the field provided information on applicable law, the student population, needs of the hearing-impaired, evaluation of the four systems, and the use of systems by other organizations.
- Library research provided the background and history of technologies that assist the hearing-impaired, explained the need, and provided the rationale for a wide-area ALS.
- Company specification sheets on specific systems provided the information to evaluate the four systems.

All the sources that were investigated are documented in the reference list in the appendix.

Overview of the report

The report includes six major sections:

1. conclusions of the study and a recommendation for the university
2. background: an explanation of the consequences of hearing

loss and how an ALS improves accessibility for hearing-impaired persons
3. a statement of the need for an ALS at the university
4. criteria for evaluation of an ALS
5. a description of the four most common ALS technologies available
6. an evaluation of those systems based on the criteria

3.0 CONCLUSIONS AND RECOMMENDATIONS

3.1 Conclusions

Conclusions are placed early in the report for the convenience of busy readers

Conclusions of this study of wide-area ALS for the hearing-impaired are grouped into three sections: the need for a system, the need for publicity, and the best system to meet the need.

The Need for an ALS

San Jose State University has a significant population of hearing-impaired regular students (30 to 50), serves another population of hearing-impaired senior citizens (20 to 30) in the Over-Sixty Program, and for lectures and other community events draws participants from the hearing-impaired population (37,000) in the San Francisco Bay Area.

The Americans with Disabilities Act (ADA), which was signed into law in July 1990 and became effective in January 1992, requires every establishment that does business with the public to make "readily achievable" changes or provide an alternate method of services to provide accessibility. For the hearing-impaired, accessibility includes listening systems for the appreciation of audio presentations.

While the university already provides personal ALS for students and has wide-area ALS for the exclusive use of employees, no provision exists for wide-area ALS for students, senior citizens, and attendees at lectures and other public gatherings.

The Need for Publicity

All segments of the hearing-impaired population who might use an ALS need to know of its availability. The ALS provided exclusively for SJSU employees has never been used. This underscores the need to publicize the existence of a wide-area ALS once the university acquires a system.

The Best System

Of the four technologies (audio induction loop, infrared, AM radio, and FM radio) the one that best satisfies the needs at the

ASSISTIVE LISTENING SYSTEMS   PAGE 3

university is the FM radio type. Systems operating on the 72–76 MHz band are preferable to those operating on the 88–109 MHz band because the FCC allows the use of higher field strength, which in turn permits a larger area of coverage. With FM technology, the reception is virtually free of electrical interference. Easy installation and operation provide the lowest cost in the long run.

Although infrared technology is popular with auditoriums (Flint Center in Cupertino expressed an interest in such a system), it does not offer the flexibility that SJSU functions would demand. In auditoriums where security from eavesdropping is important, the infrared system is best, but at SJSU, the fact that sound signals can pass through walls is not a serious consideration. One important limitation of infrared technology is that it is not functional under bright light, especially sunlight. Flexibility of use for outdoor functions as well as indoor functions is a consideration at the university.

3.2 Recommendations

Recommendations are specific, based on the conclusions, and placed early in the report

Based on the conclusions drawn from this study, we recommend the following actions:

1. **Purchase an FM radio ALS.** Preliminary studies have shown that an FM system with one transmitter and five receivers could be purchased for $2,100. We recommend that the purchase order be put out for bid, and that the quality of transmitters and receivers in bids be evaluated by audio technologists in the Instructional Resources Center and the Student Union.
2. **Obtain funds through a joint purchase** by Disabled Student Services (DSS), the Student Union (SU), and the Instructional Resources Center (IRC).
3. **Share use and responsibility** among DSS, the SU, and IRC. DSS, with its extensive student contact and newsletter, should take responsibility for publicizing the availability of an ALS. The SU should routinely schedule its use for all noninstructional activities under its control, both those in the SU itself and those at the Rec Center. IRC should store the system components, maintain and repair them, and schedule their use for instructionally related activities.

4.0 BACKGROUND

4.1 Hearing Impairment in Our Society

Details in the report body support the conclusions

Hearing impairment is the most common disability in America today, affecting 22 million Americans. In the San Francisco Bay Area,

ASSISTIVE LISTENING SYSTEMS   PAGE 4

more than 37,000 people have significant hearing loss, while only
2 percent of these people are part of the Deaf community that relies
on sign language. In this report, hearing impairment refers to the loss
of hearing after speech has been acquired. The ability to understand
speech is affected: certain sounds become distorted, and others are
lost entirely. These losses are severe enough to affect the person's
ability to understand what is being said, especially in auditoriums
and meeting rooms. Many factors can cause serious hearing loss, including disease, medicines, exposure to noise, and aging.

Straining to hear what is said, and to "get it" in social or work situations, depletes energy. Therefore, hearing-impaired people tend to
become exhausted more easily than people with normal hearing.
Tension and frustration from concentrating intensely add to their
stress. It is not uncommon for people with severe hearing loss to stop
attending social functions, church services, and cultural events.
Aware of what they are missing, and unable to control situations in
which they are at a disadvantage, hearing-impaired people often become frustrated and withdrawn as their vulnerability increases.

4.2 The Development of Technology

New technology made electronic hearing aids available in the
1930s. European manufacturers began installing telecoils in hearing
aids, and audio induction loop systems came on the European market
in the late 1950s. These features were not available in America until
later.

Wide-area listening systems based on radio frequencies came
next. In 1972, Phonic Ear petitioned the Federal Communications
Commission (FCC) for separate wide-band frequencies for use of
hearing-impaired people in educational settings. Later, an industry-
wide petition from Telex, Comtek, Earmark, Williams Sound, and
Phonic Ear requested an extension of FCC rules for FM systems so
that the systems could be used by hearing-impaired persons in auditoriums, theaters, places of worship, and amusement parks. The FCC
granted this rule change in 1982.

Infrared systems were introduced in Europe in 1976. First intended for home entertainment enhancement, within two years the
manufacturer realized the potential for infrared use in theaters and
auditoriums for both hearing and hearing-impaired persons. Introduced in the United States by Horst Anckerman and Cornelis Hofman
in 1979, the infrared systems helped hearing-impaired people over-
come background noise and distance barriers. Infrared systems are
used today in theaters, courtrooms, and international meetings for
use with language translation.

Background information supports the feasibility of the solution

4.3 Awareness of ALS Availability

In 1979, Howard E. (Rocky) Stone established an organization to help the hearing-impaired, Self Help for Hard of Hearing People, Inc. (SHHH). One of SHHH's major projects has been the installation of ALS in public meeting places. A 1986 survey by the group indicated that 12,000 large-room ALS were in operation at that time.

SHHH advocates Assistive Listening Systems and Devices. It distributes information to members; uses the personal wide-area FMs at meetings; and sponsors area workshops where ALS manufacturers display and demonstrate their products. In the *SHHH Journal,* the organization's magazine, articles on ALS appear frequently.

Without organizations like SHHH, most hearing-impaired people would not know about ALS. Hearing-aid dispensers often have no knowledge of them, and fail to make an effective case for their use to hearing-impaired people. Part of the implementation of any ALS would be an advertising campaign to publicize the system. It is up to hearing-impaired people to request and demand the access to public buildings that they are entitled to by law.

In the San Jose area, Deaf Counseling Advocacy and Referral Agency (DCARA) has worked with the San Jose Center for the Performing Arts to install ALS at the Center. The Bay Area Center for Law and the Deaf (BACLAD) is involved in getting legislation passed to install ALS in California courtrooms.

5.0 THE NEED FOR ALS

Needs are stated specifically

5.1 Laws Mandating Reasonable Access

Federal and state regulations mandate reasonable access for the hearing-impaired. The Americans with Disabilities Act of 1990 requires auxiliary aids and services to ensure effective communication with individuals with hearing impairments. In addition, Title 24 of the California Administrative Code supports the rights of hearing-impaired people to communication access in public buildings.

Americans with Disabilities Act of 1990 (Public Law 101-336)

The Americans with Disabilities Act (ADA) was designed "to provide a clear and comprehensive national mandate for the elimination of discrimination against individuals with disabilities." The act requires that auxiliary aids and services be provided "including qualified interpreters or other effective methods of making aurally delivered materials available to individuals with hearing impairments" (Title V, Sec. 3).

Details of ADA requirements for the hearing-impaired are contained in the ADA Accessibility Guidelines. Section A4.33.7 discusses Types of Listening Systems, saying that "A listening system that can be used from any seat in the seating area is the most flexible way to meet this specification." In addition, Section A4.30.7 requires symbols of accessibility for different types of listening systems. In other words, ALS must not only be provided, but an "appropriate message should be displayed with the international symbol of access for hearing loss, since this symbol conveys general accessibility for people with hearing loss." This symbol is shown in Figure 1.

Figure 1. International Symbol of Access for Hearing Loss

Title 24 of the California Administrative Code

On March 31, 1988, an opinion by state Attorney General John K. Van de Kamp concluded that state law did not require public buildings to contain an ALS. On the other hand, the law authorized the State Architect to promulgate regulations requiring ALS in new buildings and those being altered, repaired, or expanded.

Under Title 24 of the California State Building Standards Code, the State Architect requires buildings built or remodeled effective January 1, 1990, and which have a permanently installed public address system to provide ALS for people with hearing impairments.

5.2 The Hearing-Impaired University Population

Students

The first program providing support services for disabled students at SJSU began in 1972. The Disabled Students Support Program (DSSP) provides help for students whose needs cover a wide spectrum of physical and learning disabilities. DSSP attempts to work with each student for that student's maximum benefit.

The number of Deaf and hearing-impaired students matriculated at the university has increased in recent years. At the present time there are about 30 Deaf and hearing-impaired students registered with DSSP. To assist these students, DSSP employs skilled staff who are proficient in sign language and who understand Deaf culture.

Employees

SJSU purchased a wide-area FM ALS for the exclusive use of university employees as part of the "reasonable accommodations" requirement for disabled employees. The chancellor's office provided special funds for the purchase of the system. To date the system has never been used. No employee has requested it. One problem is that few people know of the existence of the system.

Senior Citizens

The Over-Sixty Program offers senior citizens the opportunity to take classes at the university free of charge. The Re-entry Program refers applicants who have disabilities of any type to Disabled Student Services. For those over sixty who are hearing-impaired, the use of an ALS would mean taking advantage of a golden opportunity for education and intellectual stimulation in their later years.

Future Demand

Hearing loss in general is on the increase. Excessive noise levels, certain types of medicinal drugs, aging, and birth defects can all cause hearing impairment. To meet this growing need, SJSU should begin now to acquire the electronic technology to make educational and cultural resources available to hearing-impaired people.

6.0 CRITERIA FOR EVALUATION OF ALS

Criteria supply a framework for evaluation of systems

The wide-area ALS adopted at the university should meet the following criteria:

1. **Portability.** The system should be usable at any location on campus so it can be shared by various campus organizations.
2. **Range of reception.** The system should allow the hearing-impaired freedom of movement or seating preference without loss of signal clarity.
3. **Flexibility.** The system should be usable outdoors or indoors.
4. **Components.** The system should consist of minimal components for ease of use, transportation, installation, and maintenance.

5. **Personal receivers.** The receiver should be tunable and compatible with a variety of hearing aid styles.
6. **Expandability.** The system should allow for future expansion as needed.
7. **Cost.** Due to budget constraints, the cost should be as low as possible. Preliminary studies indicate a range from $1,500 to $3,000.

7.0 DESCRIPTION OF THE FOUR COMMON ASSISTIVE LISTENING SYSTEMS

Each system is described in detail

Four types of ALS are available: audio induction loop, infrared, AM radio, and FM radio. These systems all have one thing in common: they are designed to bypass the acoustic space between the user and the sound source, thus reducing undesirable background noise and producing greater intelligibility.

This section describes how each type of system works.

7.1 Audio Induction Loop

The audio induction loop, the oldest technology this study considers, consists of a wire placed around the perimeter of the room or space to be served either at floor level or ceiling level (see Figure 2). The wire has both ends attached to the sound source speaker, thus creating a "loop." The electric current that flows through the loop radiates a harmless magnetic field that can be picked up by a hearing aid equipped with a telephone coil (T-switch). The T-switch is simply a tiny coil with hundreds of turns of wire that is able to pick up an electromagnetic field in close proximity to the hearing aid. All the

Graphics help readers understand each system

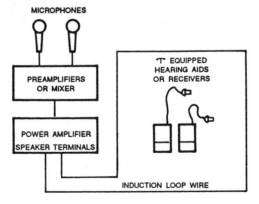

Figure 2. Audio Induction Loop Assistive Listening System

hearing-impaired person needs to do is switch the hearing aid to the telephone or "T" position (which generally turns off the normal hearing aid microphone) to hear only what is broadcast by the loop system. Users who do not have a T-switch or who do not wear a hearing aid must use a special induction receiver. Signal strength (hearing) is excellent and consistent as long as the listener is properly oriented to the loop. Extensive building metal, too small telecoils, specific location within the loop, and head angle can affect the signal.

7.2 Infrared

Infrared technology uses invisible light beams to carry information from a remote transmitter to an infrared receiver (see Figure 3). The transmitter unit is plugged into the sound system amplifier, and emits a signal consisting of light-waves, which spread throughout the room but are outside the range of human sight. The light waves carry the message from the sound source and are picked up by lightweight receivers that "see" the infrared signal. Unlike the other types of systems, the infrared type broadcasts by line-of-sight and cannot penetrate physical barriers. Reception from the receiver by the ear is through a lightweight stethoscope-like unit (similar to airline sound

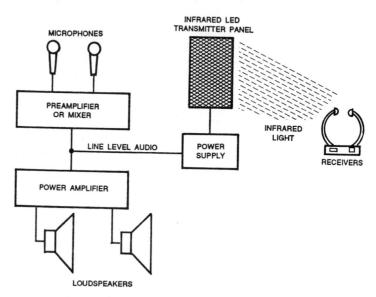

Figure 3. Infrared Assistive Listening System

systems), a T-switch (with neck induction loop), or direct audio input. Since light-waves do not pass through walls, transmission is confined to the room containing the sound source. Infrared systems are not affected by other nearby radio signals, but clear transmission may be affected by large amounts of incandescent light or sunlight.

7.3 AM Radio

Radio frequency systems for people who are hearing-impaired are of two basic types: AM (Amplitude Modulation) and FM (Frequency Modulation). Figure 4 shows how an AM or FM system works.

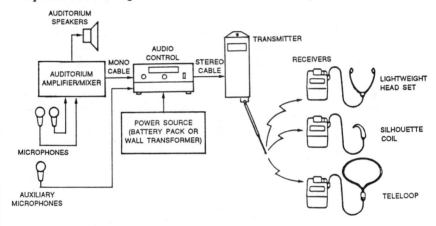

Figure 4. AM or FM Assistive Listening System

The difference in performance of the two radio systems is analogous to the difference between AM and FM radio broadcasting. The AM system typically operates at low frequencies—those of the radio broadcast band (550 to 1600 KHz) or below—and, consequently, is vulnerable to environmental interference.

The AM system consists of a transmitter connected to the existing sound source and one or more receivers set to the same frequency. The sound is conveyed from the transmitter to the individual handset receivers.

7.4 FM Radio

The FM radio system operates like the AM radio system, except that it typically operates at higher frequencies, either on the 72 to 76 MHz band designated in the United States for auditory assistance devices or on the FM broadcast band between 88 and 108 MHz.

As in the AM system, a transmitter is connected to the existing

ASSISTIVE LISTENING SYSTEMS   PAGE 11

sound source, and receivers are set to the same frequency. The sound is conveyed from the transmitter to the individual wireless handset receivers. See Figure 4.

Modes of transmission that transmit sound to hearing-impaired people include the following: (1) earphones, (2) T-Switch, in which a neck loop must be plugged into the receiver, or (3) direct audio input in which the receiver is connected to the hearing aid by a cord.

8.0 EVALUATION OF THE FOUR SYSTEMS

Evaluation of each system is by the point-by-point method

To determine which of the four ALS will best meet the needs of the university, this section evaluates each system by the seven criteria explained in section 6.0. Table 1 compares each system in detail.

Table 1. Comparison of Assistive Listening Systems

System	Advantages	Disadvantages
Induction Loop Transmitter: Transducer wired to induction loop around listening area. Receiver: Self-contained induction receiver or personal hearing aid with telecoil.	Cost-effective Low maintenance Easy to use Unobtrusive May be possible to integrate into existing public address system. Some hearing aids can function as receivers.	Signal spills over to adjacent rooms. Susceptible to electrical interference. Limited portability Inconsistent signal strength. Head position affects signal strength. Lack of standards for induction coil performance.
Infrared Transmitter: Emitter in line-of-sight with receiver. Receiver: Self-contained. Or with personal hearing aid via DAI or induction neckloop and telecoil.	Easy to use Insures privacy or confidentiality Moderate cost Can often be integrated into existing public address system.	Line-of-sight required between emitter and receiver. Ineffective outdoors Limited portability Requires installation

| AM Radio
Transmitter: Worn by speaker.
Receiver: Personal hearing aid or earphones. | No seating restriction
Less cost than FM radio | Coverage limited to 15,000 sq. ft.
Poor sound and static.
Requires 10 ft. antenna |
| FM Radio
Transmitter: Worn by speaker.
Receiver: With personal hearing aid via DAI or induction neck-loop and telecoil or self-contained with earphone(s). | Highly portable
Different channels allow use by different groups within the same room.
High user mobility
Variable for large range of hearing losses. | High cost of receivers
Equipment fragile
Equipment obtrusive
High maintenance
Expensive to maintain
Custom fitting to individual user may be required. |

Source: *Rehab Brief,* National Institute on Disability and Rehabilitation Research, Washington, DC, Vol. XII, No. 10. (1990).

8.1 Portability

The two systems that meet the portability test are the AM and FM radio systems. The audio loop system restricts use to those rooms that actually have a loop, while the infrared system cannot be used outdoors.

8.2 Range of Reception

Coverage of the systems ranges from 500 square feet to 750,000 square feet. The FM radio system has the greatest potential range; both AM and FM radio systems allow for free seating preference. The audio loop system restricts seating to within the loop, and the infrared system has a line-of-sight limitation.

8.3 Flexibility (use indoors and outdoors)

The two systems that best meet the flexibility criterion are FM radio and audio loop. Infrared technology is not functional in bright sunlight, and AM radio requires at least a 10-foot antenna. The audio loop system works outdoors if the wire loop is installed on the perimeter of the assembly area (on poles or otherwise elevated).

8.4 Components

All four systems require the following components:

- amplifier
- microphones

- some type of transmitter
- receivers
- charger (for battery-operated receivers)

In addition, AM radio requires a 10-foot antenna; FM radio requires a short antenna; audio loop requires a 50- to 200-foot wire loop; and the infrared system requires emitter panels. The audio loop system has the advantage of not requiring a receiver if the individual's hearing aid has a T-switch.

8.5 Personal Receivers

All systems require some type of receiver, though the configuration varies. No one system has a particular advantage in tuning or hearing aid compatibility.

8.6 Expandability

All systems are expandable. AM and FM systems simply add receivers. The audio loop system expands by enlarging the loop. The infrared system expands by adding panels and receivers.

8.7 Cost

If cost is figured on coverage of 15,000 square feet with five receivers (a typical room configuration and service to five hearing-impaired individuals), the four systems would cost at the minimum the following:

Audio Loop	$1,350
Infrared	$6,400
AM radio	$1,400
FM radio	$2,100

However, the audio loop system is limited to 5,000 square feet of coverage.

APPENDIX A
LIST OF REFERENCES

Americans with Disabilities Act. Public Law 101–336. 1990. EEOC-BK-19.

Blake, Ruth, Cecelia Torpey, and Patricia M. Wertz. 1987. *Effects of auditory trainers on attending behaviors of learning disabled children.* Minneapolis, MN: Telex Communications. A pamphlet that describes the finding of a study on student and teacher response to the use of FM wireless systems.

Cutler, William B. 1978. *Large room listening systems for the hard of hearing.* Unpublished manuscript. Self Help for Hard of Hearing People, Inc., Bay Area Members. A strong advocate of installing Assistive Listening Systems in all possible locations, Cutler explains the legal and technical material concerning ALS in terms the layman can understand.

Duffy, J. Trey. 1989. Disabled Student Services Coordinator. San Jose State University. (Mar.-Apr.) Dr. Duffy provided good background information on the situation at San Jose State University. He favors purchase of an ALS.

Flexer, Carol, Denise Wray, and JoAnne Ireland. 1989. Preferential seating is not enough: Issues in classroom management of hearing-impaired students. *Language, Speech, and Hearing Services in Schools.* (January): 11–21. Highlights issues of understanding the nature of hearing and consequences of hearing loss and the use of technology to enhance the signal-to-noise ratio.

Hawkins, David B. and James F. Fluck. 1982. A flexible auditorium amplification system. *ASHA,* 24.4: 263–267. Description of a rationale for and implementation of an auditorium amplification system in a university setting to improve accessibility for the hearing-impaired.

Hermann, Judy. Interview with Artemisa Valle. San Jose, CA, 17 March 1989. Hermann is the public relations director of the Student Union and provided information about student use.

Hodge, Marie. 1986. Hear what you've been missing. *50 Plus,* (November): 50–66. Hodge explores hearing loss and the advantages of Assistive Listening Systems and Devices for those who lose their hearing later in life.

Lanza, Nicholas. Staff Attorney, Bay Area Center for Law and the Deaf. Interview with Anne McLaughlin, Artemisa Valle, and Wendi Velez. 20 March 1989. Nick Lanza knows a great deal about the legal aspects of the ALS issue. As a supervisor at DCARA (Deaf Counseling Advocacy and Referral Agency), Lanza is in contact with hearing-impaired people and knows their needs.

Palomares, Luis. 1984. Kennedy center installs infrared systems. *Sound and Video Contractor.* (January): 22–28. Palomares describes infrared listening systems that were installed in the Kennedy Center theaters.

Phillips, Gene. Interview with Artemisa Valle. San Jose, CA, 27 March 1989. Phillips is an audio technician in the Instructional Resources Center and discussed the requirements of the campus public address system.

Vaughn, Gwenyth R. 1986. Assistive listening devices and systems (ALDS) come of age. *Hearing Instruments,* 37.7: 12–14. This article is an excellent summary of the need for ALS and the way it can be publicized.

____ . ALDS pioneers: Past and present. *Hearing Instruments,* 38.2: 12–14. This informative article is a good overview of how we reached the present point with assistive listening systems.

Annotated references help the reader understand the value of each source

APPENDIX B
GLOSSARY

AM (Amplitude Modulation) an ALS that uses AM radio to transmit sounds.

Amplifier a circuit or electronic apparatus for increasing the strength of electrical signals.

Assistive Listening Device any device that assists a hearing-impaired person to hear better.

Assistive Listening System (ALS) a sound enhancement system that bypasses the acoustic space between the sound source and the listener.

Audio Induction Loop System an ALS that uses a loop of wire creating a magnetic field to transmit sound.

Deaf not perceiving sound; totally or partially unable to hear. Also, a term that describes people who use sign language and live in the Deaf community.

FM (Frequency Modulation) System an ALS that uses FM radio to transmit sound.

Hearing-impaired denotes loss of the sense of hearing whether mild, moderate, or severe. A euphemism used to encompass the spectrum from hard of hearing to deaf.

Hz (hertz) the international unit of frequency, equal to one cycle per second.

Infrared System an ALS that uses the infrared wave length to transmit sound.

KHz (kilo hertz) a unit of frequency, equal to one thousand hertz.

MHz (mega hertz) a unit of frequency, equal to one million hertz.

PA (Public Address) an electronic amplification system, used in auditoriums, theaters, or the like, so that announcements and music can be easily heard by a large audience.

Receiver the component of an ALS system that receives the signal from the transmitter.

T-Switch a switch on a hearing aid that changes frequency so that the telecoil in the hearing aid inducts sound or speech without background noise.

Transmitter the component of an ALS that sends the sound to the receiver.

APPENDIX C
SPECIFICATION SHEETS

[To save space, the specification sheets are omitted, but information from the following companies is included in this appendix.]

Audio Enhancement
932 Spoede Rd. N.
St. Louis, MO 63146
USA   314/567-6141

COMTEX
357 West 2700 South
Salt Lake City, UT 84115
USA   801/466-3463

Phonic Ear Inc.
250 Camino Alto
Mill Valley, CA 94941
USA   415/383-4000

TELEX Communications, Inc.
9600 Aldrich Ave. So.
Minneapolis, MN 55420
USA   612/884-4051

Williams Sound Corp.
5929 Baker Road
Minnetonka, MN 55345-5997
USA   612/931-0291

The Process of Writing a Formal Recommendation Report

Like proposals and research reports, recommendation reports are long, complex, and composed of many parts. One challenge in writing such a report is the sheer volume of the material; another is keeping the parts intelligible by themselves but also smoothly integrated into the whole document. Before you undertake a feasibility, comparative, or recommendation report, it will pay you to review Chapters 1 through 7. You will want to refresh your memory about details that will be mentioned only briefly here.

Planning

If you are asked to write a recommendation report in industry, your primary purpose is already set, for the purpose dictates the choice of this form. The primary purpose is to interpret, evaluate, judge, or recommend—and often all of these. Secondary purposes might be to define, describe objects or procedures, or explain. If you are assigned to write a recommendation report in the college classroom, you also have a general purpose and a form, but you may not have a topic. To refine your purpose and your topic, you must think through what you intend to analyze and how your conclusions and recommendations can come from that analysis.

Recommendation reports can be written for specialists, managers, or generalists, so you must first determine your primary reader and all secondary readers. Often you will write these reports for some level of management, using your time and expertise to help managers make decisions. You could also write for generalists like consumers. As section 1.2 points out, you need to evaluate potential readers carefully, so that you can meet their needs with useful information, understandable vocabulary, and logical report organization. Knowing your readers will also help you decide what material to place in front matter, in the report body, and in back matter.

For a major project like this, you need to do both long-term and short-term scheduling. Use a planning sheet like that in section 1.6, and schedule the *whole* project first, allowing as many days or weeks for each step as seem necessary. Then schedule tasks *within* each step so that you can work on the analysis and the report writing a little at a time. My students almost always underestimate the time they need for gathering information, for writing, and for designing the document; you should add contingency time to these categories. Your instructor can give you advice about time allotments based on the length of the term and your specific writing requirements.

Gathering Information

Before you start gathering information, review section 2.1 and spend some time brainstorming to see what you already know, where the gaps are, and how the material groups itself in your mind. Then plan your search strategy. Establish tentative criteria to help guide your research. Library research can help you understand basic descriptions and procedures and is a good place to begin. But in gathering data for analysis, you are more likely to be doing field research of some kind:

- writing to or interviewing specialists
- gathering specifications and descriptions from brochures and data sheets not likely to be in a library
- conducting surveys
- observing processes
- examining places or objects
- experimenting with options, methods, or products

It's important to keep careful notes as you proceed in order to have all the needed information when you are ready to write. As you proceed, you may need to revise some criteria or eliminate some options. Do make changes as you learn more about the subject, because those changes will move you toward the organizing step. At some point, you will be schedule-driven to stop gathering information and begin organizing, but if necessary, you can go back later to fill in gaps.

Considering Legal and Ethical Implications

Once you have gathered your information, you will have a good idea if there are legal constraints or ethical implications. Avoid potential plagiarism problems by careful documentation and by using your own words to summarize the writing of others. Be sure that you observe copyright, trademark, and patent restrictions. Contact the legal department of your company if you have questions on legal issues. Clarify any ethical issues in your own mind (and perhaps with your manager). Review Chapter 2 for more details.

Organizing and Reviewing

As you begin to organize your data, don't worry about the front and back matter of the final report. Those sections can be tackled after the report itself is written. In fact, you may want to ignore the introduction as well—just don't lose sight of the report's purpose (which should be stated in the introduction). Concentrate your first efforts on organizing the body of the report—that central section where you present data and judge it by established criteria.

Read through all your information in order to settle on a main point. Then sort the data and shuffle the bits and pieces into a tentative arrangement. Choose a method of organization that will help you achieve your purpose. In a recommendation report, your organization probably will involve some kind of hierarchy, such as comparison-contrast or problem-cause-solution. These methods of organization and others are explained in Chapter 4. Remember that data can almost always be organized more than one way, so try two or three possible organizations, checking each one to see if it fulfills your purpose and meets your readers' needs. Once you have the body organized, choose material for the introduction and create an outline for that.

Determine what the conclusions and recommendations will cover and note the main points. (Or write the conclusions and recommendations first and then the introduction.) Put all the parts together and see if they fit. Adjust sections as necessary.

Be aware as you organize that you may have to find more information to support an important point. You may also have to eliminate information that does not directly apply. Electronics engineer Craig Johnson says, "It's sometimes difficult to emphasize an important point that has little supporting detail and likewise de-emphasize an unimportant point that happens to have a lot of supporting detail." In other words, don't let the *amount* of information determine your major points. Instead, go back to your purpose and let that purpose control the organization.

When you have a rough outline, review it by applying the questions in Chapter 4, and then, if possible, have three of four other people review your organization. You will welcome their perspective, which can be more objective than yours.

If you are writing this report as part of a team, meet with other team members to examine your organization relative to theirs. Remember that while each of you may be writing separate segments, the segments must interlock to produce a coherent whole.

Writing

Before you begin writing, review Chapter 5 on organizing the writing process. That first plunge into writing is scary, so begin with easy material that you know well. That might be definitions, a description, or a procedure—all concrete sections for which you probably have plenty of information. Once you've completed that material, you might turn to an explanation of your criteria. Then you can apply the criteria to your options, and so move on to conclusions and recommendations. Next you might want to write the introduction (though you could also try it earlier). Note places in the manuscript where graphics would be helpful. Later you can find or create the best graphics for those spots.

Only after you've written the main report should you worry about front and back matter. By then much of the work will be partly done. You might have created most of the glossary with the definitions you wrote to warm up; now you only need to finish and alphabetize it. If you kept track of source materials as you researched and wrote, creating your reference list will be a simple matter. The outline will convert to a table of contents with only a bit of polishing (though the page numbers can't be added until the report is typed). That leaves only the abstract, executive summary, and letter of transmittal, which must be written last since they are based on the main report. Thus, if you take one small step at a time, you can finish the draft of your report in an organized way.

Reviewing between Editing, Formatting, and Assembling

A long report needs to be reviewed several times:

- Before the draft is formatted, to check for accuracy of content, organization, grammar, syntax, and punctuation. This is called copyediting.

- After the final copy is formatted, to check for correctness and agreement with the draft. This is called proofreading.
- Before the graphics are put in final form to ensure accuracy of content and details.
- After the graphics are incorporated, to make sure they are where they're supposed to be and are properly titled and referenced in the text.
- After the pages are punched and assembled between the covers. (Nothing is more unprofessional or distracting to a reader than a page inserted upside down or backwards.)

All those reviews should ensure a report that is designed and presented as well as it is written. To create a professional-looking document, follow these suggestions:

1. Use a word processor if possible. It will speed up changes and revisions in the text.
2. Allow an extra inch for the left margin so that the pages can be bound without covering up any text.
3. Follow the suggestions in Chapter 12 for designing the document: use white space, type sizes, and features to show relationships among parts of the document.
4. Allow plenty of time to format the final copy. Even the title page, the table of contents, and the list of illustrations can take more than an hour to create.
5. Schedule time for all the levels of review discussed in Chapter 7.

Checklist for Writing a Formal Report

1. What is my purpose?
2. What is my topic?
3. Who is my primary reader?
4. Who are my secondary readers?
5. In the front matter of my report, have I included the following elements?
   ___ letter or memo of transmittal
   ___ abstract (and executive summary if I have nontechnical readers)
   ___ title page
   ___ table of contents
   ___ list of figures
   ___ list of tables
6. In the report itself, do I have these sections?
   ___ an introduction
   ___ a body (titled with key content words and organizational indicators)
   ___ a conclusion
   ___ a recommendation
7. Should I combine the conclusion and recommendation or place them in separate sections? Should I move the conclusions and recommendations before the body?

8. In back matter, do I have these sections?
   ___ a list of references
   ___ a glossary
   ___ appendixes

 EXERCISE

Study the following tables of contents and lists of illustrations and determine if they contain the major elements for an effective report and if the subheadings give enough information to help the reader. Be prepared to discuss in class.

A.

CONTENTS

	Page
Abstract	1
Introduction	2
Selection of Skis	3
Snow Conditions—Criteria Not Used	3
Skiing Ability—Choice of Skis	4
Ski Length	4
Flex Pattern	6
Torsional Rigidity	6
Thickness	8
Sidecut	9
Camber	9
Conclusion	11
Recommendation	12
Appendix A	
Glossary	13
Adult Ski Sizing Chart	15
Appendix B	
Original Proposal	16
Sample Letter of Inquiry	18
Bibliography	19

TABLES AND FIGURES

Table 1	Comparison and Rating of Ski Length	5
Table 2	Comparison and Rating of Flex Pattern	6
Table 3	Comparison and Rating of Torsional Rigidity	8

B.

CONTENTS

LIST OF TABLES AND ILLUSTRATIONS

C.

CONTENTS

▤ WRITING ASSIGNMENT

Following your instructor's specific suggestions, write a comparative, feasibility, or other recommendation report. Writing assignments in earlier chapters will help you prepare.

If this is a collaborative project, set deadlines for individual sections well in advance of the due date. Individual assignments might include responsibility for several of the following sections:

    executive summary
    introduction
    definitions
    technical description
    procedures
    analysis of options
    graphics
    conclusions and recommendations
    front matter (abstract, table of contents, list of illustrations, title page)
    back matter (list of references and appendixes)

Appoint one or two people to edit the individual sections and coordinate them so that the document reads as smoothly as though one person had written it.

REFERENCES

Leonard, David. 1985. *An evaluation of Nd:YAG surgical lasers for Allied Hospital.* Unpublished document. San Jose State University.

National Institute on Disability and Rehabilitation Research. 1990. *Rehab brief.* Washington, DC. XII, 10.

Chapter 23

Research and Project Reports

When you are working in the laboratory, the field, or the office, how can you best communicate information? How can you tell other people:

- What you have been doing?
- How you did it?
- What you have learned?
- What's important to them?

You could telephone your information or share it in a speech, but people often forget significant points or miss details when they only hear the information. Most often, therefore, you communicate information by writing some kind of report that can be read, studied, and referred to later.

If your work is research in the laboratory or field, you will write a research, scientific, or engineering report. If your work is project-oriented (building a bridge or designing a mass storage system, for example), you will write a project or completion report.

The purpose of research and project reports is usually not to evaluate, recommend, or choose from alternatives, but to present information, suggest new interpretations of data, or record accomplishments. Thus, factual and verifiable information forms the central content and makes this kind of report different in purpose and form from the comparative, feasibility, and recommendation reports discussed in Chapter 22.

23.1 Definitions of Terms

As noted in Chapter 22, the word *report* comes from a word meaning "carry back." In a report, then, you carry back or present information to the reader, and to do that, you must become an expert in the subject. Who your readers are, how much they know, and what questions they have will determine the type of report you will write.

The ultimate way to communicate research is to write a scientific or technical report and publish it in a professional journal, such as the *Journal of Bacteriology* or the *Journal of Automotive Engineering*. These documents are called *research reports, journal articles*, or *scientific papers*, and they are written by specialists for specialists. They usually follow a prescribed form, which is explained in section 23.3. Research reports that are written for use within an organization are written in the same form.

The *laboratory reports* you write as a student are an adaptation of research reports. As a student in the laboratory, you "learn by doing." In the laboratory, you conduct experiments and carry out research under controlled conditions; you learn what happens, how it happens, and what it means. Then in a laboratory report, you communicate that information to others.

Another kind of report written to share information is the *technical* or *trade article*: a report that describes applied research or development, typically giving information about new techniques or products. For example, electronics engineer Craig Johnson and technical writer Doug Klopfenstein shared information with their peers about a new procedure in a 1988 article titled "Isolating Faulty RAM Components in Record Time" published in *Evaluation Engineering*.

More often, though, you will communicate information on the job through *project* or *completion reports*, either internally for managers and specialists or externally for clients. When these reports are long or complex, they are written as formal reports. They too follow a prescribed form, though that form is less rigid than is the research paper form.

23.2 Ethical Concerns

Students writing laboratory reports often find themselves in a difficult position. Perhaps the experiment didn't go the way it was "supposed" to go. Sometimes data is missing, or results are not clear. Despite these problems, students are expected to write a coherent, complete report that is factual and verifiable.

In these circumstances, it's tempting to make up missing data, change the results slightly, or borrow information from another experiment—all minor changes to help in constructing a laboratory report that meets stated objectives and makes sense. After all, as a student you know you will be graded on your completed report.

The problem, of course, is ethical. The scientific method is built upon verifiability, and it presumes that in a research or laboratory report readers can (1) assess observations, (2) repeat experiments, and (3) evaluate intellectual processes. Those small changes you make in a report may go undetected until the reader tries to replicate the research. If the research is unreliable or if information is plagiarized, the very basis of the scientific method crumbles.

Research scientists and engineers face the same kinds of ethical issues as do students. These professionals are pressured to obtain research funds from corporations, governments, and other interest groups. The funding sources want prompt answers to their questions, influence on the research design, and access to the data. In a way, professionals are "graded" just as students are. One professional puts it this way, "If science's reputation as an impartial search for truth is to be preserved, scientists need to confront ethical dilemmas, not dig in their heels and react to them as another externally imposed distraction" (Greenberg 1991).

23.3 Research and Laboratory Reports: Organization and Examples

A research report is structured to answer specific questions posed by readers in technical and scientific fields:

- What was the problem?
- How did you study the problem?
- What did you find?
- What do these findings mean?

To answer these questions, research reports are divided into the following sections:

- title
- abstract
- introduction, including literature review
- materials and methods
- results
- discussion
- references

These parts can be grouped into front matter (the title and abstract), the report itself, and back matter (the references and supporting information like calculations). Since research reports are often brief, these divisions may or may not be set off by subheadings; each journal or organization has a slightly different style. Note that even though the parts are *presented* in this order, they need not be *written* in this order. For example, in writing the draft, you might begin with the materials and methods section, which deals with specifics and might be easier to write than the introduction.

Title

The *title* should contain both key content words and function words so that it accurately reflects the contents and can be usefully indexed. See section 6.1 for more on writing good titles. Reporting on field research, for example, biologist Brian Simmons (1990) titled his paper *A Stress Induced Die-off of Rocky Mountain Bighorn Sheep*. This paper was written for internal use by a state agency.

Abstract

The *abstract* can range from 50 to 300 words but should be proportional in length to the article. You may want to write an abstract to help you in planning the purpose, content, and organization of your report, but you should rewrite it when the report is finished and the form and content are fixed. An abstract serves several functions: (1) it can be sent as part of a query to see if journal editors are interested in your projected report; (2) it can be filed in an index of abstracts to provide condensed information about the main report to potential readers; and (3) it can precede the article, allowing readers to preview the article and decide if they want to read it. Abstracts may be either descriptive or informative; Chapter 20 gives details on writing abstracts. Here is the abstract for the field research report:

Subject of the research

Methods

An indigenous, low-elevation herd of bighorn sheep in Waterton Canyon, Colorado, was studied in 1979–1980, coincident with the construction of the Strontia Springs Dam. This small, healthy population of sheep grew steadily until fall 1980, when an outbreak of bronchopneumonia resulted in the loss of 77 percent of the herd of 78 sheep. Evidence

*Condensed
discussion*

for stress-induced disease and lungworm-pneumonia complex in bighorn sheep are reviewed and compared to conditions in Waterton Canyon. Stressors involved in the Waterton die-off include limited range, high social density, natural burdens of low-pathogenic bacteria and lungworms, human disturbance, and airborne dust.

Introduction

The *introduction* states the purpose and the main point or thesis of the report. It gives background information, including a literature review (a survey of what has already been written on the subject). This literature review establishes your mastery of the subject and helps set up a rationale for the study. In order to preview for the reader what will follow, an introduction should clearly define the problem and indicate the method of investigation. It should also define any unusual terms or symbols. Finally, it should briefly summarize the results. See section 6.5 for more on introductions. The first two paragraphs of the bighorn sheep research report show how the study is introduced.

*Background and
literature review*

Rocky Mountain bighorn sheep herds (*Ovis canadensis canadensis*) throughout North America have experienced irregular, precipitous die-offs for the last century, but especially since the 1920s (Packard 1946, Moser 1962, Post 1962, Stelfox 1971, Feuerstein et al. 1980). Die-offs have been attributed to diseases that have typically involved the lungs such as hemorrhagic septicemia, lungworm pneumonia, and pasteurellosis (Potts 1937, Marsh 1938, Packard 1946, Hunter and Pillmoore 1954, Moser and Pillmoore 1954, Buecher 1960, Moser 1962, Post 1962).

*Overview of the
study*

The Waterton Canyon bighorn herd is one of seven remaining low-elevation indigenous herds in Colorado. Behavior and ecology of the herd was studied during 1979–1980, concurrent with the construction of the Strontia Springs Dam. The Waterton herd suffered a disastrous die-off October–December 1980. The ongoing study enabled an unprecedented record of the growth and decline of a bighorn herd in the presence of human disturbance and disease.

Materials and Methods

The central section of a report, *materials and methods*, is much like a procedure: it discusses the equipment and materials used and then presents the procedure itself sequentially. This section should be specific enough so that another researcher can duplicate the procedure. The ability of others to duplicate results and verify conclusions will validate your research. See Chapter 19 for more on writing procedures.

Following a description of the study area, including location, temperature, precipitation, vegetation, and water sources, the methods section of the sheep study appears in three paragraphs.

Behavioral and population studies on the Waterton herd were conducted throughout 1979–1980 concurrent with construction activity on the Strontia Springs Dam (Simmons 1982; Risenhoover 1981, 1985). Sheep were trapped and tagged in January and December 1979. Five sheep were fitted with radio-collars; 34 others received numbered collars or ear tags.

Sheep were observed daily. Data recorded were group size, group composition, activity, vegetation type, slope, and proximity to the road. Monthly population estimates were calculated from highest one-day counts and non-duplicated sightings. A die-off period was identified during October–December 1980, based on sightings of sick sheep and discovery of carcasses. Because of drastically declining numbers, population estimates during the die-off were based on the greatest number of animals known to be alive at the end of each month.

Methods of research

Once the die-off was identified, all sheep were located as often as possible to record the progress of the outbreak. I attempted to find all carcasses of sheep thought to have died. The lower canyon was searched by 10 biologists on 25 October 1980. All carcasses found during the die-off were sexed and aged. Seventeen carcasses were transported to the Wild Animal Disease Center or to the Colorado State University Veterinary Teaching Hospital in Fort Collins for necropsy within 48 hours postmortem. Five diseased sheep were collected for necropsy. Necropsies followed the procedure described by Feuerstein et al. (1980). Gross external examinations were conducted for general condition, lesions, and parasites. Internal organs were examined grossly and were sectioned for histopathological analysis. Tests were performed for bacteria and viruses. Blood samples from fresh carcasses enabled serological and hematological tests.

Results

In the *results* section, you present the data gathered when you followed the method explained in the previous section. Often these results can be presented in a table or graph. Although they may be presented in a brief form, results actually provide the most important information in the report. The results of the bighorn sheep study are in the following paragraphs. The two supporting tables are not included here.

Summer–fall 1980 was among the driest seasons on record for Waterton Canyon. Seasonal, intermittent water supplies dried up early in 1980 because of hotter summer temperatures (22 percent greater than the June–September 50-year norm) and subnormal precipitation (92 percent less in June and 62 percent less June–October, compared with 50-year norm).

Results of the research

Known-minimum population estimates were consistent among months and within years, except during the die-off (Table 1). Lincoln-Peterson estimates never exceeded the highest known-minimum population in either year.

The outbreak was swift and severe. Dead sheep were found less than three weeks after the first sheep was observed coughing. All sex-age classes were affected and overall mortality was 77 percent (Table 1). Most sheep died of acute to sub-acute pneumonia during October–November. Necropsies of diseased sheep revealed adrenal hypertrophy, thymic involution, and lymphic hyperplasia (Table 2).

Discussion

Since the *discussion* interprets the results, sometimes you can combine these two sections. The discussion section is like the conclusions section of a feasibility or comparative report; you should tell what the results mean and why they are important. Often the results suggest the need for further testing or experiments, and in the discussion section you should explain and support that need.

The bighorn sheep study includes an extensive discussion section covering 15 pages of the report. To give you an idea of the material included in the discussion, here is an outline of the major sections.

> Natural Stressors
>> Parasites
>> Social Density
> Man-Related Stressors
>> Construction/Traffic
>> Dust
>> Research
>> Interactions
> The Waterton Canyon Die-off

The discussion section closed with the following conclusions. These conclusions serve both to summarize the discussion and extend the implications of this study to larger concerns of the Colorado State Division of Wildlife.

General summary

All-age die-offs with the involvement of bronchial pneumonia seem to be a natural but irregularly occurring event in the life history of Rocky Mountain bighorn sheep. Etiological factors include dust, lungworm, bacteria, and human disturbance, among others. The interaction of these factors, or stressors, induce a stress response in sheep. The importance of any one stressor in the onset and course of the disease varies among die-offs and individuals.

Specific conclusions

Lungworms and bacteria are normally present in low quantities in most bighorn herds; this was the case in Waterton Canyon. The presence of these organisms may induce a condition of chronic, low-grade stress in many individuals. Human disturbance was at its greatest modern level at the time of the die-off due to the construction of the Strontia Springs Dam. Construction activity during drought conditions produced excessive quantities of airborne dust. Dust seemed to be the most acute stressor involved in the Waterton die-off.

Implications for the agency

If similar die-offs are to be avoided in other bighorn herds, we should recognize that seemingly benign stressors may produce cumulative and interacting effects to create a serious problem, especially when limited habitat reduces bighorn options for avoiding stressful situations. It also would be prudent to avoid exposing bighorn sheep to abundant and persistent dust, as this was the precipitating factor in the Waterton Canyon die-off.

The Waterton Canyon bighorn herd has unique recreational values due to its proximity to a major population center; most extant herds are located in areas that are difficult to access. The herd also possesses unique biological value as one of the seven remaining low-elevation Colorado bighorn herds. As discussed by Simmons (1982), this herd retains unique adaptations for low elevation which have been lost from the gene pool of most herds. For these reasons the herd deserves all the attention and resources of the Colorado Division of Wildlife.

References

In the *reference* section, you list the published sources you have cited in your literature review and elsewhere in the article. The documentation method depends on the journal or organization for which you are writing, so obtain a copy of the guidelines for authors and study previous documents. Also see Chapter 13 for information on documenting sources. To save space, the reference section of the bighorn sheep study is not included here, but it is extensive, listing 67 citations.

Figure 23-1 shows a brief research report published in the *Review of Scientific*

A Low Energy Electron Source for Mobility Experiments*

James A. Jahnke and M. Silver

Department of Physics, University of North Carolina, Chapel Hill, North Carolina 27514
(Received 11 December 1972; and in final form, 16 January 1973)

Thin film $Al-Al_2O_3-Au$ diodes were used as sources of low energy electrons (~ 1 eV) in a drift velocity spectrometer. These diodes were found to offer several advantages over other sources used for electron mobility studies in dense gases and liquids.

The electronic states of liquids and dense gases have increasingly been studied by the injection of excess electrons. The measurement of the drift velocities of excess electrons by mobility experiments has proven to be one of the simplest and most useful methods for investigating the conducting states of gases and liquids. In our studies in dense helium gas,[1] we have used a cold-cathode emitter as a source of low energy electrons and have found that it provides several advantages over more conventional methods.

Radioactive sources and field emission tips have been most commonly used,[2-4] particularly at low temperatures. They have specific disadvantages however, since the emitted electrons have a high initial energy of several kilovolts or more. The possibility that excited species or impurity ions may be produced by these high energy electrons has continually added a dimension of uncertainty in the interpretation of drift velocity data throughout the literature. A radioactive source also has several practical experimental problems associated with it, namely, low current levels, the inability to easily turn the source off, and the procedural problems of dealing with radioactive materials themselves. Field emission tips are quite useful at liquid densities,[4] but they tend to have a short lifetime at lower gas densities where they are easily blunted. Also, at high current levels, a bubble of ionized gas tends to form around the tip, leading to increased noise and measuring problems in the experiment.[5] Photoelectric sources have proven useful in drift and other experiments.[6,7] They are, however, most easily constructed in glass systems, which makes it difficult to study pressure and temperature effects over a range of fluid densities.

The cold-cathode emitter used in the present experiments has previously been used to study the current–voltage characteristics of gases and liquids.[8] These emitters inject electrons at energies of about 1 to 2 eV[9] and easily can be operated to provide currents over the range of 10^{-10} to 10^{-12} A. The $Al-Al_2O_3-Au$ diode has been found to be one of the easiest emitters to construct and one of the most reliable.[10] An aluminum strip about 2000 Å thick is first evaporated unto a clean glass substrate and then oxidized to a depth of about 100 Å by wet electrolysis. Transverse gold strips are then evaporated onto the oxide layer to a thickness of approximately 100 Å. It is important that the gold layer be not too thick since higher voltages are then needed for emission. In the past few years, techniques in fabricating thin film diodes have become more clearly defined[10] and by carefully following these techniques we have been able to obtain satisfactory results from better than 50% of those prepared.

For the drift velocity spectrometer, two diodes were constructed on a 2.5 cm square glass substrate which was then clamped between two Plexiglas plates as shown in Fig. 1. An accelerator grid was separated from the diode by a 0.4 mm Teflon spacer. In the normal operation of the spectrometer, the system was cooled to liquid nitrogen temperature since these diodes tend to have a short lifetime at temperatures above 200 K.[10] Emission occurred when a potential difference of about 8 to 9 V was applied between the aluminum and gold layers. It has been found necessary to approach this voltage slowly and then to slowly "age" the diode at incremental voltage increases after the first emission. In the spectrometer system, the diode was first aged by floating it at -100 V above the accelerator grid which acted as the collector and biasing the diode with a 45 V battery subdivided with a 500 kΩ Helipot. A Keithley 602 electrometer was used to monitor the current on the accelerator. After a stable current was obtained, the electrometer was then connected to the collector of the spectrometer which was held at ground potential. Measurements were performed at fields of from 50 to 200 V/cm over the 2 cm length of the spectrometer. Since the diode was kept at -100 V above the accelerator, it was therefore floated at from -300 to -400 V above the collector.

In vacuum, currents on the order of 10^{-9} to 10^{-10} A were obtained between the diode and accelerator grid at diode voltages of from 10 to 11 V. Current levels decreased less than an order of magnitude when the current was passed through the grid system to the collector. The addition of gas or liquid to the system further decreased the current level by approximately an order of magnitude, so that

FIG. 1. Diode holder and drift velocity spectrometer grid system.

Figure 23-1 A Published Research Report

currents on the order of 10^{-12} A were normally measured during an experiment. The collected current is of course dependent on both the gas density and the accelerator voltage. Previous papers on injection experiments consider this in greater detail.[8,9] We have found it more practical here to maintain the accelerator voltage constant and vary the current when necessary by changing the diode voltage. In this manner, greater detail over the drift velocity spectrum may be obtained by increasing the current level to 10^{-11} or even 10^{-10} A. Driving a diode to higher current levels, however, greatly reduces its working lifetime, so a balance must often be found between problems of spectral analysis and experimental lifetime. At moderate current levels, we have had diodes operate for more than 24 h. Since the diodes do operate well over long periods at low gas densities, as well as at higher densities, they offer advantages over field emission tips in addition to avoiding some of the problems associated with radioactive sources.

*Research supported by the Advanced Research Projects agency of the Department of Defense and monitored by the U. S. Army Research Office, Durham, under Grant no. DA-AROD-31-124-72-G80, and by the Materials Research Center, UNC under contract DAHC 15-67-C-0223 with the Advanced Research Projects Agency.

[1] J. A. Jahnke and M. Silver, Chem. Phys. Lett. (to be published).
[2] F. Reif and L. Meyer, Phys. Rev. 119, 1164 (1960).
[3] H. Schnyders, S. A. Rice, and L. Meyer, Phys. Rev. 150, 127 (1966).
[4] J. A. Jahnke, L. Meyer, and S. A. Rice, Phys. Rev. A 3, 734 (1971).
[5] B. Halpern, and R. Gomer, J. Chem. Phys. 51, 1031 (1969).
[6] B. Halpern, J. Lekner, S. A. Rice, and R. Gomer, Phys. Rev. 156, 351 (1967).
[7] J. L. Levine and T. M. Sanders, Jr., Phys. Rev. 154, 138 (1967).
[8] M. Silver, P. Kumbhare, P. Smejtek, J. Chem. Phys. 52, 5195 (1970); D. G. Onn and M. Silver, Phys. Rev. 183, 295 (1969); and M. Silver, D. G. Onn, and P. Smejtek, J. Appl. Phys. 40, 2222 (1969).
[9] D. G. Onn and M. Silver, Phys. Rev. 3, 1773 (1971).
[10] D. G. Onn, P. Smejtek, and M. Silver, Proceedings of the Low Temperature Physics Conference, Boulder (1972).

Figure 23-1 Continued

Instruments. Notice how the various parts of the article blend together. Notice too that this article is written for fellow specialists, so it assumes common knowledge of terms and procedures. Therefore, unless you have experience in this field, the report will be difficult to understand—even though the writing itself is clear.

Outline of a Sample Laboratory Report

Figure 23-2 shows in outline form the detailed requirements for a laboratory report for materials engineering students. If you must write a research or laboratory report, you may want to consult these recommendations or adapt them to meet your specific requirements.

Experiment #

The Title of the Experiment

Your name
Due date
Course name and section
Instructor's name

Figure 23-2 Outline of a Sample Laboratory Report

Title of the Experiment
I. Objectives
List the given objectives or those you have formulated.
  A. Objective 1
  B. Objective 2
  C. Objective 3

II. Methods and Materials
  A. Modifications to the procedure
     1. If there were any changes to the procedure, state them here. Was different equipment used? Were different materials used? Were temperatures, concentrations, or conditions changed?
     2. If the experiment was completed exactly as outlined in the given instructions, state that here.
  B. Equipment
     1. List all equipment used in the experiment. List model numbers and exact names.
     2. Do not list the materials used.
     3. Do not describe how the equipment was used.
  C. Materials
     1. List all materials and specimens.
     2. If you used the results from materials and specimens generated by other experimenters, state that here.
  D. Experiment Steps
     1. List the individual steps you took to complete the experiment. Don't discuss the results, only what you did.
     2. When someone else reads this section, they should be able to duplicate your experiment exactly.
  E. Safety Precautions
     1. List any precautions stated in the given instructions.
     2. List any other precautions you think would be important to someone else performing this experiment.

Figure 23-2 Continued

III. Results
  A. Discussion
     Discuss what you discovered by performing the experiment.
     This section should objectively state the results of your exper-
     iment. Use tables and figures (especially graphs) to
     summarize the results. When you refer to these tables and
     graphs, use the figure or table number (for example, "See Fig-
     ure 3"). Don't list what you thought you should have found,
     only what you actually did find. Don't list the things you felt
     you should have done differently. If there were any experi-
     mental errors that would affect the results, they should be
     stated here.
  B. Tables
     Tables should be presented in the format shown below.

Table 1. Table Title Goes Here

Column 1 Head	Column 2 Head	Column 3 Head
123	0.099	6.6
134	0.087	5.5
101	0.077	6.5
119	0.088	6.9
140	0.066	7.3

Figure 23-2 Continued

Leave enough space around each graph so it does not appear crowded on the page.

Label each curve.

Clearly mark all data points.

Label both axes with the name of the units, the unit symbol, and the quantity divisions.

Tables and figures are numbered separately.

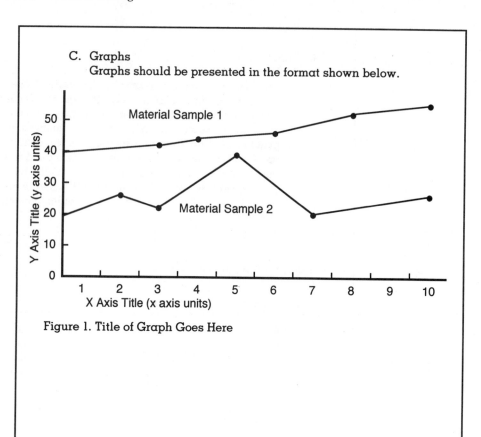

C. Graphs
    Graphs should be presented in the format shown below.

Figure 1. Title of Graph Goes Here

Figure 23-2 Continued

IV. Conclusions
   A. State how the objectives were met. Each of the objectives listed in section "I. OBJECTIVES" should be covered here in the same order. You may have met some of the objectives and not others.
   B. If an objective was not met, discuss why it was not met.
   C. Don't repeat the steps in the experiment.
   D. Don't repeat the results, except when a specific result supports your conclusion.
   E. Remember, the discussion section tells what happened in the experiment; the conclusions are based on the results.

Appendix A: Data Sheets

Copies of all your original data sheets should be included in Appendix A.

Appendix B: Sample Calculations

For each equation used for each different condition (example: for each different material), do one sample calculation. This includes equations for area and other basic information used as final or intermediate steps.

Appendix C: References

Information sources should be on this page. Note: The reference section may be the first appendix. Consult documents in your field for the correct sequence.

Figure 23-2 Continued

23.4 Project or Completion Reports: Organization and Examples

Project or completion reports also answer specific questions that specialist, managerial, or client readers have. These questions include:

- Why should they read this report?
- What was the problem (or situation)?
- What did you do?
- What was the result?
- Why is this information significant to the readers?

The form of project reports varies from organization to organization, but most reports will include the following parts:

- title
- abstract or executive summary
- introduction, including background
- methods
- factual information
- conclusions
- appendixes, including references

Sometimes project reports will add recommendations; they might also detail the next steps to be accomplished.

Title

A *title* of a project report should be useful; it should contain key content words and function words so the reader knows immediately what the report is about and what its purpose is. See Chapter 6. For example, a civil engineering project report for a new mass transit system is titled *Metro North Coast Segment: Alignment Study Beyond the Westchester Station.*

Abstract or Executive Summary

Both *abstracts* and *executive summaries* condense longer documents to save time for busy readers. Abstracts are for technical readers; they assume understanding of the situation and of the technical background. Executive summaries are for managers; they explain needed technical information and provide summaries of cost, personnel, and time considerations. See Chapter 20 for details on writing abstracts and executive summaries. In the civil engineering project report, the abstract is contained within the letter of transmittal since the client is a specialist reader.

Introduction

In a project report, the *introduction* states the background and the purpose of the work that was accomplished. It should also provide an overview of the report that will follow, either by listing the main sections or by summarizing the main points.

The civil engineering alignment study has an introduction with the subheading "Background and Purpose." This section defines the geographic area of the specific study and explains the specific contract packages that are involved. It then describes the specific problem and the purpose of the study as follows:

Background

> For a majority of its length, this alignment is adjacent to the facilities and systems of the Department of Airports (DOA) and the Federal Aviation Administration (FAA). During design for the C1001 and C1002 packages, it was decided to further study the alignment beyond the Westchester Station area until it clears the DOA/FAA area of influence. As such, a study was undertaken of the alignment along the future Westchester Parkway and turning northerly and following Lincoln Boulevard.

Purpose

> The purpose of this study is to further explore the base alignment condition of the EIR as it is impacted by the existing facilities and systems of the DOA/FAA. The alignment alternatives studied and the resultant DOA/FAA interface issues are summarized below (Bechtel 1991).

Methods

The *methods* section begins the body of a typical project report; it tells what procedures were followed on the project and how those procedures were carried out. It may discuss conditions, alternatives that were pursued, and choices that were made. It is similar to the methods section of a research report, but it may have a different title.

Factual Information

The methods followed in completing this project yielded facts, data, and information. *Factual information* is usually presented next, though sometimes methods and facts are presented in the same section (for example, this method yielded these facts, the next method yielded these facts, and so on). This section often contains tables, calculations, and figures of various kinds. Sometimes this supporting detail is part of an appendix instead of being in the main report.

Figure 23-3 shows an outline of the project report for the alignment study. Notice how the parts of the report reflect the specific content: the method and factual information are grouped under the subtitle "Summary of Alignment Alternatives Studied."

Alignment Study Beyond the Westchester Station
Prepared by Bechtel Corporation
Prepared for Caltrans and RCC
August 16, 19__

Background and Purpose
Summary of Alignment Alternatives Studied
    Conditions
    Five Alternative Alignments
Conclusions
List of Attachments
List of References

Figure 23-3 Outline of a Project Report

Conclusions

The *conclusions* section in a project report is like the discussion section of a research report: it explains to the reader what the facts mean. You must always remember that readers rely on you to help them understand, and that means interpreting results and drawing conclusions. The main thing is to be sure that your conclusions are supported by the facts in the previous section.

For the alignment study, the conclusions include these points (summarized here to exclude the language directed to specialists):

1. It is apparent that an alignment solution exists.
2. With special design efforts, the station can remain in its present location. However, any shift of the station to the north or west improves the interface between the station location and the approach slope of runway 24R.
3. The work summarized in this report is only the first phase of a more comprehensive study; details of depth, slopes, and track spacing have not been fully developed.
4. The decision on the best alignment solution should be based on more extensive analysis, including setbacks and street depressions.

Appendixes

Project reports often have extensive supporting material that would encumber the main report but still needs to be included. This material should appear in *appendixes* and can include correspondence, detailed figures, tables, and calculations. A list of references supplies documentation on sources used for information; see Chapter 13.

The list of references is often, but not always, the first appendix. In a project report, the references are likely to be previous correspondence and reports rather than books or journal articles.

23.5 The Process of Writing Research and Project Reports

The challenge in writing research and project reports is (1) to choose from the mass of information only significant facts and (2) to present them to the reader in the most efficient way. In a way, you have been the reader's representative while engaged in this project; you have conducted the experiment, completed the study, or carried out the work. Now you must allow the reader to participate in the research or project through what you write in the report.

Planning

Planning for a research or project report begins as the project itself begins. It's much easier to undertake the project knowing that one outcome will be a written document than it is to realize part way through that some crucial step should have been documented. In particular, planning requires a clear understanding of the purpose (both of the project and the report), the reader (what that reader wants), and the schedule (including responsibilities of team members).

The purpose of a research or project report is to share information, but the project itself will have very specific objectives: those must be clearly stated before the project begins, referred to as the project is underway, and included in the introduction when the report is written.

As you plan, you must also consider the potential readers. As a student, you are used to thinking of your instructor as the chief reader—a person who designed the project, probably knows more about it than you do, and will judge you on your work and your writing. Instructors, though, are not typical readers, and you need to start now to plan for real readers and think about their real needs. Readers are seeking information; they want to use that information in their own work. They may be managers, scientists, or specialists in engineering, manufacturing, or some other field, but they rely on you to give them the facts about what you did, what happened, and what's important about the result.

Scheduling is also a key element in planning. Research and project reports are often the result of teamwork. Usually, the research or the project itself demands time and effort by several people. So too, the report often must be written collaboratively. Collaborative reports demand careful planning and detailed assignment of tasks. Even more important, collaboration requires commitment by each member of the team, for only with full cooperation will each section of the report be completed properly and on time. Decide at the planning stage who will be responsible for each section of the report, who will design the graphics, and who will write or organize front and back matter. Schedule time at the end so the whole team can review the document, but assign one person to weave the parts together into a coherent whole.

Gathering Information

Research reports often require a literature review, which means that you must search carefully for *all* relevant articles, keeping notes on their content and main contributions. You may cite only a few of them in the review, but you must know about them all. Project reports may not require a literature search, but they often require nonlibrary information: laboratory or field data, surveys, interviews, calculations, and analyses. As you proceed, document your sources, because it will be much harder to reconstruct a source once you are away from it. Even if you are writing for generalists, sources and information must be accurate and complete. All readers rely on you to be the trustworthy expert on this particular topic.

Organizing and Reviewing

Research reports follow the standard organization explained in section 23.3. When readers know what to expect of each part of a report, they can process the information quickly.

Many research papers report on a project or experiment that was conducted in an organized sequence: first we did X, which resulted in Y, and then we did Z. Even though the project may have been well designed, in most cases the reader does not want to follow you through that sequence once you have explained the methods. That means in the discussion you must choose a single important point to emphasize—the results or a significant failure, for example—and organize your report to support that point.

The organization of project reports is more flexible, and you need to adapt the organization suggested in section 23.4 to the particular readers' needs. Whatever organization you use, help readers by using headings to set off major divisions. Make those headings informational with key words and organizational indicators.

Writing and Communicating Your Organization

Research reports, which describe completed actions, can often be written *almost* in the order they are presented—from introduction, to materials and methods, results, and discussion, to abstract. The introduction sets up the situation, giving background, stating the purpose, describing the method of investigation, and telling the results. The body follows a procedural organization, first telling what materials were used and what methods followed, then closing with the results, and explaining what they mean. Once written, the whole report can be summarized in an abstract, and the abstract placed in front of the article.

Students often have trouble writing research or laboratory reports because they don't distinguish between results and conclusions. You need to remember that the *discussion* tells what happened in the experiment or project; it explains the *results* that occurred. Based on those results, you draw *conclusions*, stating how the objectives were met (or not) and what the results mean. Section 23.3 shows the detailed outline developed to help materials engineering students write laboratory reports. Because it clearly differentiates the parts of the report, it will help you in constructing your own.

Project reports might be easier to write out of order. Instead of struggling with a difficult introduction, you might write the body first, since you probably know this material best. Next write the conclusions and recommendations (if any), and then try the introduction. The conclusion may even suggest an introduction. Since many methods of organization are possible, you should review section 4.2 to investigate options. Rough out graphics as you go, but don't stop to finish them until you have completed writing the draft.

Once the paper is written, write your abstract and executive summary (if needed) to reflect the actual contents of the paper. Add any documentation of references, and rewrite the title if necessary.

Reviewing, Revising, and Editing

Once the report is written, review the manuscript for technical accuracy, clarity of content and organization, and mechanics like spelling and punctuation. Focus on accuracy of graphics, labels, and titles. Double-check all calculations and entries in tables to verify numbers and placement of decimal points.

If this is a collaborative project, be sure to review each others' work for technical accuracy and clarity of presentation. Accept criticism gracefully, remembering that your ego is less important than a quality report.

Excellent research and project reports mark you as a professional. If you do your best work on those you write as a student, you can use them as samples of your writing when you interview for a job. All professions require excellent communication skills, and employers will be favorably impressed if you show that you possess those skills.

Checklist for Writing Research and Project Reports

1. What is my purpose?
2. Who is my primary reader?
3. What does that reader want from this report?
4. Have I considered ethical implications? Is my research accurate and verifiable? Have I given credit to others where appropriate?
5. Is this a research report? If so, does it include the following parts?
   ___ title
   ___ abstract
   ___ introduction, including literature review
   ___ materials and methods
   ___ results
   ___ discussion
   ___ references
6. Is this a project report? If so, does it include the following parts (or a variation that meets my readers' needs)?
   ___ title
   ___ abstract and executive summary

___ introduction, including background
___ methods
___ factual information
___ conclusions
___ appendixes, including references

7. If this is a collaborative project, do all the parts work together?
8. Have I checked the accuracy of tables, figures, headings, and calculations?
9. Have I designed the document for clarity and ease of reading?

EXERCISES

1. Find out the names of four journals in your field—two published by professional societies and two trade journals. Read through one issue of each journal that has been published within the last two years. Write a short report in which you (1) name each of the journals and summarize in a paragraph the kind of articles contained in each, and (2) explain who the intended reader group of each journal is and how you can tell.

2. Compare the laboratory report outline in section 23.3 to the requirements in your specific field. Look at textbooks or laboratory instructions for examples, or ask a professor in your field for suggestions. Write a brief report explaining where the requirements differ and where they are the same. Suggest a model outline for your field.

3. Contact a professional working in your field (an alumnus of your school, perhaps) and find out what kind of project or completion reports that person writes. Ask if you can see samples. Compare those samples with the suggested format in this chapter and report to your writing class how they are alike and how they are different.

WRITING ASSIGNMENT

Choose a subject from your major field that interests you. Conduct the field or laboratory research following the guidelines in Part 1 of this book. Then write a research paper, clearly distinguishing between results and conclusions. Target a specific reader, either a specialist at your college or university or a manager or other professional working in the field.

REFERENCES

Bechtel Corporation. 1991. Reprinted by permission of the Bechtel Corporation.
Greenberg, M. 1991. Confronting scientists' normal ethical dilemmas. *The Scientist* 12, 10 June.
Johnson, Craig and Doug Klopfenstein. 1988. Isolating faulty RAM components in record time. *Evaluation Engineering* 27, August.
Simmons, Brian. 1990. *A stress-induced die-off of Rocky Mountain bighorn sheep*. Unpublished document. San Jose State University.

Chapter 24

Speeches and Oral Presentations

Why does a book on technical writing include a chapter on speeches and oral presentations? There are two important reasons.

First, good speeches are much like good written documents. They share a basic structure of introduction, body, and conclusion. Within the body, a good speech uses a method of organization appropriate to the purpose, listeners, and content. To enhance major points, effective speakers use graphics (usually called visuals in speeches). And, in preparing a speech, you follow the same steps as in writing, except that you replace the final review with rehearsals.

Second, as a professional in a technical or scientific occupation, speaking well is important to you, whether you speak informally to a few colleagues or give a planned, practiced presentation to a large group. In fact, you may spend as much as 25 percent of your time communicating orally to clients, managers, and colleagues, and often you will succeed on the job as much or more by your ability to communicate effectively as by your technical knowledge. As one engineer put it, "The most successful people around here tend to be the ones who are able to give good oral presentations" (Almazol 1982).

Speaking and writing are different, of course, primarily because readers can go back to review what you write, while listeners must understand what you say the first time. So, because you have listeners instead of readers, a speech must be simpler, clearer, and more repetitive than a written document. You keep it simpler with short sentences, straightforward language, strong transitions from one idea to the next, good overviews, and frequent repetition of main ideas. Even the visuals for a speech must be simplified to present only one idea at a time.

24.1 Definitions and Types of Oral Presentations

A *speech* or *oral presentation* is any kind of planned spoken communication to listeners for a given purpose. *Speech* and *oral presentation* are general terms that in themselves say nothing about the content, the length, or the amount of planning involved. This chapter uses the two terms interchangeably. The focus is on planned speeches, though of course you will have to give unplanned (*impromptu*) speeches from time to time. The best preparation for an impromptu speech is thorough knowledge of your subject and a quick outline for a guide.

Planned speeches are of two types: (1) those written out in full and either memorized or read from the manuscript, and (2) those presented extemporaneously—that is, using planned and outlined main points in the form of notes but allowing for flexible wording at the time of presentation.

Written Speeches

A *written speech* has both advantages and disadvantages. One advantage is the careful polishing you can give each word and phrase. A second advantage is that when you are dealing with a highly technical subject, writing down all the details ensures that you present them accurately. A third advantage is that your speech may be published after you present it, and for publication you will need a written version.

From the listener's point of view, however, the disadvantages of a written-out speech may outweigh the advantages. The primary disadvantage is lack of listener contact: if you read a text, you must concentrate on it, and you cannot look listeners in the eye and talk directly to them. In addition, people can seldom read aloud as well as they can speak, and—in fact—most audiences dislike being read to. Memorizing a written speech is not the solution. Few speakers try to memorize a speech because of the time involved and the potential for forgetting a section. Besides, most memorized speeches sound rather mechanical because the speaker is concentrating more on remembering the next sentence than on clarifying the next point.

From watching televised speeches by the president of the United States, the secretary of state, and major television news reporters, you know that technology has made it possible for speakers to read text and still *seem* to be looking directly at the audience or the television monitor. Thus, technology may seem to be the answer to the disadvantages of the written speech. However, be aware that only a tiny percentage of speakers have access to that technology; until you become a corporation vice president or better, you will probably have to rely on a written document or an outline. If you are asked to "read a paper," at a conference for example, the best approach is to write out the speech but then to speak from an outline of that document. This is called an *extemporaneous* speech.

Extemporaneous Speeches

For most occasions, the extemporaneous speech is best. It sounds interesting because it seems spontaneous, and it enables the speaker to have eye contact with listeners, getting feedback from them and responding to the feedback. However, an extemporaneous speech is definitely a planned and practiced presentation, even though the words *impromptu, ad lib,* and *extemporaneous* are sometimes used interchangeably. An extemporaneous speech is planned to fit a designated time, organized (with main points and supporting facts written in an outline or on notecards), and rehearsed several times before it is presented. Author George Plimpton (1982) quoted Mark Twain as saying, "It takes three weeks to prepare a good ad lib speech."

This chapter concentrates on the form and the procedure for writing an extemporaneous speech. Before taking up form, though, you need to know some of the special considerations involved in speech writing.

24.2 Special Considerations in Speech Writing

Four factors make speech writing especially challenging:

1. the distraction potential
2. comprehension loss
3. time limitations
4. the group dynamic

If you understand these four factors, you can use them to your advantage, but if you neglect them, you may have a problem communicating effectively.

The Distraction Potential

When you give a speech, you rarely have an audience of only one person at a time as you do in writing. Whether your listeners number 3 or 300, speech making is a group activity, and a group means distraction. Individual listeners can be distracted by outside noises like fire sirens or passing jets or by audience noises like whispers and coughs. Their thoughts can be diverted by things they see: a fly persistently settling on a neighbor's forehead, the glare from an open window, the speaker's sway or nervous clutch at the microphone stand. And even if the room is quiet and enclosed, listeners can be led astray by their own thoughts. Thought is four times faster than speech—a fact that causes listeners to fill in the gaps between your words with all kinds of extraneous material: "That's a good point. . . . I wonder what Bill would think about it. . . . Bill sure was mad yesterday when he heard about the contract revision. . . . Well, it makes me mad too. . . . Ooops—what was that she said?"

You probably can't eliminate all external distractions, and you certainly can't control your listeners' every thought. But you do need to choose your place carefully (if you can), and arrange the room to minimize distraction and allow listeners maximum comfort, thus encouraging concentration on what you're saying. You can corral your listeners' attention by making eye contact with individuals, by responding to the signals you perceive, by organizing, and by repeating your main points.

Comprehension Loss

Add to the distraction potential the problems of limited memory and comprehension loss, and it's a wonder that speeches make an impact at all. I've frequently mentioned the important fact that people can only hold seven plus or minus two items in short-term memory at one time. That finding has enormous implications in speech writing: in fact, if you present seven or nine major points, you are probably overloading your listeners' short-term memory. In speeches you must simplify, group, and underscore each major point with graphics and with repetition. Three or four major points are probably a speech maximum.

Memory is also related to comprehension, and again the studies have been disturbing. According to financial writer Sylvia Porter (1979), "Immediately after listening to a 10-minute oral presentation, the average consumer or employee has heard, understood, properly evaluated, and retained only half of what was said. Within 48 hours, that sinks another 50 percent. The final level of effectiveness—comprehension and retention—is only 25 percent!"

If you accept those percentages as given, your task may seem impossible. Yet with careful planning and practice, you can reach your listeners with solid information and your own point of view. One reason for comprehension loss is that readers can't take breaks to mentally chew over what has been said before they go to the next point. What's more, they can't backtrack to check a fact or statistic or to review the heading of the previous subsection—all actions that careful readers take when they seek to understand. As speaker, you must take breaks for them—pausing momentarily to let a fact sink in, backtracking with summaries and repetition. The great comedians are masters of timing and the pregnant pause: Bob Hope and Bill Cosby know when not to talk. They pause, wait, and let the audience catch up.

Time Limitations

As a speaker, you almost always have a time limit; whether it's 5 minutes or 45 minutes, your listeners expect you to fit your speech in that time box. You may have page limits in writing too, but your readers can usually spend as much time as they need to read—and reread—whatever you've written. One reader can skim 10 pages (2,500 words) in 6 or 7 minutes, while another reader might study, ponder, and review the same 10 pages for 30 minutes. As listeners, however, both individuals have the same time exposure to your ideas. You can help your listeners understand by intensifying their exposure to the main ideas. You do that with repetition of key ideas, visuals for main points, handouts, even a question-and-answer period.

A second time factor is psychological: listeners expect you to respect the time limit. If they're anticipating a 15-minute speech, they will begin listening for the final point after 13 minutes. After 14 minutes they'll start to shift in their chairs and check their watches. If you pass the 15-minute mark, they'll be distracted—wondering how much longer you'll go on. What's the solution? Respect the time limit: plan your speech to fit that time box, rehearse it and time it, and then when you speak, keep track of the time.

The Group Dynamic

Listeners in a group behave differently than solo listeners, and you can use that fact to your advantage. So far the special factors in oral presentations seem to put a burden on you, but if you use the group dynamic, speaking can be easier than writing. What happens when a speaker is communicating—really reaching a listener? That listener becomes an ally, sending positive signals both to the speaker and to other listeners. Signals include sitting up and leaning forward, looking directly at the speaker, becoming still and concentrated, and even nodding, smiling, or frowning (sometimes clapping and cheering) in response to what's being said. Even though they may not realize it, other listeners are affected by those signals, and they are likely to respond in the same way. Be aware of these signals. Meet your listeners where they are and speak directly to the good ones, knowing that they can influence others.

24.3

The Form and Organization of a Speech

Like nearly every written document, a speech has an introduction, a body, and a conclusion. Each segment should be shaped by the time allotment for the whole speech. For example, if you are assigned 20 minutes to speak, the breakdown might look like this:

- Introduction   3 to 4 minutes
- Body   about 12 minutes (Perhaps 3 minutes for each major point or argument and time to summarize at midpoints. Allow about 2 minutes to present each visual.)
- Conclusion   3 to 4 minutes

For an extemporaneous speech, write an outline of these three sections, including major points and supporting details. The outline can be the same as one you'd compose in preparing to write a document, except that it may have fewer main points. If you're writing out the whole speech (perhaps for later publication), work from the same outline—estimating delivery time at approximately 100 words per minute. Your final outline can be transferred to index cards that you'll hold in your hand, or you can use standard sheets of paper, perhaps fastened in a binder so they won't be mixed up. Whatever method you use, your notes should be unobtrusive. Use them for reference, not for reading, and organize them well to avoid shuffling paper.

Figure 24-1 shows the outline for a speech by a manager of documentation at a major computer company (Dewan 1990). The audience is a group of technical writers; the subject is how to schedule writing projects.

A speech must begin by creating a friendly atmosphere and breaking the ice. You can do this by acknowledging your introduction (or perhaps introducing yourself), and by saying something complimentary to the audience. Be wary of using canned jokes to begin; they often do not fit the audience or the occasion. As soon as possible, launch into the real introduction of the speech, which should hook the listeners' attention.

In the introduction, define key terms and tell the listener where the speech is going by giving an overview. The only time you won't bluntly set forth your major point in the introduction is when you're delivering bad news. Then you should establish rapport first and give the facts that lead up to the major point—which will come at the end. This technique is further explained in Chapter 16.

The body of a speech is where you present and support your major points. Here you should apply another known fact about listeners: they pay the most attention at the beginning and again at the end of a speech. That means that whatever method of organization you choose, you should put your most important points at either the beginning or the end. The following methods of organization work well in speeches:

- *Increasing order of importance*. You start with an overview and then build your case, ending with the most important point.
- *General-to-specific*. Again you begin with the larger picture and move to details. At the end you must return to the larger picture.
- *Problem-solution*. You can make problems dramatic and therefore attention-getting, which will start you off strongly. Then you supply details in the middle and end strongly with the solution.
- *Cause-effect*. The links between causes and effects (or effects and causes) can be used as glue to bind your points together. Often cause-and-effect will

**TITLE: "I'M A GOOD WRITER SO WHY IS MY JOB SO HARD?" OR
"SCHEDULING SKILLS YOU DIDN'T LEARN IN SCHOOL"**

1. What about Me?

   - Education
   - Writing experience
   - Managing experience

2. Why Is Scheduling So Difficult?

   In general:

   - <u>Companies</u> emphasize doing, not planning
   - <u>Groups</u> (superiors, peers, subordinates) have conflicting goals
   - <u>Individuals</u> face unknown or changing information

   As technical writers: We are a service; therefore reactive, not proactive.

3. What Makes a Good Schedule?

   A good schedule:

   - Sets priorities
   - Identifies dependencies (people, information, money)
   - Anticipates problems
   - Focuses effort
   - Reflects commitment

4. How Can I Create a Good Schedule?

   - Schedule from the large to the small, the bottom to the top (example)
   - Rely on experience (other workers, work logs)
   - Balance the needs of others (negotiate)
   - Achieve a commitment from others
   - Monitor your progress and make changes

5. How Can This Advice Be Applied?

   - Preparing to write
   - Gathering information and writing
   - Reviewing

Figure 24-1 Speaker's Outline of the Major Points

develop into a narrative or chronological sequence. This sequence is effective with listeners because of its storylike approach, and you can plan a dramatic climax of the narrative near the end of your speech.

- *Specific-to-general.* Interesting examples or illustrations will attract your listeners' attention at the beginning. Then you can move into more general statements in the body, culminating with your main points and so ending on a strong note.

Chapter 4 gives details on methods of organization that you can use in planning speeches.

The conclusion of a speech almost always includes a summary, designed to reinforce the main points and increase listener comprehension. Conclusions can also propose or recommend action. Many effective speakers close with an anecdote or vivid example; some use a quotation. The kind of conclusion depends, of course, on the occasion and the type of audience. The important thing is that you *plan* for an effective conclusion. Don't just let your words dribble off with a vague "That's about it." Listeners also expect you to signal the conclusion with your voice and body. You drop your voice or slow your pace and use terms like *finally* and *in conclusion* to tie up the last ends neatly.

24.4 Visuals for Oral Presentations

Important as the sense of hearing is, we receive only about 5 percent of our information through our ears. Knowing this, good speakers supplement and underscore what they say with graphics, which in speeches are often called *visuals*. Visuals also help focus listeners' attention, and they provide variety. Figure 24-2, for example, shows the key points printed on a transparency that is projected while the speaker discusses and elaborates on these points (Wells 1990). Keeping the key points in front of the listeners helps them concentrate on the topic.

All the forms of graphics discussed in Chapter 11 could be used in a speech, but they need to be simplified. Drawings, simple tables, a list of main points, graphs, and even photographs can be effective. You can present these graphics as visuals in several different ways:

- *Chalk Board.* Because it is permanent and reusable, the chalk board will probably always be with us. If you write as you speak, your actions will draw the listeners' attention. But elaborate drawings can take too much time, and you must turn your back on your listeners to do them. If left on the board when you go on to other topics, the drawings may be a distraction.
- *Posters and Flip Charts.* Both posters and flip charts can be prepared in advance, using bright colors to highlight key points. Posters (on card stock) should be placed on an easel for better viewing and should be large—typically 2 × 3 feet. Flip charts are prepared on large sheets of paper fastened together at the top and placed on an easel. Leave the top sheet bank to prevent distraction. Then uncover each sheet as you reach that point in your speech,

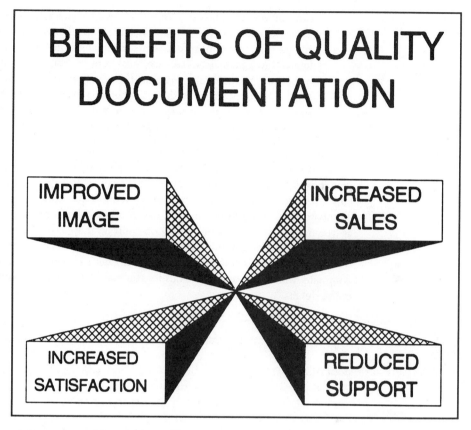

Figure 24-2 A Transparency Summarizing Key Points in a Speech

flipping the used sheet over the top as you finish. You can also create the visuals as you go with felt-tipped pens.

- *Slides*. Because they can be seen most easily, slides are often the most effective vehicle for visuals if you're speaking to a large group. Their disadvantage is that the room must be darkened, and you may lose some listeners if they can't see you.

- *Handouts*. You can put visuals and other key information directly into your listeners' hands. However, passing them out is a distraction in itself, and you'll probably lose many listeners as they switch their attention to what's directly in front of them. For this reason, limit handouts to what is needed as illustrative material during your presentation and pass them out ahead of time. Never spend your speaking time reading aloud what's on the handout, but highlight points as the readers follow along. If the handout material summarizes your speech, pass it out at the end.

- *Transparencies*. For small and medium-sized audiences, the transparency used with an overhead projector is the most effective and versatile way to present visuals. The advantages of transparencies are that you can face the listeners, leave the room lights on, and still project the visuals for easy viewing. In addition, you can prepare transparencies in advance from either typed or hand-drawn materials, because most copy machines will also produce transparencies. You can also write or draw on the surface as you talk, or you can overlay transparencies to show a process or change. Perhaps most important, you can remove the transparency or turn off the projector when you are finished, thus removing potential distraction when you go on to another point.

All visuals used in speeches must be big enough to see easily. Even on transparencies, you should use oversized type and limit the number of words. Keep drawings and graphs simpler than those you'd use in a written report, and split complex tables into sections, presenting only one section at a time. In a graph or drawing, eliminate extra lines or focus on only one portion. Whether you do them in advance or as you speak, spend enough time on your visuals to make them neat and well organized, remembering that they should show you at your professional best.

Finally, don't overburden your speech with visuals; they should enhance, not dominate, what you say. Use visuals for the main points and let your spoken words fill in the details.

24.5 The Process of Writing a Speech

Planning

As in writing, your first planning task must be to determine your purpose in giving this speech. You need to ask: "What do I want my listeners to go away with?" and "Why was I asked to speak?" Presumably, you speak for the same reasons you write—primarily to share information, perhaps to recommend or persuade. But sometimes your primary purpose in a speech will be to entertain, even if you're also providing information. Therefore, you need to evaluate the social context and the time of day of the speech. A speech during working hours to a group of professionals can be very different in purpose from a speech after a luncheon or dinner meeting—even though the general topic and the audience may be the same.

You must try to learn as much as you can about your listeners, both as a group and as individuals. How many will there be? What are their interests? How can most of them be characterized? Are they specialists, technicians, operators, managers, or generalists? Will the audience be mostly of one type or a mix of two or three types? What do they need—and want—to know? What is their anticipated attitude: that is, will they be supportive and enthusiastic, or perhaps hostile and antagonistic? What is your relationship to your listeners? You may be speaking down to subordinates, horizontally to peers, or up to your bosses. In each case, your preparation and approach will differ, for you must meet your listeners' needs.

Listener type and background will also influence your topic, or at least its level of difficulty and the amount of detail that you include. No matter what type of audience you have, though, the content of a speech must be simpler than the comparable content of a written report for the same audience.

Finally, learn about the location for your speech. Determine the size of the room, its physical makeup (and how many distractions come with it), and the availability of a lectern, lights, microphones, overhead projectors, and slide projectors. Many speakers carry their own slide projectors with them to ensure that the equipment is available and working. If your speech is one of a series, find out how many other people will be speaking and where your speech will fall in the sequence. If you are at the beginning, the middle, or the end of a sequence of speakers, you will face different kinds of expectations (and perhaps meet with different levels of interest) from your listeners.

Gathering Information

In writing a speech, you follow the same information-seeking route as in writing a report. Many times, of course, your speech results from work you have been doing and will even be accompanied by a written document. Two hints may help you in gathering information:

1. Plan to know more than you will put in the speech itself. Have a broad base of facts, figures, comparisons, and expert opinion to undergird the three or four points around which you'll build your speech. This extra information will help you select the information to present, and you can draw on it later if there's a question-and-answer period.

2. Seek interesting information and dramatic ways of presenting facts. Remember that to be effective, your points must be heard and understood. That means you must keep your listeners' attention. One way to do that is to have a theme—with variations—that you return to throughout. Another is to focus on the human side of facts and figures and especially to present facts in terms of how they will affect the listeners personally.

> *Not* Employment trends as reported by the U.S. Bureau of Labor Statistics do not substantiate the belief that high technology will provide vast numbers of new jobs, the latest estimate being that not more than 4 percent of new opportunities will be in high technology.

> *But* You can't depend on high tech for substantial gains in employment openings. According to the U.S. Bureau of Labor Statistics, only 4 of every 100 new jobs will come from what we call high technology.

A third way is to include anecdotes and human interest narratives. In planning, though, don't spend so much time looking for the human interest angle that you neglect your purpose and your primary content.

Organizing, Writing, and Reviewing

Since an extemporaneous speech is given from an outline, the outline provides both the backbone and the substance of your speech. Jim Jahnke, a scientist who frequently gives speeches on air pollution topics, says, "I never give prepared speeches because they come out too stilted. I outline the basic points I want to talk about on one sheet of paper. Then I highlight the key words I wish to emphasize and weave my talk around them" (1987).

As you sort your information, consolidate the bits and pieces into three or four major points, each of which can be strongly stated and frequently reviewed. Organize the body portion first to ensure that the topic is covered and the method of organization is dynamic—going somewhere and bringing the listeners along to the desired destination. To personalize the speech, plan to speak directly to your listeners with second-person pronouns, contractions, and relatively short sentences.

> *Not*　The agency has completed within the last month a revision of the long-range projections involving both employment and population in Madison County. The principal effect of these changes will be . . .

> *But*　We've recently revised the long-range projections of employment and population in Madison County. Here's how they will affect you . . .

As a speaker, you can emphasize words by using greater volume, higher (or lower) pitch, changes of pace, facial expressions, and physical gestures like pointing or clenching your fist. Albert Mehrebian, a communication expert at UCLA, says that up to 55 percent of a person's impact is from nonverbal signals (Almazol 1982). Since none of these features is available in writing, you have a real advantage as a speaker. But because of listeners' limited attention span and the distraction potential, your sentences do have to be less complicated than in writing. To remind yourself where you want to use transparencies, slides, or other visuals, mark those spots in your speech.

Organize the conclusion next to see where you will end. Make sure that somewhere in the conclusion you summarize the main points that you've raised. Then go back to the introduction. Perhaps now you'll have found a good way to attract your listeners' attention and focus it on your content. Don't try to be clever in the introduction; instead, let the material shape how you begin. Be sure that the introduction provides an overview of the whole speech—telling your audience where they're going. Review the organization to ensure that you've included all the key information. Then you are ready for the first rehearsal.

Rehearsing

Most people feel silly talking to themselves, so you may be tempted to skip this step. Don't. You may have the best organized speech in your county, but until you've actually used that outline to speak from, you don't know if it will work. What's more, until you actually say all the words you intend, you don't know how long it will take to say them. Some speakers go for long walks and practice the speech as they walk; some lock themselves in the bathroom with a shower running so no one will hear them.

These may be extreme measures, but do what works best for you. Follow these hints to make the practice sessions easier.

- Look up all the tricky words before you start and mark the pronunciation on your cards in big letters. Practice the pronunciation until you can do it without stumbling.
- Stand up when you practice and try to duplicate the conditions you'll have when you speak (lectern or not, overhead projector, and so on). If possible, practice in the room where you'll be making your speech, or at least visit the room beforehand so you know what it's like.
- Use your cards or outline, but make sure you're not reading from them. Practice in front of a mirror to make sure you're looking up.
- Time your presentation with a stopwatch. Listen to yourself when you speak and make sure you are not rushing.
- After the first rehearsal, adjust for timing and smooth out any places where you stumbled. Be aware that the actual presentation usually takes longer than rehearsals, so allow for that. Write out transitions if necessary. Practice two or three times more, but don't overpractice.
- Videotape or audiotape one rehearsal and review it to see how you look and sound to others.

Speaking

On the day of your speech, dress professionally for the occasion, avoiding loud or clashing colors without dressing like a mouse. For men, unless the occasion is informal, a jacket and tie are usually expected. For women, a tailored suit or dress is a good bet. Arrive at least 15 to 20 minutes before your speech to check on the room and equipment and to prepare yourself mentally and emotionally. Don't worry if you're nervous—even professional speakers usually are, and you can use your extra energy to good advantage. Instead of thinking about potential problems, think positively about what you can accomplish. Meet the person who will introduce you, or learn if you must introduce yourself. (It's even better to know this in advance.)

Start slowly when you begin to speak, deliberately keeping the pace down at the beginning. Look your listeners right in the eye; in fact, find three or four supportive people in different parts of the room and talk directly to those individuals. That will ensure that your eyes move through the audience. Don't be afraid to move and to gesture, point to your visuals, and even underline key points. Keep your watch in front of you so you can stay within your time box, adjusting your pace as you go. Don't glance at the watch constantly; instead, look at your audience. End on time and with a real ending, not a weak gasp.

If a question-and-answer session follows, repeat or summarize each question to ensure that everyone in the audience has heard it and that you understand it. This also gives you time to plan your response. If you don't catch the gist of a question, feel free to ask that it be repeated. Don't let yourself be drawn into an argument with a listener; instead, offer to speak with the person later. If a question stumps you, admit you don't know the answer—perhaps offering to find out or giving an alternate source of information. Usually you can field questions for 10 to 15 minutes before they begin to repeat, but don't be afraid to end the session if it's not productive.

Finally, try to get some feedback from your listeners. Sometimes you can chat with individuals later, and sometimes you can ask for written comments. Responses, of course, will depend on the purpose and nature of the speech as well as the type of audience, but the more you can evaluate what you did, the better prepared you'll be for the next speech you're asked to give.

In this chapter, I've concentrated on rather formal speeches, but more often you will give informal oral presentations to your colleagues or managers. Take these presentations seriously because they are an important part of your professional career. Adapt the procedure and the form to your needs, but always spend time carefully preparing and rehearsing whenever you must give a short—or long—speech or oral presentation.

Checklist for Writing a Speech

1. In preparing my speech, have I considered these elements?
   ___ the distraction potential
   ___ comprehension loss
   ___ time limitations
   ___ the group dynamic
2. Does the introduction of my speech accomplish the following?
   ___ create a friendly atmosphere
   ___ hook the listeners' attention
   ___ present an overview of what's coming
3. Does the body present and support the major points, using the most appropriate method of organization for the purpose and the type of listener?
4. Does the conclusion summarize and, if appropriate, propose action?
5. Have I included anecdotes, vivid examples, or quotations to provide interest?
6. Have I planned visuals to highlight key points?
7. Which of the following methods should I use to present my visuals?
   ___ a chalk board
   ___ posters or flip charts
   ___ slides
   ___ handouts
   ___ transparencies

EXERCISES

1. Choose a procedure from your major field that interests you and that you understand well. Plan a 5-minute extemporaneous speech in which you explain this procedure to the members of your technical writing class. In your planning, determine your classmates' level of technical knowledge so you know how technical you can be.

Use at least one kind of visual in your speech. As a member of the audience, write a brief critique of each classmate's speech, commenting on the content as well as the presentation and noting areas that were well explained to a generalist audience.

2. Near the end of your term project, plan and present a 10-minute speech to your technical writing class, giving the results and conclusions of your study. Prepare an outline of your speech to give your instructor before you begin, and include at least two visuals.

3. Attend a speech given by someone in your major field (at a technical conference, on-campus presentation, or meeting of a professional society). Take notes on the speech; then, later, analyze the speech makeup. How long did the speaker spend on each segment? How many major points were made? What kind of visuals were used? How were they presented? Report your findings to the class either orally or in an informal report.

REFERENCES

Almazol, Susan, 1982. Professionals getting the word on speaking skills. *San Jose Mercury News* PC-1, 31 January.

Dewan, Tom. 1990. I'm a good writer, so why is my job so hard? Speech at San Jose State University.

Jahnke, Jim. 1987. Letter to author, 8 September.

Plimpton, George. 1982. How to make a speech. Advertisement for International Paper Co. New York.

Porter, Sylvia. 1979. Are you listening? Really hearing? *San Francisco Chronicle.* 14 November.

Wells, Kathy. 1990. Formal document reviews. Speech at San Jose State University.

Appendix

Handbook

This handbook contains both explanations and practice exercises. It does not pretend to explain every detail of punctuation, abbreviations, numbers, capitalization, and spelling, but it will answer many questions that you may have about those subjects, and it will give you guidelines to follow when writing technical and scientific documents.

1. Punctuating Sentence Types

As a student, you've studied the rules of punctuation for years, and you already know many of the topics covered in this handbook. However, you may be unsure how to use some punctuation marks, and this course may well be your last chance to polish your punctuation skills before you enter the working world. Punctuation is important. When it is done well, it provides the reader subtle guidelines in interpreting meaning, saying "Stop here," or "Pause here and then go on," or "These two ideas are related." Readers accept those signals subconsciously, not noticing the marks themselves but concentrating on the content. But if a sentence is either unpunctuated or badly punctuated, readers do notice the punctuation because they must slow down to decipher the sentence's meaning.

This handbook covers those marks of punctuation you are most likely to use in your writing, and it provides explanations, examples, and exercises to help you practice using particular punctuation marks. You can use the handbook both as a reference source and as a workbook. When you work through exercises, use the standard copyediting symbols on the inside of the back cover to show the changes. These symbols are widely used in industry and publishing and will indicate your professionalism.

One use of punctuation marks is to help readers identify the types of sentences they are reading. This section will help you use punctuation effectively to end sentences and to separate the parts of compound and complex sentences.

Simple Sentences

The key to sentence punctuation is a clear understanding of what makes a sentence.

Definitions

- *sentence*: an independent unit of words that contains a subject and a verb and closes with a mark of end punctuation. (For identification in the examples that follow, subjects are underlined once and verbs twice.)

    Writers write.

    Sometimes the subject is understood.

    Write! ("you" understood)

- *subject*: that part of a sentence about which something is said or asked.

    Students write.

The *simple subject* is the main noun or pronoun.

> He ran.

The *complete subject* is the main noun or pronoun plus its associated modifiers.

> The man in the red jumpsuit ran.

Subjects can be multiple.

> The truck and the motorcycle collided.

- *verb*: the word or words in a sentence that show action, occurrence, or existence.

> Students write.
> Jones is in his office.

Verbs can be multiple.

> College students eat, sleep, and study.

Verbs can be made up of more than one word.

> Chemistry 101 will be offered at night.

The *predicate* is the verb plus its modifiers and *complements*. The predicate is the part of the sentence that says or asks something about the subject. A complement completes the sentence meaning. Complements can be direct objects, indirect objects, subject complements, or object complements. (In the following, the subject is underlined once and the predicate twice.)

> The fire burned 42 acres of brush and trees.

- *end marks*: the marks of punctuation that tell the reader what the sentence is doing. *Periods* end sentences that make statements. *Question marks* identify questions, and *exclamation marks* indicate strong statements.

Compound Sentences

Writers frequently combine related ideas into one sentence. If you want to join two or more ideas of equal importance, you may want to write a compound sentence. Look at the two sentences below:

> Mary graduated in June with a degree in geology.
> She now works as a computer programmer.

Suppose that you want to join those two ideas into one sentence, giving equal importance to each idea. You might say:

> Mary graduated in June with a degree in geology, and she now works as a computer programmer.

What are you implying by this combination? A simple chronology. Mary did that and now is doing this. Another choice might be this one:

> Mary graduated in June with a degree in geology; she now works as a computer programmer.

But think about the *meaning* you want to communicate. Is that job as a programmer an expected outgrowth of a geology degree? Or do you want to imply a contrast by the way you combine those sentences? For example, you might say:

> Mary graduated in June with a degree in geology, but she now works as a computer programmer.
>
> *or* Mary graduated in June with a degree in geology; however, she now works as a computer programmer.

In each of the four sentences above, the two ideas are correctly combined into a compound sentence with equal importance given to each idea. However, the implications are different in the four sentences. The change in meaning comes from the choice of the connecting word combined with the punctuation.

Definitions

Before you learn the punctuation rule for compound sentences, you need to be sure of the meanings of five key terms:

- *clause*: a group of related words that contains a subject and a verb. The subject may be understood. A short sentence is called a clause when it becomes part of a larger sentence.

  > but he examined the precipitate

- *independent clause*: a clause that can stand alone as a complete sentence.

  > the report summarized data during a six-month period

- *coordinators*: the connecting words that follow. They are also called *coordinating conjunctions. Co* implies equal, like the word *co-worker*.

and	either	yet
but	neither	so
or	nor	for

- *adverbial connectors*: adverbs that can be used to join independent clauses. They are also called *conjunctive adverbs*. They include the following:

accordingly	however	now
also	indeed	otherwise
anyhow	instead	still
anyway	meanwhile	then
besides	moreover	therefore
consequently	namely	thus
furthermore	nevertheless	too
hence		

Phrases can also serve as adverbial connectors.

as a result   for example
in fact   in addition

- *compound sentence*: a sentence containing two or more independent clauses joined in some way.

Punctuation Rule

A compound sentence can be punctuated in one of three ways:

1. *with a comma and a coordinator*:
   independent clause   (, coordinator)   independent clause

   The flywheel recovers in less than a second, *and* another power stroke begins.

2. *with a semicolon*:
   independent clause   (;)   independent clause

   The flywheel recovers in less than a second; another power stroke begins.

3. *With a semicolon and an adverbial connector followed by a comma*:
   independent clause   (; adverbial connector,)   independent clause

   The flywheel recovers in less than a second; *then*, another power stroke begins.

The adverbial connector often starts the second clause, but it can also appear elsewhere in that clause.

   The tests are not material performance tests; they are, *instead*, material quality tests.

EXERCISE

To review your understanding of key terms, answer each of the following questions. Try to answer without looking at the definitions. Then check your answers against the definitions given earlier in this section.

   What is a clause?
   What is an independent clause?
   How would you define a compound sentence?
   When should you write a compound sentence?
   What are the three ways a compound sentence may be punctuated?

Following are pairs of simple sentences. Rewrite each pair into a compound sentence using *each* of the three ways of compounding. Be sure that you write a compound sentence, even though you could, perhaps, write a simple sentence. Mark the best one with a star and be prepared to defend your choice in class.

1. Normal range for the AM-FM radio is approximately 35 miles. On flat terrain and with powerful transmitters, the range could be considerably extended.
2. A home radio has constant voltage and is rarely subjected to wide variations in temperature. The FM car radio operates under more severe conditions.
3. The tandem master cylinder is mounted on the forward end of the booster. The rear end of the booster is connected to a frame-mounted bracket.
4. Two output lines are connected to the power section. One leads to the power steering gear and the other to the reservoir.
5. Brake adjustment is required after installation of new or relined brake shoes. Adjustment is also necessary whenever excessive travel of pedal is needed to start braking action.

Complex Sentences with Subordinators

Sometimes you want to combine two or more ideas in a sentence, and you want to make each idea equally important. Then you write a compound sentence. At other times, you want to combine ideas, but you want to make one idea dominant and the other subordinate. Then you write a complex sentence. Occasionally, you want to do both. Then you write a compound-complex sentence.

Look at the following two sentences.

J. Smith designed the SF 150 module.

He was promoted to manufacturing manager.

Which of these two ideas do you, as writer, want to be the more important?

Suppose you choose the idea of promotion as more important. That sentence will become your main, or *independent* clause; the sentence about designing the module will become a subordinate or *dependent* clause. You can form a dependent clause by beginning it with a *subordinator* (or a *subordinating conjunction*).

Because J. Smith designed the SF 150 module, he was promoted to manufacturing manager.

The independent clause can appear at the beginning or the end of the sentence.

J. Smith was promoted to manufacturing manager _because_ he designed the SF 150 module.

If you chose the design idea as more important, you would begin the promotion clause with a subordinating conjunction.

J. Smith designed the SF 150 module _because_ he was promoted to manufacturing manager.

or **_Because_ he was promoted to manufacturing manager, J. Smith designed the SF 150 module.**

Do you see how the meaning has changed? We changed it by the choice of the main clause. We could also change the meaning dramatically by choosing a different subordinating conjunction. Suppose we use *after*:

> J. Smith designed the SF 150 module *after* he was promoted to manufacturing manager.
>
> *or*   *After* he was promoted to manufacturing manager, J. Smith designed the SF 150 module.

Subordinators can be divided into two classes according to how they function in a sentence. One class is called *causal/logical* subordinators because those words indicate cause and logical relationship. The second class is called *temporal/spatial* subordinators because those words indicate time or place relationship.

Definitions

- *clause*: a group of related words that contains a subject and a verb. The subject may be understood. A short sentence is called a clause when it becomes part of a larger sentence.
- *independent clause*: a clause that can stand alone as a complete sentence.
- *dependent clause*: a clause that cannot stand alone as a complete sentence. It is subordinate to an independent clause.

> **if the product is announced**

- *complex sentence*: a sentence containing one independent clause and one or more dependent clauses.
- *subordinators*: connecting words that begin dependent clauses and join them to independent clauses. Subordinators are also called *subordinating conjunctions*. *Sub* implies "under." The following words can be causal/logical subordinators:

although	even if	though
as though	if	unless
because	since	whereas

The following words can be temporal/spatial subordinators:

after	since	when, whenever
before	till	where, wherever
once	until	while

Punctuation Rule

Use these guidelines to punctuate a complex sentence when the dependent clause begins with a subordinating conjunction:

1. If the dependent clause stands first, it must be followed by a comma.

> **Before high-speed maglevs will be practical, a breakthrough in electrical generation is necessary.**

2. If the dependent clause stands after the independent clause, no punctuation is required unless the subordinator is *though, although,* or *whereas*.

> **A breakthrough in electrical generation is necessary before high-speed maglevs will be practical.**
>
> **The harbor entrance rapidly filled with sand, even though it was dredged out every year.**

☰ EXERCISE

To review your understanding of key terms, answer each of the following questions. Try to answer without looking at the definition. Then check your answers against the definition given earlier in this section.

> What is a dependent clause?
> What is a subordinating conjunction?
> When should you subordinate a clause?
> What is the punctuation rule for a complex sentence?

Using a subordinator, join the following independent clauses to make a complex sentence. Be sure to write a complex sentence, even though you could write a simple or compound sentence. Try to use a different conjunction for each one, and try to place the dependent clause in different positions. Use standard copyediting marks to insert punctuation marks and words. Be prepared to defend your choices.

1. Darwin was a fledgling naturalist in 1832. His observations of plant and animal life were precise and detailed.
2. Darwin took extensive notes on his round-the-world journey in the *HMS Beagle*. He later published the *Journal of Researches* based on his journey.
3. Darwin's observations were limited to those animals on friendly shores. Some obstreperous South American natives would not allow the crew of the *Beagle* to land.
4. Today scientists try to use only objective words in their technical descriptions. Darwin used terms like *hideous, ugly, repulsive,* and *disgusting*.
5. Students today do not appreciate the accomplishments of the nineteenth century naturalists. They realize that Darwin and Huxley were breaking new ground in both concentration and technique.
6. James Watson, an American molecular biologist, was in his early twenties. He went to Cambridge to study the structure of proteins.
7. He collaborated with English biologist Francis Crick. They discovered the chemical structure of DNA.
8. Watson became a professor of biology at Harvard. He published a popular account of their search in the book *The Double Helix*.
9. Watson and Crick were searching for the DNA model. Linus Pauling was also close to finding the answer.
10. Pauling sent a preliminary manuscript about his study from California to Cambridge. Pauling's son, Peter, was a biophysicist.

Complex Sentences with Relative Pronouns

Dependent Clauses Used as Adjectives

Complex sentences consist of dependent and independent clauses. Some dependent clauses are formed by adding subordinating conjunctions to clauses; other dependent clauses are formed by adding relative pronouns. These clauses may be used either as adjectives or as nouns. A clause used as an adjective stands after the word it modifies.

Common relative pronouns that can begin clauses are

that	which	whom
what	who	whose

For example:

> The wallet *that I lost* contained all my credit cards.
>
> The Ste. Claire Hotel, *which was rundown and dirty*, has been restored to its former splendor.
>
> The student *who caught the shoplifter* received a substantial reward.
>
> The candidate *whom we supported* was not elected.
>
> The student *whose bike was stolen* had to commute by bus.

Each dependent clause above is used as an *adjective*; that is, it modifies or explains the noun that precedes it, answering the questions "What?" or "Who?"

Dependent Clauses Used as Nouns

Dependent clauses can also be used as *nouns* in the sentence. The clauses may look the same, but they are not punctuated. For example:

> *That I lost his confidence* is painfully obvious. [subject]
>
> I know in my heart *what I am doing*. [direct object]
>
> He wrote the letter to *whom it might concern*. [object of preposition]

Punctuation Rules

Dependent clauses used as *nouns* are generally not separated from the rest of the sentence by commas. Dependent clauses used as *adjectives* sometimes must be set off by commas. When deciding whether to set off the clause with commas, follow these guidelines:

1. If the clause is necessary to establish the specific identity of the noun it follows, it is a *primary identifier*: it restricts or limits the meaning of the word it describes. Such a primary identifier is called *restrictive*, and no commas surround it.

   > The woman *whom you have just met* is in charge of the project. [restrictive]
   >
   > The book *that I am now reading* is a collection of essays. [restrictive]

2. If the clause merely adds extra information to the word that it describes, it is a *secondary identifier*: it could be dropped from the sentence without materially affecting the meaning. Such a secondary identifier is called *nonrestrictive*, and it should be surrounded by commas. For example:

> Felicia Lopez, *whom you have just met*, is in charge of the project. [nonrestrictive]
>
> *A Moveable Feast, which I am now reading*, is a collection of essays. [nonrestrictive]

3. As a general rule, use *that* in restrictive clauses (without commas) and *which* in nonrestrictive clauses (with commas).

> A carburetor *that will give better gas mileage* is badly needed. [restrictive]
>
> The Solex carburetor, *which gives better gas mileage*, is an ecological bonus. [nonrestrictive]

EXERCISES

1. Write one sentence using each clause as a restrictive or "primary" adjective modifier. Then write another sentence using the same clause as a nonrestrictive or "secondary" adjective modifier. Punctuate correctly.

1. who had seen the entire incident
2. that/which polluted the atmosphere
3. whom he admired
4. whose term expired this year
5. that/which stopped at ten o'clock
6. who was elected in 1988
7. that/which was sold yesterday

2. Underline the relative clause in each sentence below. Circle the relative pronoun.

1. In addition to the contaminants that may be present in free air, the air compressor machinery may introduce contaminants.
2. Scuba tanks often are fitted with a rubber or plastic boot that should have holes to permit draining.
3. An operator who is unaware of the problems involved with compressed air systems can endanger the diver's life.
4. Portable high-pressure compressors, which are used for filling scuba tanks, should deliver a minimum of 2 standard cubic feet per minute of air at 2,000 psi.
5. Water-lubricated compressors that do not produce carbon monoxide or other contaminants are available on a limited basis.
6. A molecular sieve, which has an extremely large surface area to enhance its capacity for adsorption, removes harmful contaminants by causing them to adhere to its surface.

7. For field use, instruments are available that provide sufficiently accurate data to determine the safety of a gas as a breathing medium.
8. Oxygen is shipped in gas cylinders that are color-coded *green*.
9. Nitrogen, which is the most commonly used diluent, is limited because of its tendency to produce narcosis.
10. Helium has a high diffusivity that allows it to leak through penetrators and into equipment.

3. Underline the relative clause in each sentence and mark it restrictive or nonrestrictive. Using standard copyediting marks, punctuate correctly, and change *that/which* if necessary.

1. Speech is one of those basic abilities that set us apart from animals.
2. We all use automatons like the dial telephone and automatic elevator which either get their instructions from us or report back to us on their operations.
3. The movements of the vocal organs generate a speech sound wave that travels through the air between speaker and listener.
4. Pressure changes at the ear activate the listener's hearing mechanism and produce nerve impulses that travel along the acoustic nerve to the listener's brain.
5. We can compare all the words of a language and find those sounds that differentiate one word from another.
6. There are some words, like "awe" and "a," which have only one phoneme, and others that are made up of 10 or more phonemes.
7. The rules that outline the way sequences of words can be combined to form acceptable sentences are called the grammar of a language.
8. Peter Denes and Elliot Pinson who wrote *The Speech Chain* are both communication scientists.
9. *The Speech Chain* is part of a science study series by scientists who write for students and laymen.
10. The neuron has an expanded part, the cell body which contains the cell nucleus.

2.

Using Punctuation within Sentences

Writers use punctuation marks *within* sentences in certain conventional ways to help readers understand how the parts of the sentences are related to one another. This section explains the most common uses for a variety of punctuation marks.

Commas in a Series

Commas are used to separate three or more individual items in a series. Series items include the following:

- dates

    The order was shipped on June 12, 1992, by U.S. Mail.

- places

  > The central manufacturing plant is located at 1066 S. Main Street, Rockford, Illinois.

- sentence elements

  > The manufacturing plant requires bulk, container, and pallet facilities.

Always use the comma before the conjunction that precedes the last item. This comma is never wrong, and it may avoid confusion.

> *Not*   The cafeteria stocks strawberry, mocha nut, vanilla and chocolate swirl ice cream.

Can you tell how many flavors the writer intended to list?

> *But*   The cafeteria stocks strawberry, mocha nut, vanilla, and chocolate swirl ice cream.

EXERCISE

Use copyediting marks to correct any comma errors in the following sentences.

1. The four most abundant elements in seawater are oxygen, hydrogen, chlorine and sodium.
2. Seawater is always slightly alkaline because it contains earth minerals such as sodium, calcium magnesium, and potassium.
3. Divers need to know about all five kinds of pressure: atmospheric hydrostatic absolute gauge and partial.
4. The second edition of *The NOAA Diving Manual* was published on December 1 1979 and is for sale by the U.S. Government Printing Office Washington DC 20402.
5. Advantages of open-circuit scuba are mobility, portability adaptability to small-boat operations and availability of training.
6. In many cases the instruments, tools and techniques were made or modified by individual scientists to meet the needs of the project.
7. A base-line study was made by Turner, Ebert, and Given at Point Loma San Diego County California.
8. The Pratt Macrosnooper is an underwater magnifying system consisting of three lenses, spacers, and plastic housing.
9. Soap mineral or fungus deposits on the Macrosnooper lenses may be removed by overnight soaking in bleach vinegar or laundry detergent.
10. Shellfish studies by divers have been directed toward the ecology of the organism its behavior relative to sampling gear and artificially implanted tags, and the efficiency of the sampling gear for capturing the organism.

Commas with Introductory and Interrupter Words and Phrases

Commas are used to set off introductory words or phrases from the rest of the sentence. A phrase is a group of related words.

Introductory words

- nouns of address

    Paul Smith, will you please give your analysis first?

- transitional words that break the continuity of the writing

    The manufacturing plant will be closed the first week in September. However, the development lab will operate as usual.

- conversational openers like *yes, no, now, well*, or *why*

    No, the shipment has not arrived.

Introductory phrases

- prepositional phrases of three words or more

    At each level of design, the review committee meets to solve problems.

- verbal phrases (made from verbs, but not acting as verbs)

    To advance the spark, change the ignition timing.
    Printing to screen, he checked the format of the document.

- short prepositional phrases that could cause misreading

    In time, all bearing surfaces will show wear.

- long modifiers out of usual order

    Safe for humans, the spray is toxic to insects.

Interrupter words and phrases

- direct address

    Call me, Dr. Howard, if you have any questions.

- appositives (a second noun that amplifies the meaning of the first noun)

    Alan Sourth, manager of publications, will be the principal speaker.

- parenthetical elements (extra or added information)

    The temperature varied, he explained, because of the inversion.

- contrasting statements

    The mechanical components, not the electrical, were causing the problem.

• transitional expressions like *for example* and *especially*

> So-called stone fruits, for example, peaches and nectarines, are technically soft tree fruits with pits.

EXERCISE

Using standard copyediting marks, edit the following sentences for comma errors.

1. Pyrotechnics the craft of making fireworks is beginning to use scientific chemical applications.
2. However the industry in the United States has been dominated for decades by small family firms.
3. To fuel an aerial firework pyrotechnists traditionally use charcoal and sulfur.
4. For the lifting charge and the bursting charge the pyrotechnist uses a mixture of gunpowder and fuel.
5. To create the colorful streaks across the sky fireworks manufacturers add "stars," for example strontium compounds for red and barium compounds for green.
6. New fuels especially magnesium and aluminum alloys can create brighter and deeper colors than were previously attainable.
7. Bill Page, manager of Celebrity Fireworks in California has perfected a new oxidizer potassium perchlorate that is more difficult to ignite.
8. Less hazardous than other powders, the new oxidizer is however harder to use in aerial shells.
9. Some fireworks accidents according to Page, are blamed on discharge of static electricity.
10. To avoid static discharge workers at the fireworks plant wear plug-in wristbands to keep themselves grounded.

Commas and Semicolons in Compound and Complex Sentences

Commas and semicolons are also used to help readers understand the relationship of clauses in compound and complex sentences. See section 1.0 for details on punctuation in different sentence types.

Colons

College students seldom use colons, perhaps because they are unsure how to use them correctly. Professional writers, however, use colons extensively, relying on them to convey subtle shades of meaning to the reader. You can use a colon in four situations:

1. To introduce a formal list or series, especially one including the word *following*.

> The following Pacific Coast bays are really drowned river valleys: San Francisco Bay, Coos Bay, Humboldt Bay, and Grays Harbor.

Note, however, that you *should not* use a colon to separate a verb from its object or complement or to separate a preposition from the object.

Not   The three most conspicuous plants in brackish water marshlands are: cattails, bulrushes, and sedges.

But   The three most conspicuous plants in brackish water marshlands are cattails, bulrushes, and sedges.

Not   Collected animals should be put in: plastic pails, plastic bags, or sloshing water.

But   Collected animals should be put in plastic pails, plastic bags, or sloshing water.

2. To introduce a word group or clause that summarizes or explains the first word group. In this case, the colon can replace *that is* or *namely*.

There is only one way to preserve collected animals: keep the new surroundings as close as possible to the original habitat.

If the clause following the colon is a complete sentence, it is often capitalized.

The coloration of the upperside of the carapace varies extensively: Sometimes it is red or purplish, but in many specimens it is gray, brown, or olive.

3. To introduce a long or formal quotation.

Kozloff's warning is plain: "Leaving a rock 'belly up' is an almost sure way to kill most of the animals that are living on its under- side, and perhaps also the animals and plants on its upper side."

4. To show distinctions or divisions according to convention.

Dear Mr.Simpson:
4:30 P.M.
"The Pacific Northwest: An Ecological Study"

EXERCISE

Using standard copyediting marks, insert colons in the following sentences if they are needed. Below the sentence, write the reason for inserting the colon. If no colons are needed, give the reason.

1. Of the DC and AC types, the AC variety is most prevalent It permits larger displays, although it's more difficult to build and requires more complex drive circuitry.

2. The main difference between AC and DC plasma panels is AC displays have dielectric layers separating the gas from the unit's activating electrodes.

3. The thin-film EL panels used in information displays usually fail in one of three ways package fracture from shock, dielectric breakdown from over-voltage, brightness decay from high temperatures.

4. The new development program could prove to be one of the most significant in the history of flat-panel displays the production of a full-color EL display.

5. Hix stated "The use of class 10 clean rooms allows production of panels with 100 percent functional pixels and no pin-hole defects."

6. VF panels basically employ the same principle as a triode vacuum tube The phosphor-coated anode emits light when struck by electrons from the cathode.

7. So far this technology has appeared in only two military applications a compass computer and a digital message device.

8. Application for AC plasma displays have included the following Trident submarines, the Harpoon missile system, and an antenna-directing system in the E-4B emergency presidential command aircraft.

Apostrophes

The apostrophe is used for three reasons:

1. to show possession
2. to show the omission of letters or numbers
3. to prevent misreading of numbers or figures used as words

To Show Possession

- Add '*s* to singular nouns and pronouns to show possession.

  a product's reliability
  John Parker's letter
  somebody's badge

- Add '*s* to singular nouns that end in *s*. If the pronunciation is awkward, you may use the apostrophe only.

  Charles's report
  this class's schedule
  Keats' study

The Government Printing Office (GPO) *Style Manual*, however, requires only the apostrophe, as do some other style guides; follow your company style guide on this point.

- Add only the apostrophe to plural words ending in *s*.

> **three hours' worth**
> **all managers' reports**

- Add *'s* to plurals that do not end in *s*.

> **men's contracts**
> **women's rights**

Personal pronouns are already possessive and do not need apostrophes.

hers	its	ours
his	theirs	whose

In particular, remember to write *its* (no apostrophe) to show possession.

> **The snake shed *its* skin.**

To Show Omission

Use the apostrophe to replace the missing letters or numbers in a contraction. Put the apostrophe where the letters or numbers would otherwise be.

she'll	she will
he's	he is
can't	can not
I'm	I am
'93	1993
it's	it is

In particular remember to write *it's* (with apostrophe) for the contraction.

> ***It's* time for work.**

To Prevent Misreading

Use the apostrophe for clarity when you are using numbers, letters, or symbols as words.

> **The chart omits all *0's* to the left of the decimal point.**
> **Technical people use a crossbar on *7's* to avoid mistaking them for *1's*.**
> **In my handwritten memo he misread my *a's* for *e's*.**

If there is no chance of confusion, the apostrophe can be eliminated.

☰ EXERCISE

Use copyediting marks to correct any apostrophe errors in the following sentences.

1. The Germans intensive bombing on December 29, 1940, caused more than 1500 fires in London.
2. Due to equipment shortages, some fire brigades engines were made from cut-down cars or trucks.

3. The Coventry-Climax engines adaptability made it widely used in portable "wheelbarrow pumps."
4. New York Citys highest fire of the 40s occurred when a Mitchell bomber crashed into the 78th and 79th floors of the Empire State Building.
5. Due to the elevators failure, firefighters had to carry equipment from the 67th floor.
6. Modern Super Pumpers have been developed to fight urban centers large fires.
7. The combined force of the Super Pumper and it's satellites can hurl up to 37 tons of water on a fire each minute.
8. Super Pumpers high pressure hose's use all synthetic materials that can withstand more than 700 psi.
9. Ships fires, even at dockside, cant always be fought by simply pouring on water.
10. It's said that an aircraft fire must be dealt with within 60 seconds of the crash.

Hyphens

Hyphen usage can be complicated, and neither dictionaries nor handbooks always agree on suggested rules. But technical writers and editors need solid information about hyphens because they are used so frequently in technical writing. Here are the most common uses of the hyphen. For further information, see Karen Judd: *Copyediting: A Practical Guide* (Los Altos: Crisp Publications, 1990), or the Government Printing Office *Style Guide*.

To Indicate Syllable Breaks

The most common use of a hyphen is marking a break in a word at the end of a line. Some word-processing programs do this automatically. If not, the only things you have to remember are these:

- Hyphenate between syllables only. If you don't know where the break is, look in the dictionary.
- Do not divide one-syllable words.
- Do not hyphenate names, contractions, abbreviations, acronyms, or numerals.
- Do not hyphenate the last word on a page or the last word in a paragraph.

In Compound Words

Authorities and dictionaries often do not agree on whether a hyphen is required in a compound word. Usually what happens is that a compound word begins as two separate words, soon is hyphenated, and eventually becomes a single word. You need to remember these things:

- Be consistent. Choose a single dictionary, handbook, or company style guide and follow it faithfully.

- Keep your reader in mind. Hyphens are supposed to make the reading easier. Sometimes you must use hyphens simply to avoid confusion. For example:

  an un-ionized compound (not *unionized*)
  re-cover it (make a new cover, not *recover* something lost)
  ball-like
  cave-in

In Compound Words Used as Adjectives

Use hyphens in the following compound words when they are used as adjectives *preceding* nouns. Do not hyphenate compound words when they follow a verb.

- numbers from twenty-one to ninety-nine written in words

  Thirty-three engineers attended the meeting.

- a number plus a unit of measure used as an adjective

  9-minute interval

- spelled-out fractions used as adjectives

  one-third cup

- an adjective or noun plus a past or present participle used as an adjective

  motion-activated alarm
  software-testing strategies

- an adjective plus a noun used as an adjective

  high-performance workstations
  dual-port design

- two colors used as an adjective

  yellow-green smoke

- words with prefixes, if the root is a proper name or a number

  pre-1980
  post-Newtonian

In Compound Words Used as Nouns

Compound words used as nouns include the following:

- all compound nouns beginning with *self*

  a self-study

- a compound made of two nouns of equal value

  owner-operator
  motor-generator

- numbers from twenty-one to ninety-nine when written out

    seventy-three

Do not use hyphens in the following situations:

- with most prefixes and suffixes

    starlike
    pretest

- with names of chemicals used as adjectives

    sodium chloride deposits

- with an adverb ending in *ly* plus an adjective or participle

    a finely tuned engine

- if the individual words modify the noun separately

    a new digital analyzer

EXERCISE

Insert or delete hyphens in the following sentences if necessary. In the space below each sentence, write the reason for your action.

1. The chemist performed the complex analysis by using his balance and finely-graduated weights.

2. During the decision making process, one needs to think through alternative actions.

3. The procedure has been used for the last twentyfive years.

4. That refinery has a long term contract with Exxon Corp.

5. The powerplant uses 16 cylinder diesel electric marine modules.

6. The power modules combine on a single skid GE's custom 8000 selfventilated generator.

7. The car buying public has responded well to front wheel drive.

8. That business requires highly skilled and technically trained management.

9. A 5.7-liter engine is used in that model.

10. Fuel mileage is rated at sixteen mpg, which is a three mpg improvement.

Dashes, Commas, Parentheses, and Brackets

Dashes, commas, parentheses, and brackets are all used to set off certain words from the rest of the sentence. Unlike commas, which give equal weight to the words they surround, dashes emphasize and parentheses de-emphasize the enclosed words. Brackets usually indicate that the enclosed words come from a different source or are not an integral part of the sentence.

What we usually call the dash is more specifically known as the *em dash*. The name comes from the width of the dash—about the same width as the capital *M*. In typing, the dash is shown by two hyphens, and no spaces are used before or after the dash. Dashes are often use as substitutes for other marks of punctuation, but they are not as precise. The dash tends to be informal, especially if it is used frequently, but it is most effective when used sparingly. Use dashes for the following reasons.

To Set Off Parenthetical Material

Commas and parentheses can also set off parenthetical material, but for the *strongest* emphasis on the parenthetical material, use dashes.

> Certain cells of the retina—the familiar rods and cones—undergo photo-chemical changes that enable you to see in various light levels.

For *equal* emphasis between the sentence and the parenthetical material, use commas.

> Certain cells of the retina, the familiar rods and cones, undergo photo-chemical changes that enable you to see in various light levels.

To *reduce* the importance of the parenthetical material, use parentheses.

> Certain cells of the retina (the familiar rods and cones) undergo photo-chemical changes that enable you to see in various light levels.

Which version do you think the author used for this sentence? If you guessed the first example, you are right.

To Show a Break in Thought

A dash can set off an emphatic idea or summarize an idea.

> Modern arithmetic was the result of a long search—"research"—for a tool that would handle complicated quantitative problems.

> The human eye is a dual organ—two eyes working together to transmit information to the brain.

For Clarity

When commas are already used within parenthetical material, use dashes to set off the major parenthetical material.

> Complex sounds—such as musical notes, spoken words, or an engine's roar—all produce a more or less definite pitch.

Brackets are like parentheses in setting off from the rest of the sentence words that are reduced in emphasis. However, brackets usually indicate that the words are not an integral part of the sentence. For example:

The printer [subject] provided [verb] ten different fonts [direct object].

 **EXERCISE**

In the following sentences, insert or delete dashes where necessary, or change inappropriate dashes to other punctuation marks. Below each sentence explain why you took the action you did.

1. If the computer is a mini or super-mini class, it should have the ability to use micros as emulators—this will prevent it from being limited to system application software.

2. This makes available to the user a large library of less expensive, already available microcomputer software—the 9816 is capable of running CPM software in addition to HP software.

3. Unlike those types of instruments whose sound source is produced by plucking, striking, or blowing, electronic instruments, the synthesizer for example, produce sound through an electrical circuit that generates an oscillating electrical current.

4. The synthesizers in this report are called hybrids, a combination of analog and digital technology.

5. Each letter of the ADSR generator represents a separate stage of its four major functions—Attack, Decay, Sustain, and Release.

6. Each envelope generator has four different time durations that can be programmed to modify a sound—these time durations control the speed and volume level of the programmed event and are expressed in rates and levels, seconds, and milliseconds.

7. When one or more of the control sources, the keyboard, LFO, or envelope generator, is triggered, the audio chain is enacted, resulting in audio output.

8. The upper mid-section of the board of an ARP Odyssey II is devoted to a panel of slide switches that control many of the gut components—the low-frequency oscillator (LFO), the digitally controlled oscillator (DCO), high-pass filter (HPF), voltage-controlled filter (VCF), voltage-controlled amplifier (VCA), and envelope generator (EG).

9. There is apparently (I'm not sure) no way to enter hard spaces or characters.

10. This "desk-top mainframe" is a 32-bit super-mini computer, which currently is the only super-mini to use a "super chip," a 32 bit processor on a single chip.

Quotation Marks

This section discusses some guidelines for using quotation marks.

To Set Off Someone's Exact Words

- Use a comma (or a colon in formal writing) after the identifier, the word or words telling who made the statement.

   > Thomas Gold said, "Oil and gas come from methane gas deep in the earth."

   > The engineer concluded by saying: "The fog horns sound whenever visibility drops below three miles."

- Use commas before and after the identifier if it divides a complete sentence.

   > "I am convinced," Gold reported, "that oil and gas can be found anywhere if you dig down far enough."

- Place commas and periods inside end quotation marks; place semicolons and colons outside. Question marks go inside or outside depending on whether the quotation is a question or is part of a question.

   > A new theory uses a three-dimensional map to show that the universe is composed of gigantic "bubbles."

   > "If we are right, these bubbles fill the universe just like suds filling the kitchen sink," said John P. Huchra of the Harvard-Smithsonian Center for Astrophysics.

   > The observations "pose serious challenges for current models for the formation of large-scale structure"; conventional explanations say that gravity played a dominant role.

   > Are the stars and galaxies gathered on the surface of these "gigantic bubbles"?

- Use single quotes to enclose a quotation within a quotation.

   > Cruikshank reported last week: "We think the asteroid-meteorite link has been strengthened considerably by the recent discovery of '1982 RA,' which has an orbit that crosses earth's orbit."

- Do not, however, use quotation marks around indirect quotations.

   > He reported that a telescope was not required.

To Identify New Technical Terms or Words Discussed as Words
(Sometimes, such words are printed in italics instead.)

Gold was an originator of the "steady state" theory, which held that the universe was unchanging.

To Identify Parts of a Larger Published Work

Winters reported on Chapter 7, "Direct Redistributional Issues."

EXERCISE

Using standard copyediting marks, edit the following sentences for the punctuation associated with quotation marks.

1. "The biological theory of petroleum arose when all assumed that complex compounds of carbon could be formed only biologically" a science writer reported.
2. "Now it is clear he continued, that hydrocarbons are present everywhere in the solar system".
3. "If the earth started with a quantity of carbon compounds comparable to that on other planets, the question must be "Where is it all?"" the science writer said.
4. Geologist Gold contended that most of the methane is deeper than people have looked. "I expect the region between 15,000 and 30,000 feet below the surface to be very rich in methane, he said.
5. The Swedish government announced plans to drill a deep test hole in granite bedrock. Geologist Michel Halbouty wrote: The Siljan Ring area represents one of the best prospects in the world to test the abiogenic gas theory"
6. Gerard Demaison, however, recently said "There are people that are overreacting to Dr. Gold and think he's wasting everybody's time, and some who believe everything he says. I think both are extreme positions."
7. Gold countered that several kinds of evidence already established provided as firm a proof as is ever possible in science.
8. "Analysis of petroleum and gas from adjacent fields in the Middle East Gold contended, "shows that the oil is very similar chemically".
9. "However," he continued, "the three fields span totally different geological regions: folded mountains in Iran, the Tigris River valley in Iraq, and flat-bedded sediments in Saudi Arabia.
10. Gold concluded Detailed studies of the chemistry of oil and gas fields in Oklahoma and Texas that occur in geologically different formations provide additional insurmountable evidence for the theory."

Ellipses

An ellipsis is a series of spaced periods inserted in quoted text to show that words have been left out. If the ellipsis appears in the middle of a sentence, it consists of three spaced periods; if it appears at the end of a sentence, it consists of four spaced periods—one for the end of the sentence. For example:

> According to professor Richard A. Lanham, "Prose style . . . does not finally come down to a set of simple rules about clarity, brevity, and sincerity. . . . People who tell you prose style is simple are kidding you."

The three dots in the first quoted sentence indicate that a word or words have been omitted; the four dots between sentences show that a sentence (or more) has been omitted.

3.   Abbreviating Words, Names, and Titles

An abbreviation is a shortened form of a word; like a contraction, an abbreviation is formed by omitting some of the letters of the word. In many abbreviations, a period replaces the omitted letters, but in special types of abbreviations, no periods are used. A good general guideline is this: If the abbreviation is in lowercase letters, periods are usually used. If the abbreviation is in capital letters, usually no periods are used.

Common Abbreviations That Usually Require a Period
(Notice that a space follows the period in initials preceding a name.)

- a person's initials
    E. T. Jones
    Robert P. Waring

    But   FDR and JFK

- designation of streets, avenues, and buildings in addresses

    4320 Mission Blvd.
    Bldg. 12

- a title preceding or following the person's name

    Ms. Shirley Agnew
    Maj. Alexis Prudhomme
    Atty. Roger Hewlett
    Lawrence Tegren, Jr.

    But   William Jones III

- an academic degree

    B.S.   Ph.D.
    M.D.   M.B.A.

- a company name (Always use the abbreviation if it is part of the official company name; otherwise, use the abbreviated form only in addresses, lists, and bibliographies.)

  Smith Bros.
  Chrysler Corp.
  A. B. Price, Inc.

- a date or a time

  Jan 3, 1993
  3:00 P.M.
  345 B.C.

- a part of a book in a cross reference, table, or figure

  Ch. 14
  p. 372

- a country name when used as an adjective

  U.S. Department of State
  But also common: USA

- measurement notations that could be confused with words

  in.   = inch
  tan.   = tangent
  at.wt. = atomic weight

Abbreviations That Do Not Require a Period

These abbreviations also do not need a space between letters.

- measurements (Use the abbreviation with a unit of measure only, and only when it cannot be confused with an existing word; see above. Otherwise, write out the word.)

  3000 kc   175 Hz   450 mm

- postal abbreviations for states (See Chapter 16 for a complete list of postal abbreviations.)

  VA   WI   WA

- SI (Systéme International) units

  m   = meter
  kg = kilogram
  s   = second
  A   = ampere
  K   = kelvin
  mol = mole

cd = candela
min = minute
d   = day
h   = hour
L   = liter

Acronyms and Initialisms

An acronym is formed from the first letter or letters of several words. Acronyms are pronounced as words. Initialisms are also formed from the first letter of each word, but the letters are pronounced individually.

radar   radio detecting and ranging
laser   light amplification by stimulated emission of radiation
scuba   self-contained underwater breathing apparatus
NATO   North Atlantic Treaty Organization
RAM   random access memory
IRS   Internal Revenue Service
UFW   United Farm Workers

The first time you use an acronym or initialism, put the identifying words first, then write the acronym in parentheses. Always do this unless you are sure that *all* your readers will recognize the shortened term.

 **EXERCISE**

Using standard copyediting marks, edit the following sentences for errors in the use of abbreviations, acronyms, and initialisms.

1. Cubic Corp developed a mine detector for the US Army that uses radio waves to find buried explosives of any kind.
2. The vehicle-mounted road mine detector (V.M.R.M.D.) can operate up to 8 mph to clear roads and communication lines of metal and plastic mines.
3. The device uses a Texas Instrument Co microprocessor.
4. Radio signals are transmitted and received by antennas on a search head placed 16 ft in front of a tank or other vehicle.
5. Signals are analyzed by the T.I. microprocessor and displayed on a dash-mounted control unit.
6. The unit contains a 4 in square T.V. displaying vertical red lines that correspond to each antenna.
7. Antennas ride 2 in above the surface for maximum sensitivity.
8. Older detectors could only find metal mines, according to Sgt GF. Frannis.
9. The VMRMD was announced in New York on Sept 1, 1979; a hand-held unit with similar technology is also being developed by Cubic Corp.
10. R Keller, product manager at Cubic, says that the VMRMD also has potential non-military applications.

4. **Using Numerals and Words for Numbers**

As a technical writer, you must use numbers of all kinds: counting numbers, ordering numbers, fractions, measurements, and percentages. You need to establish some simple guidelines so that you know when to use numerals, or numerical symbols (1, 7, 360), as opposed to word equivalents (one, seven, three hundred sixty), and always treat numbers consistently.

No single standard has been accepted in industry or publishing; therefore, you must use common sense in setting up guidelines for number use. When you must write about numbers:

1. Determine your company's accepted style and follow it.
2. If your company has no accepted style or you are working on your own, follow these guidelines, which have been adapted from the Government Printing Office *Style Manual*. The assumption throughout this section is that while words may make smoother sentences, numerals are easier for readers to understand.
3. As a general rule, write out one to nine, but use numerals for 10 and above.

Numerals

Use numerals in the following situations:

- in counting 10 items or more

    14 agricultural workers

- in ordering items from the 10th on

    the 14th floor

- in sentences containing 2 or more numbers, when 1 number is 10 or more

    Yesterday we shipped 15 orders, but today we only managed 8.

- in all units of time, measurement, or money

    His shift ends at 4:00 P.M.
    The diameter is 250 mm.
    Total cost of the repair was $1,452.37.
    They forecast a 20 percent rise in sales over the next 5 years.
    If the diameter is 2.469 inches or less, replace the shaft and both rotors.

If a sentence contains both numbers under nine and units of time, measurement, or money, you would still spell out the numbers one to nine and use numerals for the units.

Each of the six volunteers was tested for 45 minutes.

- for very large round numbers (In this case, use a numeral followed by a word.)

    2 billion years
    $4 million

- when two numbers appear together (In this case, spell out the lower number and use numerals for the higher number.)

    14 three-foot sections
    six 100-lb. bags of flour

Remember to use a hyphen between a number and a measurement when the two are used together to modify a noun.

Words

Always use words:

- in counting up to nine items

    three maps
    seven undergraduates

- in ordering items up to the ninth

    the third building on the left
    Fifth Avenue

- to begin a sentence

    Forty-three geographical areas have been identified.
    Six examples are included in the interim report.

If the numeral is more than two words, it's better to reorder the sentence.

Not   **Six hundred thirty-three patents were processed.**

But   **The patent office processed 633 patents.**

- to indicate general approximations

    during the eighties

But use numerals when the approximation is used with a modifier.

    almost 70 errors
    approximately 300 experiments

- if two numbers occur together (In this case, spell out the lower number and use numerals for the higher number.)

    three 12-inch rods
    144 nine-ounce containers

● to indicate fractions, unless the fraction is used as a modifier

> **three-fourths of an inch**
>
> *But*   **3/4-inch pipe**

When possible, change the fraction to a decimal.

> *Not*   **2 1/2 million**
>
> *But*   **2.5 million**

EXERCISE

Evaluate each of the following sentences to see if it conforms to the rules for number use. Make necessary changes, using standard copyediting marks.

1. Florida has the nation's highest concentration of millionaires: nineteen for every 1,000 people.
2. Most millionaires are ordinary people in their early sixties; they have worked hard for 30 years, six days a week.
3. Jupiter, the largest of the planets, has a diameter only 1/10 of the sun.
4. Venus, though similar to Earth in size and mass, has a dense atmosphere and a surface temperature of over 400 degrees C.
5. The Earth makes a complete revolution around the sun every 365 days, five hours, forty-eight minutes, and forty-six seconds.
6. Two and one-half million dollars were earned by the Spacki Corp. during fiscal 1987.
7. Some employees now work 4 10-hour days instead of 5 8-hour days.
8. These employees frequently begin work as early as 6:00 A.M.; however, they may begin work at eight in the morning and leave as late as seven in the evening.
9. Araxco's offices are on the ninth floor at 153 W. Tenth Street, Topeka, Kansas.
10. Lake Texoma, created by the damming of the Red River, covers more than ninety-three thousand acres and borders on the two states of Texas and Oklahoma.

5. Capitalizing and Spelling Effectively

Capitalization

Like good punctuation, good capitalization should be invisible; that is, it should help the reader understand subtle distinctions in meaning without drawing attention to itself.

The first rule in capitalization is to follow your company style sheet; however, if no style sheet exists, follow these guidelines to set up your own style guide.

General Rules
  Capitalize the following.

- Names of:
    people
    organizations
    structures
    vehicles
    nationalities
    languages
    religions
    months
    days of the week
    holidays (but not seasons like winter)
    geographic areas (but not directions like north)
    courses or studies (if they are official titles)
- Titles of:
    people
    books
    articles
    chapters

Specific Rules

- In lists that complete a sentence, do *not* capitalize the list items.

    Preflight procedures include the following:

    1. visual inspection of the exterior
    2. draining of fuel sumps
    3. checking of oil
    4. visual inspection of the engine

- In titles and headings, capitalize all major words. Capitalize a preposition, article, or conjunction if it is the first or last word of the title or heading.

    Roll around a Point
    A Manual of Style

  Special rules apply to listing titles in reference lists. See Chapter 13.

- Capitalize trade names, even those trade names that have taken on common usage. See Chapter 2 for more on trade names.

    | Kleenex | Teflon | Plexiglas |
    | Formica | Scotch tape | |

- Capitalize references to figures, tables, and chapters when they are followed by a number or letter. Do not capitalize them when you make a general reference.

> See Table 1.
> All the tables that follow are from the U.S. Census Bureau.

Spelling

For a technical writer, one of the advantages of a word processor may be its spelling checker. Spelling checkers generally have a large vocabulary of common words and will flag misspellings and typos so they can be corrected. However, as the writer you still have to make final judgments, and you can't always count on having access to word processing. Therefore, it will benefit you to pay attention to your spelling.

No magic formula for good spelling exists, but here are some guidelines to help you become a better speller.

- Buy a good dictionary, like *Webster's Third International Dictionary*, and look up every word you're not sure of. Even better, get your company to buy you one as an investment in your writing. Use it all the time.
- Keep a list of your own problem words, either on index cards, on the wall by your desk, or on your computer terminal. Use it as a quick reference. Few people have more than 20 problem spelling words, and a list will save you time. Here are some typical problem words. Do you recognize yours?

accessible	comparative	definitely
competitive	separate	judgment
accommodate	receive	occurrence
consistent	descendant	liaison

- When you write a document, proofread for spelling by reading it backward, one sentence at a time: read the whole last sentence; then read the next-to-last sentence, and so on. This will force you to pay attention to individual words.
- Ask someone whose judgment you trust to read your documents to check for spelling. In return, offer to read their work.

≡ EXERCISE

Edit the following sentences for capitalization and spelling errors. Consult a dictionary. Use standard copyediting marks to make any necessary changes.

1. After flight 854 to Dallas was cancelled, vice-president R.A. Jones and the sales manager booked a flight to Houston.
2. With each occurence of failure, the clock stops.
3. The court's judgement is that school district 109 already complys with the desegregation order.

4. Plant no. 10 in Oshkosh replaced all its fluoresent bulbs with sodium vapor bulbs during the Summer.

5. Sales Manager Shockman serves as liason with the university on the cooperative education program.

6. When we receive acknowledgment of the benefit increase, we will send you a xerox copy of the original.

7. Professor Haloran diagramed the process on the board, but the students in chemistry 101 still didn't understand why naptha was used as the solvent.

8. If you mispell any words on your résumé, you will not be contacted by our personel office.

9. Mortality From Principle Types Of Accidents (heading)

10. "Comunicating Ideas through Writen Language" (title)

Acknowledgments (continued from p. IV)

Figure 16-6 reprinted by permission of Michael S. Clark, U.S. Suzuki Motor Corporation.

Figure 16-7 reprinted by permission of Paula Bell, Manager, Technical Publications, Silicon Compiler Systems.

Figure 16-8 reprinted by permission of Douglas A. Mendoza © 1992 PDP Systems.

Chapter 19

Figure 19-2 reprinted with permission of Jonathan Neher.

Chapter 20

Exercise: "Groundwater Management Study of Pajaro Basin Completed," *AMBAG Update*, Aug. 1984, Vol. 10. The preparation of this report was financed in part through an Areawide Waste Treatment Management Continuing Planning Program Grant from the U.S. Environmental Protection Agency, Region IX, under the provisions of Section 208 of the Federal Water Pollution Control Act, as amended. This does not signify that the contents necessarily reflect the views and policies of the U.S. Environmental Protection Agency. Reprinted by permission.

Chapter 21

Figure 21-2. Reprinted by permission of the City of San Jose, 1992.

Chapter 22

Leonard, David S. Portions of "An Evaluation of Nd:Yag Surgical Lasers for Allied Hospital." Reprinted with permission.

Chapter 23

Portions of "A Stress-Induced Die-off of Rocky Mountain Bighorn Sheep" reprinted by permission of Brian Wesley Simmons.

Figure 23-1 reprinted by permission. "A Low Energy Electron Source for Mobility Experiments," by James A. Jahnke and M. Silver in *Review of Scientific Instruments*, Vol. 44, No. 6, June 1973, pp. 776–777.

Reprinted by permission of Bechtel Corporation.

Index